Key Technologies on New Energy Vehicles

Key Technologies on New Energy Vehicles publishes the latest developments in new energy vehicles - quickly, informally and with high quality. The intent is to cover all the main branches of new energy vehicles, both theoretical and applied, including but not limited:

- Control Technology of Driving System
- Hybrid Electric Vehicle Coupling Technology
- Cross Disciplinary design optimization technology
- Single and Group Battery Technology
- Energy Management Technology
- Lightweight Technology
- New Energy Materials and Device
- Internet of Things (IoT)
- Cloud Computing
- 3D Printing
- Virtual Reality Technologies

Within the scopes of the series are monographs, professional books or graduate text-books, edited volumes, and reference works purposely devoted to support education in related areas at the graduate and post-graduate levels.

Hui Zhang · Rongrong Wang · Junmin Wang

Robust Gain-Scheduled Estimation and Control of Electrified Vehicles via LPV Technique

華中科技大學出版社
Huazhong University of Science and Technology Press

Springer

Hui Zhang
Beihang University
Beijing, China

Rongrong Wang
School of Mechanical Engineering
Shanghai Jiao Tong University
Shanghai, China

Junmin Wang
The University of Texas at Austin
Austin, TX, USA

ISSN 2662-2920 ISSN 2662-2939 (electronic)
Key Technologies on New Energy Vehicles
ISBN 978-981-19-8508-9 ISBN 978-981-19-8509-6 (eBook)
https://doi.org/10.1007/978-981-19-8509-6

Jointly published with Huazhong University of Science and Technology Press
The print edition is not for sale in China (Mainland). Customers from China (Mainland) please order the print book from: Huazhong University of Science and Technology Press.

This Springer imprint is published by the registered company Springer Nature Singapore Pte Ltd.
The registered company address is: 152 Beach Road, #21-01/04 Gateway East, Singapore 189721, Singapore

Foreword

The past two decades have witnessed the research and development (R&D) and the industrialization of China's new energy vehicles. Reviewing the development of new energy vehicles in China, we can find that the "Tenth Five Year Plan" is the period when China's new energy vehicles began to develop and our nation started to conduct organized R&D of the electric vehicle technology on a large scale; the "Eleventh Five Year Plan" is the period when China's new energy vehicles shifted from basic development to demonstration and examination as the Ministry of Science and Technology carried out the key project themed at "energy saving and new energy vehicles"; the period of the "Twelfth Five Year Plan" is the duration when China's new energy vehicles transitioned from demonstration and examination to the launch of industrialization as the Ministry of Science and Technology organized the key project of "electric vehicles"; the period of the "Thirteenth Five Year Plan" is the stage when China's new energy vehicle industry realized the rapid development and upgrading as the Ministry of Science and Development introduced the layout of the key technological project concerning "new energy vehicles".

The decade between 2009 and 2018 witnessed the development of China's new-energy automobile industry starting from scratch. The annual output of new-energy vehicles developed from zero to 1.27 million while the holding volume increased from zero to 2.61 million, each of which occupied over 53% in the global market and ranked 1st worldwide; the energy density of lithium-ion power batteries had more than doubled and the cost reduced by over 80%. In 2018, 6 Chinese battery companies were among the top 10 global battery businesses, with the 1st and the 3rd as China's CATL and BYD. In the meanwhile, a number of multinational automobile businesses shifted to develop new-energy vehicles. This was the first time for China to succeed in developing high-technology bulk commodities for civic use on a large scale in the world, also leading the trend of the global automobile development. The year of 2020 marked the landmark in the evolution of new-energy automobile. Besides, this year was the first year when new energy vehicles entered families on a large scale and the watershed where new-energy vehicles shifted from policy-driven to market-driven development. This year also saw the successful wrapping up of the mission in the *Development Plan on Energy Saving and New Energy Vehicle Industry (2012–2020)* and the official release of *Development Plan on New Energy Vehicle Industry (2021–2035)*. At the end of 2020, in particular, president Xi Jinping proposed that China strove to achieve the goal typified by peak carbon dioxide emissions by 2030 and carbon neutral by 2060, so as to inject great power into the sustainable development of new energy vehicles.

Looking back to the past and looking forward to the future, we can see even more clearly the historical position of the current development of new energy vehicles in the energy and industrial revolution. As is known to us all, each and every energy revolution started from the invention of power installations and transportation vehicles. On the other hand, the progress of power installations and transportation vehicles contributed to the development and exploitation of energy and led to industrial revolution. In the first energy revolution, steam engine was used as the power installation, with coal and train as energy and transportation respectively. As for the second energy revolution, internal combustion engine was taken as the power installation, oil and natural gas as energy, gasoline and diesel as energy carrier, and automobile as the transportation vehicle. At the current stage of the third energy revolution, all kinds of batteries are power installation, the renewable energy as the subject of energy and electricity and hydrogen as energy carrier, and electric vehicles as the means of transportation. In fact, the first energy revolution enabled the UK to outperform Netherland while the second energy revolution made the USA overtook the UK, both in terms of the economic strength. The present energy revolution may be the opportunity for China to catch up with and surpass other nations. How about the fourth industrial revolution? In my opinion, it is the green revolution based on renewable energy and also the smart revolution on the basis of digital network.

From the perspective of energy and industrial revolution, we can find three revolutions closely related to the new energy vehicles: electrification of power—the revolution of electric vehicles; low-carbon energy—the revolution of new energy; systematic intelligence—the revolution of artificial intelligence (AI).

First, electrification of power and the revolution of electric vehicles.

The invention of lithium-ion battery triggered the technological revolution in the area of storage battery over the past 100 years. Viewed from the development of power battery and power electronic device, the involvement of high specific energy battery and high specific power electric drive system would contribute to the platform development of electric chassis. The volume power of the machine controller based on new-generation power electric technology has more than doubled to 50 kw. In future, the volume power of the high-speed and high-voltage machine can be nearly doubled to 20 kw and the power volume of the automobile with 100 kw volume power could be no more than 10 L. With the constant decline of the volume of the electric power system, the electrification will lead to the platform and module development of chassis, which will lead to a major change in terms of vehicle design. The platform development of electric chassis and the lightweight of body materials will bring about the diversification and personalization of types of vehicles. Besides, the combination of active collision avoidance technology and body lightweight technology will result in a significant change in automobile manufacturing system. The revolution of power electrification will promote the popularity of new energy electric vehicles, and will eventually contribute to the overall electrification of the transportation sector. China Society of Automobile Engineers proposed the development goals of China's new energy vehicles in the *2.0 Technology Road Map of Energy Saving and New Energy Vehicles*: the sales of new energy vehicles would reach 40% of the total sale of vehicles by 2030; new energy vehicles would become the mainstream by 2035 with its sale accounting for over 50% of the total sale of vehicles. In the foreseeable future, electric locomotives, electric ships, electric planes and other types will become a reality.

Second, low-carbon energy and new energy revolution.

In the Strategy on Energy Production and Consumption Revolution (2016-2030) jointly issued by National Development and Reform Commission and National Energy Administration, a target was proposed that the non-fossil energy would account for around 20 percent of total energy consumption by 2030 and over 50% by 2050. Actually, there are five pillars aimed to realize the energy revolution: firstly, the transformation of renewable resources and the development of photovoltaic and wind power technologies; secondly, the energy system is shifting from centralized to distributed development, turning every building into a micro-power plant; thirdly, the storage of intermittent energy by way of relevant technologies such as hydrogen and battery; fourth, the development of energy (electric power) Internet technology; fifth, enabling electric vehicles to become the end of energy usage, energy storage and energy feedback. In fact, China's photovoltaic and wind power technologies are fully qualified for large-scale distribution, but energy storage remains a bottleneck which needs to be solved by way of battery, hydrogen and electric vehicles. With the large-scale promotion of electric vehicles, along with the mixture of electric vehicles and renewable energy, electric vehicles will become the "real" new energy vehicles utilizing the entire chain of clean energy. In so doing, it could both solve the pollution and carbon emission problems of the vehicle itself, but could also be conducive to the carbon emission reduction of the entire energy system, thus bringing about a new energy revolution for the entire energy system.

Third, intelligent development of system and AI revolution.

Electric vehicles have three attributes of travel tools, energy devices and intelligent terminals. Intelligent and connected vehicles (ICVs) will restructure the industrial chain and value chain of vehicles. Software defines vehicles while data determine value. The traditional vehicle industry will be transformed into a high-tech industry leading the AI revolution. In the meanwhile, let's take a look at the Internet connection and the feature of sharing regarding vehicles, among "four new attributes", from the perspectives of both intelligent travel revolution and the new energy revolution: for one thing, the connotation of the Internet pays equal attention to the Internet of vehicle information and the Internet of mobile energy. For another, the connotation of sharing lays equal emphasis on sharing travel and energy storage information. And both stationery and running electric vehicles can be connected to the mobile energy Internet, finally realizing a full interaction (V2G, Vehicle to Grid). As long as the energy storage scale of distributed vehicles is large enough, it will become the core hub of intelligent transportation energy, namely, the mobile energy Internet. Intelligent charging and vehicle to grid will meet the demand of absorbing renewable energy fluctuations. By 2035, China's inventory of new energy vehicles will reach about 100 million. At that time, the new energy vehicle-mounted battery power will reach approximately 5 billion KWH (kilowatt-hours) with 2.5 billion – 5 billion KWH as the charging and discharging power. By 2035, the maximum installed capacity of wind power and photovoltaic power generation will not surpass 4 billion KW. The combination of Vehicle-mounted energy storage battery and hydrogen energy could completely meet the demand of load balance.

All in all, with the accumulation of experience over the past two decades, since 2001, China's electric vehicles have "shifted to another path and led in the sector of new energy vehicles" worldwide. At the same time, China could build its advantage in terms of renewable resources with AI leading the world. It can be predicted that the period between 2020 and 2035 will be a new era when the revolution of new energy electric vehicles, the revolution of renewable energy and the revolution of artificial intelligence will leapfrog and develop in an coordinated manner and create a Chinese miracle featuring the strategic product and industry of new energy intelligent electric vehicles. Focusing on one strategic product and sector, such three technological revolutions and three advantages will release huge power, which could help realize the dream of a strong vehicle nation and play a leading role in all directions. With the help of such advantages, China will create a large industrial cluster with the scale of the main industry exceeding 10 trillion yuan and the scale of related industries reaching tens of trillions of yuan. The development of new energy vehicles at a large scale will result in a new energy revolution, which will bring earthshaking changes to the traditional vehicle, energy and chemical industry, thus truly embracing the great changes unseen in a century since the replacement of carriages by vehicles.

The technology revolution of new energy vehicle is advancing the rapid development of related interdiscipline subjects. From the perspective of technical background, the core technology of energy saving and new energy vehicles—the new energy power system technology, remains the frontier technology at the current stage. In 2019, China Association for Science and Technology released 20 key scientific and engineering problems, 2 of them (electrochemistry of high energy and density power battery materials, and hydrogen fuel battery power system) belonging to the scope of new energy system technology; The report of *Engineering Fronts* 2019 published by Chinese Academy of Engineering mentioned the power battery 4 times, fuel battery 2 times, hydrogen energy and renewable energy 4 times as well as electricity-driven/hybrid electric-driven system 2 times. Over the past two decades, China has accumulated plenty of new knowledge, new experience, and so many methods during the research and development regarding new energy vehicles. The "research series of key technologies on energy saving and new energy vehicles" are based on Chinese practice and the international frontier, aiming to review China's research and development achievements on energy saving and new energy vehicles, meet the needs of technological development concerning China's energy saving and new energy vehicles, reflect the key technology research trend of international energy saving and new energy vehicles, and promote the transformation and application of key technologies as regards China's energy saving and new energy vehicles. The series involve four modules: vehicle control technology, power battery technology, motor driving technology as well as fuel battery technology. All those books included in the series are research achievements with the support of National Natural Science Foundation of China (NSFC), major national science and technology projects or national key research and development programs. The publish of the series plays a significant role in enhancing the knowledge accumulation of key technologies concerning China's new energy vehicle, improving China's independent innovation capability, coping with climate change and promoting the green development of the vehicle industry. Moreover, it could contribute to China's development into a strong vehicle nation. It is hoped that the series could build a platform for academic and technological communication and the author and readers could jointly make contributions to reaching the top in the international stage concerning the technological and academic level in terms of China's energy saving and new energy vehicle.

January 2021

Ouyang Minggao
Academician of Chinese Academy of Sciences
Professor of Tsinghua University (THU)
Beijing, China

Preface

Vehicle control is one of the key technologies in modern society. The challenges in vehicle control arise from the nonlinearities of dynamics and actuators, needs to adapt to various road conditions and drivers, new configurations of power and transmission systems, network-induced issues, and multi-objective optimization. With the rapid development of control methodologies, we have frequently felt that a textbook which summarizes the recent research results from the control theory and the vehicular applications would be beneficial.

Though there are various textbooks on vehicle dynamics and control, they do not reflect the new techniques for handling the aforementioned challenges. This book attempts to present some techniques such as the robust control and nonlinearity approximation using linear-parameter-varying (LPV) techniques. Meanwhile, the control of independently driven electric vehicles and autonomous vehicles is introduced. Specifically, this book covers a comprehensive literature review, robust state estimation with uncertain measurements, sideslip angle estimation with finite-frequency optimization, fault detection of vehicle steering systems, output-feedback control of in-wheel motor-driven electric vehicles, robust path following control with network-induced issues, and lateral motion control with the consideration of actuator saturation.

It is hoped that this textbook can serve as a useful reference source to researchers from both industry and academia. It can also be used as a textbook for undergraduate- or graduate-level courses on vehicle dynamics and control.

Beijing, China
Shanghai, China
Austin, USA
February 2022

Hui Zhang
Rongrong Wang
Junmin Wang

Acknowledgements

Dr. Hui Zhang's initial research is focused on robust control with Prof. Yang Shi for the doctor degree. It is very grateful to Prof. Yang Shi for introducing Hui to the field. Hui Zhang started to work on vehicle dynamics and control when he collaborated with Prof. Junmin Wang as a postdoctoral researcher in 2012. During this period, Hui Zhang worked together with Rongrong Wang who was a Ph.D. student supervised by Prof. Junmin Wang. After joining new positions, the authors continued to work on vehicle dynamics and control, particularly for electric vehicles. Without the collaborations, the authors could not have finished this book.

Most chapters of this book are from previous published papers. The authors would like to express their gratitudes to Panshuo Li from Guangdong University of Technology, Anh-Tu Nguyen from Université Polytechnique des Hauts-de-France, Haiping Du from University of Wollongong, Yan Wang from Ford Motor Company, Guoguang Zhang from the Ohio State University, Hui Jing and Nan Chen from Southeast University, Kaijiang, Chuan Hu and Fengjun Yan from McMaster University, Mohammed Chadli from University of Picardie Jules Verne, and Yue Liu from Beihang University.

Finally, the authors would like to thank their families for the support.

Committe of Reveiwing Editors

Contents

1 Polytopic LPV Approaches for Intelligent Automotive Systems: State of the Art and Future Challenges 1
1.1 Introduction 1
1.2 Polytopic Linear Parameter-Varying Systems 3
1.2.1 A Motivating Automotive Application 3
1.2.2 Polytopic LPV System Description 5
1.2.3 Lyapunov-Based Stability of Polytopic LPV Systems 7
1.2.4 Gain-Scheduled Control Laws for LPV Systems 11
1.2.5 Gain-Scheduled Control of Polytopic LPV Systems 13
1.2.6 Multiple Convex Summation Relaxation 16
1.2.7 Observer Design for LPV Systems 16
1.2.8 Polytopic LPV Models and Takagi–Sugeno Models 18
1.3 Applications to Vehicle Dynamics Control 20
1.3.1 Vehicle Dynamics 20
1.3.2 Choices of Scheduling Parameters for LPV Control 25
1.4 Applications to Autonomous Vehicles 30
1.5 Applications to Vehicular Powertrain Systems 31
1.5.1 Internal Combustion Engines 31
1.5.2 Electric Vehicles 33
1.5.3 Aftertreatment Systems 35
1.6 Future Research Trends and Challenges 38
1.6.1 LPV Complexity Reduction 38
1.6.2 Fault Detection and Fault-Tolerant Control 38
1.6.3 Limited Capacities of Perception and Motion Planning 39
1.6.4 Driver-Automation Shared Driving Control 39
1.7 Concluding Remarks 40
References 40

2 $\mathcal{H}_\infty$ Observer Design for LPV Systems with Uncertain Measurements on Scheduling Variables: Application to an Electric Ground Vehicle ... 51
2.1 Introduction ... 51
2.2 Problem Formulation and Preliminary ... 53
2.3 Observer Design ... 57
2.4 Application to an EGV ... 64
2.5 Conclusion ... 71
References ... 71

3 Sideslip Angle Estimation of An Electric Ground Vehicle Via Finite-Frequency $\mathcal{H}_\infty$ Approach ... 75
3.1 Introduction ... 75
3.2 Problem Formulation and Preliminary ... 77
3.2.1 Introduction of the Electric Ground Vehicle ... 77
3.2.2 System Modeling and Identification ... 78
3.2.3 Model Transformation and Problem Formulation ... 82
3.2.4 Design Objectives ... 85
3.3 Observer Design ... 87
3.4 Experimental Results ... 90
3.5 Conclusion ... 93
References ... 93

4 Active Steering Actuator Fault Detection for an Automatically Steered Electric Ground Vehicle ... 97
4.1 Introduction ... 97
4.2 System Introduction and Problem Formulation ... 98
4.2.1 Acquisition System and Steering Actuator of EGV ... 98
4.2.2 EGV System Modeling ... 100
4.3 Main Results ... 109
4.3.1 Stability Analysis and Observer Design ... 110
4.3.2 $\mathcal{H}\_$ Performance and Observer Design ... 114
4.3.3 $\mathcal{H}_\infty$ Performance and Observer Design ... 120
4.3.4 Mixed $\mathcal{H}\_/\mathcal{H}_\infty$ Observer Design ... 124
4.4 Experiment-Based Simulation Results ... 125
4.5 Conclusions ... 128
References ... 128

5 Robust $\mathcal{H}_\infty$ Output-Feedback Yaw Control for In-Wheel Motor-Driven Electric Vehicles with Differential Steering ... 131
5.1 Introduction ... 131
5.2 System Modeling and Problem Formulation ... 133
5.2.1 Vehicle Dynamics with Differential Steering ... 133
5.2.2 Vehicle Modeling with Parameter Uncertainties ... 137
5.2.3 Problem Statement ... 139
5.3 Robust Controller Design ... 140

5.4 Simulation Results 145
5.4.1 J-Turn Simulation 146
5.4.2 Double-Lane Change 148
5.5 Conclusion 150
References 151

6 Robust $\mathcal{H}_\infty$ Path Following Control for Autonomous Ground Vehicles with Delay and Data Dropout 153
6.1 Introduction 153
6.2 System Modeling and Problem Formulation 155
6.2.1 Path Following Model 155
6.2.2 Vehicle Model 156
6.2.3 Path Following with Delay and Data Packet Dropout 158
6.2.4 Problem Statement 160
6.3 Robust $\mathcal{H}_\infty$ Controller Design with Delay and Data Dropout 160
6.4 Simulation Results 165
6.4.1 Single-Lane Change Maneuver 165
6.4.2 Double-Lane Change Maneuver 168
6.5 Conclusion 169
References 170

7 Robust Lateral Motion Control of Four-Wheel Independently Actuated Electric Vehicles with Tire Force Saturation Consideration 173
7.1 Introduction 173
7.2 System Modeling 175
7.2.1 Vehicle Model 175
7.2.2 Vehicle Model Considering Parameter Uncertainties 177
7.3 Control System Design 179
7.3.1 Higher-Level Controller Design 179
7.3.2 Lower-Level Controller Design 184
7.4 Simulation Studies 190
7.4.1 Single Lane Changing 190
7.5 J-Turn Simulation 195
7.6 Conclusion 200
References 200

Appendix: Fundamentals of Robust $\mathcal{H}_\infty$ Control 203

Acronyms

ABS	Anti-lock braking system
ACC	Adaptive cruise control
AFS	Active front steering
AHS	Automated highway systems
APU	Auxiliary power unit
DOC	Diesel oxidation catalyst
DPF	Diesel particulate filter
DYC	Direct yaw-moment control
EGV	Electric ground vehicle
ESP	Electronic stability program
EV	Electric vehicle
FWIA	Four-wheel independently actuated
GPS	Global positioning system
HEV	Hybrid electric vehicle
IAS	Intelligent automotive system
LFT	Linear fractional transformation
LMI	Linear matrix inequality
LPV	Linear parameter varying
LQG	Linear quadratic Gaussian
MPC	Model predictive control
NCS	Networked control system
PID	Proportional-integral-derivative
PMI	Proportional multiple-integral
SCR	Selective catalytic reduction
SISO	Single-input single-output
SMC	Sliding mode control
TRC	Transportation research center
TRFC	Tire-road friction coefficient

Chapter 1
Polytopic LPV Approaches for Intelligent Automotive Systems: State of the Art and Future Challenges

Abstract With more and more stringent requirements on driving comfort, safety, and fuel economy, polytopic LPV approaches have become popular in intelligent automotive control systems due to their merits in dealing with the complex nonlinearities. This survey starts with a review on control theory of polytopic LPV systems. Stability analysis and controller design are provided with techniques in obtaining less conservative results. Then, some key applications in vehicle dynamics control are provided. Several LPV models concerning the vertical dynamics, lateral dynamics, and integrated dynamics are introduced. Different polytopic LPV control designs are summarized taking various settings on time-varying parameters into account. Moreover, polytopic LPV approaches in vehicle path following and powertrain control are concluded, including the applications in internal combustion engines, electric vehicles, and aftertreatment systems. Finally, from recent advances on polytopic LPV control theory and automotive applications, future research directions and related challenges are discussed.

1.1 Introduction

Linear-parameter-varying (LPV) technique has been demonstrated as an effective approach to deal with complex nonlinear systems [1]. Depending on the type of trajectories of the parameters, the LPV framework can be used to represent various classes of nonlinear systems, including LPV systems with slowly or arbitrarily fast-varying parameters [2], switched systems [3], hybrid dynamical systems [4], and periodic systems [5]. The main interest of LPV approaches is to make possible the extension of some linear concepts to the case of nonlinear systems such as $\mathcal{H}_\infty$ control, sensitivity shaping, $\mathcal{D}-$stability, etc. Consequently, LPV approaches have been exploited for modeling, estimation, and control in a variety of engineering applications [1, 2, 6], especially automotive nonlinear systems [7]. One of the major challenges of LPV approaches arises from deriving less conservative optimization-based solutions with affordable computational load for complex high-dimensional nonlinear systems.

H. Zhang et al., *Robust Gain-Scheduled Estimation and Control of Electrified Vehicles via LPV Technique*, Key Technologies on New Energy Vehicles,
https://doi.org/10.1007/978-981-19-8509-6_1

With the rapid development of modern automotive industry, increasingly stringent requirements on passenger comfort, fuel economy, vehicle safety, and pollutant emissions have been proposed. The intelligent automotive systems (IASs) emerge to copy with those various requirements, of which a major task is the effective control and estimation algorithm design. Equipped with an increasing number of sensors, actuators, and other integrated devices, modern IASs involve more and more nonlinearities and uncertainties. Although kinds of new technologies have been adopted in modern vehicles [8], the complex nonlinearities and uncertainties in IASs give rise to great challenge in the controller and estimator synthesis. Due to the legacy control system and the highly nonlinear and highly complex vehicle systems, most vehicle control features are still heavily relying on feedforward PID-based controls with gain scheduling as a way to address the nonlinearity. In addition, most of the controller loops are designed for SISO which leads to significant optimality deficiency when SISO loops are highly coupled. The usual way to deal with these interactions is to detune individual control loop performances. Despite a solid theoretical foundation, LPV methodology is still not widely used in industrial applications, especially in automotive industry. Therefore, it is timely to conduct a review on the use of LPV approaches for IAS applications to show both the interests and the recent advances which have been achieved in this topic.

This chapter first provides in Sect. 1.2 a concise overview on LPV approaches. The goal is not to comprehensively review the literature which already includes several monographs [1, 2, 7, 9, 10], surveys [6, 11, 12], and numerous technical articles therein. Instead, a selective list of notable references is given to describe LPV approaches, especially the mainstreams to reduce the related design conservatism. Our goal is to provide the insights of LPV technique which will be useful for the control design of real-world applications. Moreover, the presented methods to reduce the design conservatism can be exploited as practical guidelines to improve the control performance for complex engineering systems. Among various existing LPV approaches [2], e.g., polytopic LPV design, linear fractional transformation (LFT) LPV design, gridding-based LPV design, here the emphasis is put on polytopic LPV framework without loss of generality. Note that systems with other parameter dependencies can be transformed into polytopic LPV models [9]. Moreover, polytopic LPV approaches have become very popular, at least from the viewpoint of the number of related publications [6]. In Sect. 1.3, we discuss the polytopic LPV approach in vehicle dynamics control. Vehicle dynamic systems can be modeled as polytopic LPV systems because of the time-varying parameters. Several LPV models concerning vehicle vertical dynamics, lateral dynamics, and integrated dynamics are briefly introduced. According to different time-varying parameters considered, various gain-scheduled controllers to improve the driving comfort and safety are concluded. Since some parameters vary independently, methods on shrinking the polytopic to obtain less conservative results are provided as well. Autonomous vehicles have been regarded as a promising future for complex IASs. The successful application of polytopic LPV to the control of autonomous vehicles is reviewed in Sect. 1.4. Besides the autonomous vehicles, the powertrain control has gained considerable attentions as well during the past decades. The polytopic LPV approaches for intelligent powertrain systems are reviewed in Sect. 1.5. In Sect. 1.6, we present our viewpoint on the current research trends and challenges. This chapter aims to complement other surveys on LPV theory [1, 11, 12] and applications [6] with a special focus on polytopic LPV approaches for IASs [13, 14]. Note that recent comprehensive surveys on some mainstream techniques for path following control of autonomous vehicles, including pure pursuit, feedback linearization, MPC control, and many others, can be also found in [15, 16].

Notation

$\mathcal{I}_N$ denotes the set $\{1, 2, \ldots, N\}$. $\mathbb{R}$ is the field of real numbers. For a vector $x \in \mathbb{R}^n$ and $i \in \mathcal{I}_n$, x_i denotes the ith entry of x, and $\|x\|_\infty = \max_{i=1,\ldots,n} x_i$ denotes the infinity norm of x. I denotes the identity matrix of appropriate dimension. For a matrix X, $X^\top$ indicates its transpose. For any square matrix X, $X \succ 0$ indicates a symmetric positive-definite matrix, and $\text{He}X = X + X^\top$. $\text{diag}(X_1, X_2)$ denotes a block-diagonal matrix composed of X_1, X_2. The symbol $\star$ represents matrix blocks that can be deduced by symmetry. The time dependency of the variables is omitted when convenient.

1.2 Polytopic Linear Parameter-Varying Systems

This section first provides an automotive application to motivate the need of LPV control in improving the practical performance of IASs. Then, after a brief description, we discuss some key points related to the stability analysis and synthesis of polytopic LPV systems.

1.2.1 A Motivating Automotive Application

Adaptive cruise control (ACC) is an advanced driver-assistance system for road vehicles that automatically adjusts the vehicle speed to maintain a safe distance from vehicles ahead [14]; see Fig. 1.1. Due to its important role for the safety issue of vehicles in the same lane, these ACC systems have been the research and development focus of the automotive industry [17]. Hereafter, we consider the multi-objective control problem of an ACC system to motivate the need for an LPV control framework in IAS applications.

The longitudinal dynamics of autonomous vehicles can be described by the following simplified nonlinear differential equations [19]:

$$\begin{aligned} M_v \dot{V}_x &= F_l - F_d, \\ \tau \dot{F}_l &= -F_l + u, \end{aligned} \tag{1.1}$$

where V_x is the longitudinal speed, F_l is the longitudinal force realized on the wheels, M_v is the total mass of the vehicle, τ denotes the time constant of the longitudinal

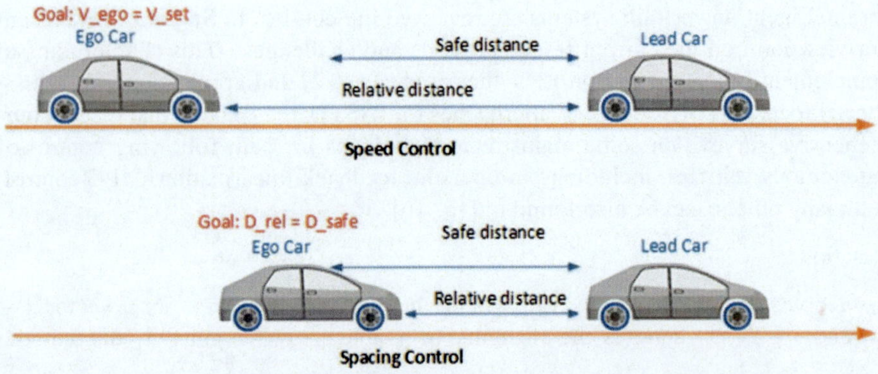

Fig. 1.1 Illustration of vehicle ACC systems [18]. $V_$ego is the ego vehicle velocity, $V_$set is the ego vehicle velocity setpoint, $D_$rel is the distance relative between lead vehicle and ego vehicle, and $D_$safe is the safe distance between lead vehicle and ego vehicle

actuators. The control input u represents the required longitudinal control force. The longitudinal disturbance force F_d aims at taking into account the aerodynamics and the road conditions as [20]

$$F_d = C_a V_x^2 + C_r M_v g \cos\phi + M_v g \sin\phi, \tag{1.2}$$

where ϕ is the road slope, C_a and C_r are the vehicle parameters related to the aerodynamics and rolling resistances.

From (1.1) and (1.2), the longitudinal dynamics can be represented in the following state-space form:

$$\dot{x} = A(V_x)x + Bu + Ew(\phi), \tag{1.3}$$

where $x = \begin{bmatrix} V_x & F_l \end{bmatrix}^\top$ is the state vector, $w(\phi) = C_r \cos\phi + \sin\phi$ is the system disturbance, and

$$A(V_x) = \begin{bmatrix} -\frac{C_a}{M_v} V_x & \frac{1}{M_v} \\ 0 & -\frac{1}{\tau} \end{bmatrix}, \quad B = \begin{bmatrix} 0 \\ \frac{1}{\tau} \end{bmatrix}, \quad E = \begin{bmatrix} M_v g \\ 0 \end{bmatrix}. \tag{1.4}$$

Remark that the dynamic matrix $A(V_x)$ of system (1.3) *explicitly* depends on the vehicle speed, which is time-varying and bounded as follows [14]:

$$V_x \in [5, 30]\ \text{m/s}, \quad \dot{V}_x \in [-3.5, 2.5]\ \text{m/s}^2. \tag{1.5}$$

Note that classical linear time-invariant approaches, e.g., linear quadratic regulator (LQR), LQR with preview of the disturbance, model predictive control, loop-shaping robust control, etc., can be applied to the LPV system (1.3) for related control problems. A notable survey on these linear control approaches for vehicle dynamics applications can be found in [13]. Moreover, MPC technique has shown large potential for use in automotive applications [21]. However, when the system to be dealt with is highly nonlinear and uncertain, linear MPC technique may not provide a satisfactory control performance. Robust and/or nonlinear MPC control schemes should be required in this situation. Unfortunately, there are still some fundamental challenges to overcome when using robust/nonlinear MPC techniques [13, 21]. First, the calibration effort for MPC controllers can be costly in many cases. Second, the control design of nonlinear MPC controllers are still too computationally complex, and low-complexity explicit control laws or fast optimization algorithms are necessary. Third, guaranteeing the stability of MPC *a priori*, without increasing excessively the algorithm complexity is still widely open. Finally, several applications have significant nonlinearities and uncertainties, as the usual case of intelligent automotive systems, which complicate MPC design and implementation. LPV control technique could be an effective alternative to overcome these major drawbacks for such complex nonlinear systems. In particular, taking into account the information on the time-varying parameters such as in (1.5), the LPV design conservatism can be significantly reduced to further improve the practical control performance [1].

Hereafter, we review some basic features of LPV control to highlight its ability to take into account the *time-varying* nature of system parameters in the design procedure for performance improvements. The mainstream of current research on LPV control theory, with a focus on polytopic approaches, is also discussed. These discussions aim at providing different theoretical directions that could be performed to further improve the practical performance of real-world LPV control systems such as in the case of IAS applications.

1.2.2 Polytopic LPV System Description

For generality, we consider the LPV system (1.3) in its general state-space form as

$$\begin{aligned} \dot{x} &= A(\theta)x + B(\theta)u + E(\theta)w, \\ y &= C(\theta)x + F(\theta)w, \end{aligned} \tag{1.6}$$

where $x \in \mathbb{R}^{n_x}$ is the state of the system, $u \in \mathbb{R}^{n_u}$ is the control input, $w \in \mathbb{R}^{n_w}$ is the disturbance input, $y \in \mathbb{R}^{n_y}$ is the system output, and $\theta \in \mathbb{R}^p$ is the vector of unknown *time-varying* parameters whose measurement is available in real time for gain scheduling control. Assume that the parameter $\theta = \begin{bmatrix}\theta_1 \dots \theta_p\end{bmatrix}^\top$ and its unknown rate of variation $\dot{\theta}(t)$ are smooth and respectively valued in the hypercubes

$$\begin{aligned} \mathcal{S}_\theta &= \{(\theta_1, \dots, \theta_p)^\top : \theta_j \in [\underline{\theta}_j, \overline{\theta}_j], \ j \in \mathcal{I}_p\}, \\ \mathcal{S}_\upsilon &= \{(\dot{\theta}_1, \dots, \dot{\theta}_p)^\top : \dot{\theta}_j \in [\underline{\upsilon}_j, \overline{\upsilon}_j], \ j \in \mathcal{I}_p\}, \end{aligned} \tag{1.7}$$

where $\underline{\theta}_j \leq \overline{\theta}_j$ (respectively $\underline{\upsilon}_j \leq \overline{\upsilon}_j$) are *known* lower and upper bounds on θ_j (respectively $\dot{\theta}_j$), for $j \in \mathcal{I}_p$. We assume that the time-varying matrices $A(\theta)$, $B(\theta)$, $C(\theta)$, $E(\theta)$, and $F(\theta)$ of system (1.6) are continuous on the hypercube $\mathcal{S}_\theta$. Note that condition (1.7) of the time-varying parameter θ is a *generalized* version of (1.5).

Remark 1.1 Several methods have been proposed to obtain the LPV model in (1.6) from a nonlinear system, for instance Jacobian linearization [22], function substitution [23], state transformation [24]. More details on analytical LPV modeling and experimental LPV modeling can be found in [7, Ch. 1] and [10], respectively. Note that for a given nonlinear system, its LPV representation is not unique and different models yield different properties with respect to stability analysis and/or control performance. Especially, an LPV model can suffer from *overbounding*, i.e., the parameters are related to each other by inherent couplings, which can increase the numerical complexity and the conservatism of the design results [25]. The so-called *parameter set mapping* approach in [26] can be used to obtain a less conservative LPV representation.

Using the sector nonlinearity approach [27, Ch. 2], the LPV model (1.6) can be *equivalently* rewritten in the polytopic form

$$\begin{aligned} \dot{x} &= \sum_{i=1}^{N} \eta_i(\theta)(A_i x + B_i u + E_i w), \\ y &= \sum_{i=1}^{N} \eta_i(\theta)(C_i x + F_i w), \end{aligned} \tag{1.8}$$

with $N = 2^p$. The constant matrices of appropriate dimensions A_i, B_i, C_i, E_i, and F_i, for $i \in \mathcal{I}_r$, represent the set of N local linear sub-models, which are defined as

$$\Pi_i = \Pi(\theta)|_{\eta_i(\theta)=1}, \text{ for } \forall \Pi \in \{A, B, C, E, F\}.$$

The weighting functions $\eta_i(\theta)$, for $i \in \mathcal{I}_N$, are continuously differentiable and belong to the simplex, defined as

$$\Xi_\theta = \left\{ \eta(\theta) \in \mathbb{R}^N : \sum_{i=1}^{N} \eta_i(\theta) = 1, \ \eta_i(\theta) \geq 0, \ \forall \theta \in \mathcal{S}_\theta \right\}.$$

Since $(\theta, \dot{\theta}) \in \mathcal{S}_\theta \times \mathcal{S}_\upsilon$, with $\mathcal{S}_\theta$ and $\mathcal{S}_\upsilon$ defined in (1.7), then the lower bound ϕ_{i1} and the upper bound ϕ_{i2} of $\dot{\eta}_i(\theta)$ can be easily computed from the analytical expression of $\eta_i(\theta)$ as

$$\dot{\eta}_i(\theta) \in [\phi_{i1}, \phi_{i2}], \quad \phi_{i1} \leq \phi_{i2}, \quad i \in \mathcal{I}_N. \tag{1.9}$$

Remark 1.2 The sector nonlinearity approach [27] allows deriving an *exact* polytopic form of the generic LPV system (1.6). The weighting functions $\eta_i(\cdot)$, for $i \in \mathcal{I}_N$, can capture the parameter nonlinearities, i.e., they can be a *nonlinear* function of components of θ. Hence, stability and control methods based on the polytopic LPV model (1.8) can be applied to a larger class of parametric dependencies than, e.g., linear, affine, rational, or linear fractional transformation (LFT).

1.2.3 Lyapunov-Based Stability of Polytopic LPV Systems

For stability analysis, we consider an autonomous polytopic LPV system (1.8) of the form

$$\dot{x} = \sum_{i=1}^{N} \eta_i(\theta) A_i x. \tag{1.10}$$

In the sequel, we provide an overview on Lyapunov-based stability analysis of the polytopic LPV system (1.10).

1.2.3.1 Quadratic Stability

The quadratic stability analysis of system (1.10) is a direct extension of the result for linear time-invariant systems in [28]. Consider a quadratic Lyapunov function of the form

$$V(x) = x^\top P x, \quad P \succ 0. \tag{1.11}$$

The following quadratic stability result is readily obtained.

Theorem 1.1 *The equilibrium of the LPV system in* (1.10) *is globally asymptotically stable if there exists a common positive-definite matrix P such that*

$$A_i^\top P + P A_i \prec 0, \quad i \in \mathcal{I}_N. \tag{1.12}$$

Note that condition (1.12) is represented in the form of linear matrix inequalities (LMIs). Hence, the stability analysis can be effectively performed with available numerical solvers [28]. This particular feature has sparked a growing interest in LPV-based approaches for nonlinear control theory and its real-world applications. The insights on the state of the art of prominent results in the early of this century are discussed in [11, 12], whereas more recent theoretical developments of LPV systems and control can be found in [1, 2, 10]. Reference [6] also provides an application-oriented survey on this research topic.

For stability analysis, a common Lyapunov matrix P is used to check the stability for all local linear sub-systems. Moreover, the weighting functions of the LPV system (1.10) are not involved in the stability condition. For these reasons, quadratic stability analysis makes no difference between time-invariant parameters, slowly varying parameters, and arbitrarily fast-varying parameters. Despite its simplicity, this type of stability may lead to overconservative results [2]. Therefore, a considerable research effort has been devoted to the conservatism reduction issue [1, 29–32]. The current mainstream to reduce the conservatism of LPV model-based approaches is based on the choices of different families of Lyapunov functions. These choices can be accompanied with an $S-$variable approach [33] to introduce slack variables for relaxation purposes [34]. Some insights of this research mainstream are discussed hereafter.

1.2.3.2 Poly-Quadratic Stability

The conservatism of using a single quadratic Lyapunov function (1.11) for LPV stability analysis is illustrated in [2]. A natural way to overcome this major drawback consists in constructing Lyapunov functions that are explicitly parameter dependent. To illustrate the idea of using slack variables for relaxations purposes, consider the following poly-quadratic Lyapunov functional candidate:

$$V(x) = x^\top \mathcal{P}(\theta)x, \tag{1.13}$$

where $\mathcal{P}(\theta) = \sum_{i=1}^{N} \eta_i(\theta)(P_i + X)$, and $P_i + X \succ 0$, for $i \in \mathcal{I}_N$. The time derivative of the Lyapunov function (1.13) is given by

$$\dot{V}(x) = \begin{bmatrix} x \\ \dot{x} \end{bmatrix}^\top \begin{bmatrix} \dot{\mathcal{P}}(\theta) & \star \\ \mathcal{P}(\theta) & 0 \end{bmatrix} \begin{bmatrix} x \\ \dot{x} \end{bmatrix}.$$

Remark 1.3 Note that the quadratic Lyapunov function (1.11) is directly recovered from (1.13) with $P_1 = \cdots = P_N = P$ and $X = 0$. Then, function (1.11) is only a special case of the poly-quadratic Lyapunov function (1.13).

Since $\eta(\theta) \in \Xi_\theta$, it follows that $\sum_{i=1}^{N} \dot{\eta}_i(\theta)X = 0$, for any matrix X. Then, the term $\dot{\mathcal{P}}(\theta)$ can be rewritten in the following form [35]:

$$\begin{aligned} \dot{\mathcal{P}}(\theta) &= \sum_{k=1}^{N-1} \dot{\eta}_k(\theta)(P_k + X - P_N) + \dot{\eta}_N(\theta)X \\ &= \sum_{k=1}^{N-1} \sum_{l=1}^{2} \left[\omega_{kl}(\theta)\phi_{kl}\mathcal{X} + \frac{1}{N-1}\omega_{Nl}(\theta)\phi_{Nl}X \right], \end{aligned} \tag{1.14}$$

with $\mathcal{X} = P_k + X - P_N$, and

$$\omega_{k1}(\theta) = \frac{\phi_{k2} - \dot{\eta}_k(\theta)}{\phi_{k2} - \phi_{k1}}, \quad \omega_{k2}(\theta) = \frac{\dot{\eta}_k(\theta) - \phi_{k1}}{\phi_{k2} - \phi_{k1}}. \tag{1.15}$$

The bounds ϕ_{kl}, for $(k,l) \in \mathcal{I}_N \times \mathcal{I}_2$, are given in (1.9). Note that $\omega_{kl}(\theta) \geq 0$, $\sum_{l=1}^{2} \omega_{kl}(\theta) = 1$, for $k \in \mathcal{I}_N$.

Using expression of $\dot{\mathcal{P}}(\theta)$ in (1.14) and Finsler lemma [28], the following theorem provides a poly-quadratic stability analysis for LPV systems.

Theorem 1.2 *Consider the LPV system* (1.10) *with* $(\theta, \dot{\theta}) \in \mathcal{S}_\theta \times \mathcal{S}_v$ *and* $\eta(\theta) \in \Xi_\theta$. *If there exist symmetric matrices* $X \in \mathbb{R}^{n_x \times n_x}$, $P_i \in \mathbb{R}^{n_x \times n_x}$, *for* $i \in \mathcal{I}_N$, *and matrices* $M \in \mathbb{R}^{n_x \times n_x}$, $G \in \mathbb{R}^{n_x \times n_x}$ *such that the following linear matrix inequalities are satisfied:*

$$P_i + X \succ 0, \quad i \in \mathcal{I}_N, \tag{1.16}$$

$$\begin{bmatrix} \Psi - MA_i - A_i^\top M^\top & \star \\ P_i + X + M^\top - GA_i & G + G^\top \end{bmatrix} \prec 0, \quad i \in \mathcal{I}_N, \tag{1.17}$$

with $\Psi = \phi_{kl}(P_k + X - P_N) + \frac{1}{N-1}\phi_{Nl}X$, *for* $k \in \mathcal{I}_{N-1}$ *and* $l \in \mathcal{I}_2$. *Then, the LPV system* (1.10) *is asymptotically stable.*

Note that condition (1.16) guarantees that function (1.13) is a proper candidate of Lyapunov functions for LPV stability analysis purposes. The proof of Theorem 1.2 is adapted from [36, Theorem 1.1] for the polytopic LPV setup.

Remark 1.4 Due to the presence of $\dot{\mathcal{P}}(\theta)$, the time derivatives of the weighting functions $\dot{\eta}_i(\theta)$, $i \in \mathcal{I}_N$, appear *explicitly* in the expression of $\dot{V}(x)$. Hence, in contrast to quadratic stability, exploiting the information on the time-varying parameter and its rate of variation plays a key role for poly-quadratic stability of LPV systems. Indeed, the bounds ϕ_{kl}, for $k \in \mathcal{I}_N$ and $l \in \mathcal{I}_2$, defined in (1.9), are directly involved in the stability condition (1.17).

Remark 1.5 For relaxation purposes, the decision matrices X, M, and G are introduced in the stability condition (1.17) in Theorem 1.2 as slack variables. Indeed, imposing $X = 0$, $P_i = P \succ 0$, for $\forall i \in \mathcal{I}_N$, and $M = -P$, we can prove that the result of Theorem 1.2 precisely includes that of Theorem 1.1; see also Remark 1.3. With an appropriate selection of the slack variables, we can also theoretically prove that the stability condition in Theorem 1.2 leads to less conservative results than the classical poly-quadratic stability result in [2, Ch. 2]. Moreover, note that with the use of the slack decision variable X, the Lyapunov-related decision matrices P_i, for $i \in \mathcal{I}_N$, are not required anymore to be positive definite as most of existing parameter-dependent Lyapunov-based stability results [35].

1.2.3.3 Other Choices of Lyapunov Functions

Apart from poly-quadratic Lyapunov functions, other classes of Lyapunov functions have been proposed in the literature to overcome the conservatism issue of quadratic stability analysis.

□ Polyhedral Lyapunov Functions

Based on the ∞−norm of the state, this class of Lyapunov functions is constructed as follows [37, 38]:

$$V(x) = \|P^\top x\|_\infty, \tag{1.18}$$

where $P \in \mathbb{R}^{n_x \times s}$ is a full row rank matrix. Although necessary and sufficient stability conditions can be derived using polyhedral Lyapunov functions [39], this class of Lyapunov functions is only suitable for LPV systems with arbitrarily fast-varying parameters. Moreover, it leads to a nonconvex control framework, which induces numerical difficulties.

□ Piecewise Quadratic Lyapunov Functions

This class of Lyapunov functions is defined as

$$V(x) = \max_{i=1,\ldots,n} \{x^\top P_i x\}, \tag{1.19}$$

where $P_i \succ 0$, for $i \in \mathcal{I}_n$. Due to the presence of multiple matrices P_i, the piecewise Lyapunov function (1.19) can yield more relaxed stability results compared to the quadratic Lyapunov function (1.11). However, same to polyhedral Lyapunov functions, this class of Lyapunov functions leads to a nonconvex framework for stability analysis and control design [40, 41].

□ Homogeneous Lyapunov Functions

Homogeneous polynomially parameter-dependent quadratic Lyapunov functions are constructed as

$$V_m(x, p) = x^\top P_m(p)x, \tag{1.20}$$

where $P_m(p) \in \mathbb{R}^{n_x \times n_x}$ is a homogeneous matricial form of degree m, i.e., matrix whose entries are (real q−variate) homogeneous forms of degree m. As an extensions of poly-quadratic Lyapunov functions (1.13), the effectiveness in terms of conservatism reduction with respect to (1.13) has been demonstrated in [42, 43]. However, this class of Lyapunov functions suffers two major drawbacks for LPV stability analysis: (i) the parameters should vary arbitrarily fast, (ii) the exponential increase of the computational burden as the degree of the homogeneous polynomial increases [44].

1.2.4 Gain-Scheduled Control Laws for LPV Systems

Extension of stability results to the control design is the main goal in LPV-based framework for nonlinear systems. As for the stability analysis, robust control theory plays a key role in the theoretical developments of LPV control. However, the parameter of LPV systems is real-time available for control design, which is not the case of uncertain/unknown parameter in robust control theory [33]. The incorporation of the time-varying parameter in the control laws leads to the concept of LPV gain-scheduled controllers [45, 46]. We review below the four most common gain-scheduled control laws, in terms of both theoretical developments and real-world applications, existing in the literature for the LPV system (1.6).

□ Gain-Scheduled Static State-Feedback Control

The gain-scheduled state-feedback control is the most simple control law, whose structure is given by

$$u = K(\theta)x, \tag{1.21}$$

where the parameter-dependent gain $K(\theta) \in \mathbb{R}^{n_u \times n_x}$ is to be determined. Despite its easy design, this control structure requires full-state information for real-time implementation, which may not suitable for many practical situations [1].

□ Gain-Scheduled Static-Output-Feedback Control

The gain-scheduled static-output-feedback (SOF) is another simple control structure, defined as

$$u = K(\theta)y, \tag{1.22}$$

where the parameter-dependent gain $K(\theta) \in \mathbb{R}^{n_u \times n_y}$ is to be determined. The great advantage of this control law consists in its simplicity for implementation. However, the related control formulation is *inherently* nonconvex, which leads to a major challenge in deriving design conditions that can be efficiently solved with available numerical solvers [47]. Based on $S-$variable approach [33], some numerically tractable designs of gain-scheduled SOF control have been proposed in the literature [48–50].

☐ Gain-Scheduled Observer-Based Feedback Control

For many engineering applications, the system state is not fully available for state-feedback control and an SOF control cannot provide a viable control solution. In this situation, we can consider a gain-scheduled observer-based control scheme, whose structure relies on a full-order Luenberger observer of the form

$$\begin{aligned} \dot{\hat{x}} &= A(\theta)\hat{x} + B(\theta)u - L(\theta)(y - \hat{y}), \\ \hat{y} &= C(\theta)\hat{x}, \end{aligned} \tag{1.23}$$

where $\hat{x} \in \mathbb{R}^{n_x}$ is the estimated state, and $L(\theta) \in \mathbb{R}^{n_x \times n_y}$ is the observer gain. Then, the gain-scheduled control law is now defined as

$$u = K(\theta)\hat{x}, \tag{1.24}$$

where $K(\theta) \in \mathbb{R}^{n_u \times n_x}$ is the control gain. Note that a simultaneous design of both the observer gain $L(\theta)$ and the control gain $K(\theta)$ may not admit a convex formulation [2]. However, a separate observer-control design always results in LMI synthesis conditions [51–53].

☐ Gain-Scheduled Dynamic-Output-Feedback Control

Dynamic output feedback (DOF) control can be used to overcome the drawbacks of the two above control laws. The structure of a gain-scheduled DOF control law is given by

$$\begin{aligned} \dot{x}_c &= A_c(\theta)x_c + B_c(\theta)y, \\ u &= C_c(\theta)x_c + D_c(\theta)y, \end{aligned} \tag{1.25}$$

where $x_c \in \mathbb{R}^{n_c}$ is the state of the controller, and the matrices of appropriate dimensions $A_c(\theta)$, $B_c(\theta)$, $C_c(\theta)$, $D_c(\theta)$ are to be designed. This important class of LPV controllers has been largely studied in LPV control framework, especially from theoretical viewpoint; see [54–58] and related references. An interesting feature of DOF scheme is that in many control setups, convex solutions can be achieved with a full-order controller (1.25), i.e., $n_c = n_x$. However, due to the presence of a rank constraint, the design of reduced-order DOF controllers ($n_c < n_x$) is inherently nonconvex [59, 60].

It is important to note that the gain-scheduled SOF controller (1.22) and observer-based controller (1.23)–(1.24) are special cases of the DOF controller (1.25). Indeed, if $A_c(\theta) = 0$, $B_c(\theta) = 0$, $C_c(\theta) = 0$, and $D_c(\theta) = K(\theta)$ in (1.25), the SOF control law (1.22) can be directly recovered. Moreover, the DOF controller (1.25) becomes the observer-based control law (1.23)–(1.24) with $x_c = \hat{x}$ and

$$\begin{aligned} A_c(\theta) &= A(\theta) + B(\theta)K(\theta) + L(\theta)C(\theta), \quad & C_c(\theta) &= K(\theta), \\ B_c(\theta) &= -L(\theta)C(\theta), & D_c(\theta) &= 0. \end{aligned}$$

A recent survey focusing on gain-scheduled DOF law for three LPV control setups (polytopic LPV synthesis, gridding-based LPV synthesis, and multiplier-based LFT LPV synthesis) is given in [6].

1.2.5 Gain-Scheduled Control of Polytopic LPV Systems

General speaking, the procedure to design LPV controllers can be described with the following steps.

- Step 1: Select the form of the gain-scheduled controller, i.e., state-feedback control, SOF control, observer-based control, DOF control.
- Step 2: Define the closed-loop LPV system with its performance specifications.
- Step 3: Apply a set of Lyapunov-based stability conditions, verifying the predefined specifications, to the closed-loop system.
- Step 4: Transform the stability conditions in Step 3 into numerically tractable design conditions, using for instance robust control tools [28, 33].

For simplicity without loss of generality, we consider here in Step 1 a state-feedback law with $\mathcal{L}_2-$gain performance to illustrate step-by-step the control design of polytopic LPV systems. Within the polytopic control setup, the gain-scheduled controller (1.21) takes the form

$$u = \sum_{i=1}^{N} \eta_i(\theta) K_i x. \tag{1.26}$$

Note that controller (1.26) and the system in (1.8) share the same weighting functions. For Step 2, we define the closed-loop LPV system from the expressions of system (1.8) and controller (1.26) as

$$\dot{x} = \sum_{i=1}^{N} \sum_{j=1}^{N} \eta_i(\theta) \eta_j(\theta) ((A_i + B_i K_j) x + E_i w). \tag{1.27}$$

For Step 3, we distinguish two specific cases for illustrations: quadratic control design and poly-quadratic control design.

1.2.5.1 Quadratic Gain-Scheduled Control Design

The following result is readily obtained for the closed-loop system (1.27) using the quadratic Lyapunov function (1.11).

Theorem 1.3 *[2, Ch. 3] Consider the LPV system* (1.8) *and the gain-scheduled control law* (1.26)*. If there exist a positive-definite matrix $Q \in \mathbb{R}^{n_x \times n_x}$, matrices Y_i, for $i \in \mathcal{I}_N$, and a positive scalar γ such that the following LMI conditions hold:*

$$\Phi_{ii} \prec 0, \quad i \in \mathcal{I}_N, \tag{1.28}$$

$$\Phi_{ij} + \Phi_{ji} \prec 0, \quad i, j \in \mathcal{I}_N, \ i < j, \tag{1.29}$$

where

$$\Phi_{ij} = \begin{bmatrix} \mathrm{He}(A_i Q + B_i Y_j) & \star & \star \\ E_i^\top & -\gamma I & \star \\ C_i Q & F_i & -\gamma I \end{bmatrix}. \tag{1.30}$$

Then, the closed-loop LPV system (1.27) *is quadratically stable and the $\mathcal{L}_2-$gain of the transfer $w \to y$ is smaller than γ, for all $\theta \in \mathcal{S}_\theta$. Moreover, the control feedback gains in* (1.26) *are defined as*

$$K_i = Y_i Q^{-1}, \quad i \in \mathcal{I}_N.$$

The control result in Theorem 1.3 comes from the bounded-real lemma [28]. Indeed, substituting the closed-loop system (1.27) into this well-known lemma, it follows that

$$\begin{bmatrix} \mathrm{He}[P(A(\theta) + B(\theta)K(\theta))] & \star & \star \\ E(\theta)^\top P & -\gamma I & \star \\ C(\theta) & F(\theta) & -\gamma I \end{bmatrix} \prec 0. \tag{1.31}$$

Note that inequality (1.31) is *nonconvex* due to the nonlinear coupling between the Lyapunov matrix P and the control gains K_j, for $j \in \mathcal{I}_N$. Then, Step 4 is necessary for the control design. To this end, applying a congruence transformation [28] to (1.31) with $\mathrm{diag}(Q, I, I)$, followed by the change of variable $Y_j = K_j Q$, with $Q = P^{-1}$, it follows that

$$\sum_{i=1}^{N}\sum_{j=1}^{N} \eta_i(\theta)\eta_j(\theta)\Phi_{ij} \prec 0, \tag{1.32}$$

where Φ_{ij} is defined in (1.30). Then, from the convexity property, it is clear that conditions (1.28)–(1.29) guarantee (1.32).

1.2.5.2 Poly-Quadratic Gain-Scheduled Control Design

To overcome a conservative design due to the use of quadratic Lyapunov functions in Theorem 1.3, the following theorem provides design conditions based on quadratic Lyapunov functions and $S-$variable approach.

Theorem 1.4 *Consider the LPV system* (1.8) *and the gain-scheduled control law* (1.26). *If there exist symmetric matrices* $W \in \mathbb{R}^{n_x \times n_x}$, $Q_i \in \mathbb{R}^{n_x \times n_x}$, *and matrices* $M \in \mathbb{R}^{n_x \times n_x}$, $Y_i \in \mathbb{R}^{n_u \times n_x}$, *for* $i \in \mathcal{I}_N$, *and positive scalars* τ, γ *such that the following LMIs are satisfied:*

$$Q_i + W \succ 0, \; i \in \mathcal{I}_N, \tag{1.33}$$

$$\Gamma_{\rm iikl} \prec 0, \; i \in \mathcal{I}_N, \; k \in \mathcal{I}_{N-1}, \; l \in \mathcal{I}_2, \tag{1.34}$$

$$\Gamma_{\rm ijkl} + \Gamma_{\rm jikl} \prec 0, \; i, j \in \mathcal{I}_N, \; i < j, \; k \in \mathcal{I}_{N-1}, \; l \in \mathcal{I}_2, \tag{1.35}$$

where

$$\Gamma_{\rm ijkl} = \begin{bmatrix} \Upsilon_{kl} - {\rm He}\mathcal{A}_{\rm ij} & \star & \star & \star \\ Q_i + M - \tau \mathcal{A}_{\rm ij} & \tau(M + M^\top) & \star & \star \\ E_i^\top & \tau E_i^\top & -\gamma I & \star \\ C_i M^\top & 0 & F_i & -\gamma I \end{bmatrix},$$

$$\mathcal{A}_{\rm ij} = A_i M^\top + B_i Y_j, \; \Upsilon_{\rm kl} = \phi_{\rm kl}(Q_k + W - Q_N) + \phi_{\rm Nl} W.$$

Then, the closed-loop LPV system (1.27) *is poly-quadratically stable and the* $\mathcal{L}_2-$*gain of the transfer* $w \to y$ *is smaller than* γ, *for all* $(\theta, \dot{\theta}) \in \mathcal{S}_\theta \times \mathcal{S}_\upsilon$. *Moreover, the control feedback gains in* (1.26) *are defined as*

$$K_i = Y_i M^{-\top}, \quad i \in \mathcal{I}_N.$$

Note that condition (1.33) guarantees a proper choice of poly-quadratic Lyapunov function candidates. Conditions (1.34)–(1.35) guarantee the poly-quadratically stability of the closed-loop LPV system (1.27) and the $\mathcal{L}_2-$gain performance. The proof of Theorem 1.4 follows similar steps as for the quadratic control results in Theorem 1.3. Here, the key difference is that the $S-$variable approach is exploited in Theorem 1.4 to introduce the slack decision variable M into the design, enabling the decoupling between the Lyapunov matrices $P_i + X$ and the control gain matrices K_i, for $i \in \mathcal{I}_N$. This allows not only convexifying the design conditions but also reducing further the design conservatism.

Remark 1.6 As for the LPV stability analysis, in terms of conservatism relaxation, we can theoretically prove that the control result in Theorem 1.4 includes precisely that in Theorem 1.3. Indeed, this can be done by selecting $W = 0$, $M = Q_i = Q$, for $i \in \mathcal{I}_N$, and a *sufficiently* small scalar $\tau > 0$.

Remark 1.7 One of the main sources of conservatism for the control result in Theorem 1.4 is that the slack variable M is *parameter-independent*. This can be solved by ying the control law (1.26) as

$$u = K(\theta)M(\theta)^{-1}x, \tag{1.36}$$

where $M(\theta) = \sum_{i=1}^{N} \eta_i(\theta)M_i$ and $M_i \in \mathbb{R}^{n_x \times n_x}$. Compared to the widely-used control structure (1.26), controller (1.36) allows avoiding some special structure of matrix decision variables, leading to less conservative design conditions [48, 61]. However, note that such a nonpolytopic control law requires a real-time inversion of a parameter-dependent matrix, which may induce numerical difficulties for control implementation, especially when the number of LPV sub-models N becomes significantly large.

1.2.6 Multiple Convex Summation Relaxation

For polytopic LPV control, in contrast to the stability analysis, the design conditions usually involve solving multiple convex summations, for instance the double convex summation (1.32). Due to the presence of the weighting functions, inequality (1.32) is an *infinite* LMI condition, which cannot be directly solved by numerical solvers. To convert (1.32) into a *finite* set of LMI constraints, the following usual decomposition can be performed [62]:

$$\begin{aligned}
&\sum_{i=1}^{N}\sum_{j=1}^{N} \eta_i(\theta)\eta_j(\theta)\Phi_{\text{ij}} \\
&= \sum_{i=1}^{N} \eta_i^2(\theta)\Phi_{ii} + \sum_{i=1}^{N}\sum_{j>i}^{N} \eta_i(\theta)\eta_j(\theta)(\Phi_{\text{ij}} + \Phi_{\text{ji}}).
\end{aligned} \tag{1.37}$$

Since $\eta(\theta) \in \Xi_\theta$, it is clear from (1.37) that conditions (1.28)–(1.29) are *sufficient* to guarantee (1.32). The *sufficiency* of multiple convex summation-based conditions may cause additional conservatism to the control design. Various approaches have been proposed to reduce this source of conservatism, for instance without introducing slack variables [62, 63], and with the use of slack variables [64, 65]. Among these approaches, Pólya's theorem-based relaxation [65] provides *asymptotically necessary and sufficient* LMI-based conditions to check the definite positiveness of multiple convex summations. However, such a relaxation result is only meaningful from the theoretical viewpoint since the computational burden *exponentially* increases with respect to the homogeneous degree of the summations [66].

1.2.7 Observer Design for LPV Systems

In many cases, the system states are unmeasurable or the measurement cost is not acceptable. As a result, the observer technique is developed, and the observer design for LPV systems is discussed in the section. Consider an LPV system given as follows:

$$
\begin{aligned}
\dot{x}(t) &= A(\rho)x(t) + B(\rho)u(t) + E(\rho)w(t) \\
&= \sum_{i=1}^{n} \theta_i(\rho(t))(A_i x(t) + B_i u(t) + E_i w(t)), \\
y(t) &= C(\rho)x(t) \\
&= \sum_{i=1}^{n} \theta_i(\rho(t))C_i x(t),
\end{aligned}
\tag{1.38}
$$

where $x(t), u(t)$, and $w(t)$ are the state vector, system input, and external disturbance, respectively. $y(t)$ is the measurable output. $A(\rho)$, $B(\rho)$, and $C(\rho)$ are the system matrix, input matrix, and output matrix, respectively. $E(\rho)$ is the given matrix with appropriate dimension. Based on the LPV system, a simple Luenberger observer is introduced:

$$
\begin{aligned}
\dot{\hat{x}}(t) &= A(\rho)\hat{x}(t) + B(\rho)u(t) + L(\rho)(y(t) - \hat{y}(t)), \\
\hat{y}(t) &= C(\rho)\hat{x}(t),
\end{aligned}
\tag{1.39}
$$

where $\hat{x}(t)$ and $\hat{y}(t)$ are the observed state vector and system output. $L(\rho)$ is the observer matrix needed to be determined. Defining the observation error as $e(t) = x(t) - \hat{x}(t)$, the observation error system could be obtained:

$$
\begin{aligned}
\dot{e}(t) &= \hat{A}(\rho)e(t) + E(\rho)w(t), \\
z(t) &= Fe(t),
\end{aligned}
\tag{1.40}
$$

where

$$
\begin{aligned}
\hat{A}(\rho) &= A(\rho) - L(\rho)C(\rho) = \sum_{i=1}^{n}\sum_{j=1}^{n} \theta_i(\rho(t))\theta_j(\rho(t))(A_i - L_j C_i), \\
F &= I_{n\times n}.
\end{aligned}
$$

Here, $z(t)$ is the controlled output.

Theorem 1.5 *Given a positive scalar γ, the observation error system (1.40) is asymptotically stable and the $\mathcal{H}_\infty$ performance is satisfied if there exist positive-definite symmetric matrix P, matrix K_j with appropriate dimension satisfying the following inequality:*

$$
\Theta_{\mathrm{ij}} + \Theta_{\mathrm{ji}} < 0, \quad 1 \le i \le j \le n, \tag{1.41}
$$

where

$$
\Theta_{\mathrm{ij}} = \begin{bmatrix} A_i^{\mathrm{T}} P + P A_i + C_i^{\mathrm{T}} K_j^{\mathrm{T}} + K_j C_i & P E_i & F^{\mathrm{T}} \\ * & -\gamma^2 I & 0 \\ * & * & -I \end{bmatrix}. \tag{1.42}
$$

Moreover, the gain-scheduled observer gain could be solved through $L_j = P^{-1}K_j$.

Proof Based on the Lyapunov stability theory, a candidate Lyapunov function is defined as follows:

$$\begin{aligned} V(e(t)) &= e^{\mathrm{T}}(t)Pe(t) \\ \dot{V}(e(t)) &= \dot{e}^{\mathrm{T}}(t)Pe(t) + e^{\mathrm{T}}(t)P\dot{e}(t) \\ &= \xi^{\mathrm{T}}(t)\Theta\xi(t), \end{aligned} \tag{1.43}$$

where

$$\begin{aligned} \xi(t) &= \begin{bmatrix} e(t) \\ w(t) \end{bmatrix}, \\ \Theta &= \begin{bmatrix} (A(\rho) + L(\rho)C(\rho))^{\mathrm{T}}P + P(A(\rho) + L(\rho)C(\rho)) & PE(\rho) \\ * & 0 \end{bmatrix}. \end{aligned}$$

Here, the Lyapunov function matrix P is a positive-definite symmetric matrix. It can been seen that the observation error system (1.40) is stable if $\dot{V}(e(t)) < 0$ is satisfied. In addition, an $\mathcal{H}_\infty$ performance is guaranteed if the following condition holds:

$$\begin{aligned} J &= \dot{V}(e(t)) + z^{\mathrm{T}}(t)z(t) - \gamma^2 w^{\mathrm{T}}(t)w(t) < 0 \\ &= \xi^{\mathrm{T}}(t)\Phi\xi(t) < 0, \end{aligned} \tag{1.44}$$

where

$$\Phi = \begin{bmatrix} (A(\rho) + L(\rho)C(\rho))^{\mathrm{T}}P + P(A(\rho) + L(\rho)C(\rho)) + F^{\mathrm{T}}F & PE(\rho) \\ * & -\gamma^2 I \end{bmatrix}. \tag{1.45}$$

By using Schur complement and the summation properties, the condition in (1.41) can guarantee that $J < 0$ is satisfied.□

It can be found from Theorem 1.5 that the $\mathcal{H}_\infty$ performance index γ indicates an attenuation level. Lower value of the performance index ensures better disturbance attenuation performance. As a result, the following corollary is introduced:

Corollary 1.1 *The minimum $\mathcal{H}_\infty$ performance index γ^* can be obtained by computing the following optimization problem:*

$$\begin{aligned} \gamma^* &= \min \gamma \\ &\text{s.t. } (1.41) \end{aligned}$$

1.2.8 Polytopic LPV Models and Takagi–Sugeno Models

Takagi–Sugeno (T-S) fuzzy model-based approaches have been known as an effective framework to deal with nonlinear systems [27, 66]. T-S fuzzy modeling, first proposed in [67], is expressed by fuzzy IF- THEN rules which represent local dynamics of nonlinear systems as

$$\begin{aligned}\text{Rule } R_i: \quad & \text{If } z_1 \text{ is } \mathcal{M}^i_1 \text{ and } \cdots \text{ and } z_p \text{ is } \mathcal{M}^i_p, \\ & \text{Then } \dot{x}(t) = A_i x + B_i u + E_i w,\end{aligned} \tag{1.46}$$

where R_i denotes the ith fuzzy inference rule, N is the number of inference rules, $\mathcal{M}^i_j$, with $i \in \mathcal{I}_N$ and $j \in \mathcal{I}_p$, are the fuzzy sets, and (A_i, B_i, E_i) the state-space matrices of appropriate dimensions of the ith local linear sub-model. The vector of premise variables is defined as $z = [z_1 \cdots z_p]^\top$. Using the center-of-gravity method for defuzzification, the T-S fuzzy model (1.46) can be rewritten in the compact form

$$\dot{x} = \sum_{i=1}^{N} h_i(z)(A_i x + B_i u + E_i w), \tag{1.47}$$

where the membership function $h_i(z)$ is defined as

$$h_i(z) = \frac{\omega_i(z)}{\sum_{i=1}^{N} \omega_i(z)}, \quad \omega_i(z) = \prod_{j=1}^{p} \mu^i_j(z_j), \quad i \in \mathcal{I}_N.$$

The grades of membership of the premise variables in the corresponding fuzzy sets $\mathcal{M}^i_j$ are given as $\mu^i_j(z_j)$. Note that the normalized membership functions satisfy the convex sum property as [27]:

$$0 \le h_i(z) \le 1, \quad \sum_{i=1}^{N} h_i(z) = 1, \quad \sum_{i=1}^{N} \dot{h}_i(z) = 0. \tag{1.48}$$

From their respective expressions that polytopic LPV system (1.8) and T-S fuzzy system (1.47) share some analogies [68]. Specifically, the scheduling parameter θ of LPV system (1.8) corresponds to the premise variable z of T-S system (1.47). Moreover, these systems are constructed by merging linear sub-models together with the weighting functions $\eta_i(\theta)$ for LPV systems or the membership functions $h_i(z)$ for T-S fuzzy systems. Hence, T-S fuzzy systems can be considered as polytopic quasi-LPV systems. Perhaps the major difference between polytopic LPV and T-S fuzzy systems consists in their historical backgrounds, i.e., robust control theory in LPV case and fuzzy theory in T-S case.

Due to the strong analogies, a large number of theoretical tools can be applied to both polytopic LPV systems and T-S fuzzy systems, especially Lyapunov method in conjunction with LMI techniques [28]. However, note that within T-S fuzzy framework the premise variables generally depend on the state vector. Then, the time derivatives of the membership functions, depending on the time derivative of the state, are generally not available for control design of system (1.47). This implies much more numerical and theoretical challenges when using poly-quadratic Lyapunov functions, also called fuzzy Lyapunov functions [69], for stability analysis and control design of continuous-time T-S fuzzy systems. Indeed, most of results are formulated using local analysis settings with different degrees of conservatism; see [66] for a recent discussion.

1.3 Applications to Vehicle Dynamics Control

This section provides some polytopic LPV-based results on vehicle dynamics control. Different suspension models, descriptions of vehicle lateral dynamics, and integrated vehicle models used for control design are briefly discussed. Then, polytopic LPV controller design methods with various settings of scheduled parameters are summarized.

1.3.1 Vehicle Dynamics

1.3.1.1 Vehicle Suspensions

The main functions of a well-designed suspension system can be summarized as follows.

- *Ride quality*. A vehicle suspension has a function of providing an isolation by decreasing forces transmitted from the vehicle axle to the vehicle body. The accelerations of the sprung mass are used as a performance indicator.
- *Suspension deflection limit*. An excessive suspension bottoming should be avoided. Hence, the suspension stroke should be constrained to a prescribed level.
- *Road holding*. The wheels and the road should contact uninterruptedly. Moreover, the dynamical tire load should not exceed the static one.

For the design of suspension systems, three categories of suspension models, including full-car models [70–74], half-car models [75, 76], and quarter-car models [77–85], have been widely used. According to different control strategies, the designed suspensions can also be divided as active suspensions [71, 72, 75–77, 79, 81, 82, 86–88] and semi-active suspensions [73, 74, 78, 80, 83, 84, 89, 90].

Active Suspension

Let us consider the half-car model depicted in Fig. 1.2. Assume that the structure of the vehicle is symmetrical, the pitch angle is small, and all the springs and dampers have linear dynamics. As a result, the half-car model can be described as follows [75]:

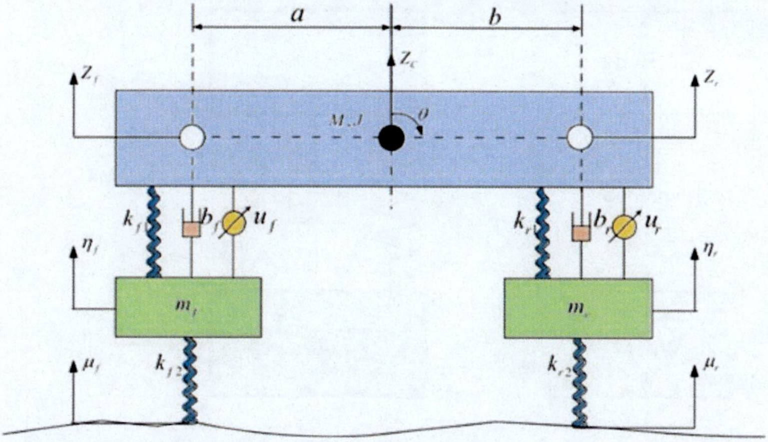

Fig. 1.2 Schematic diagram of a half-car active suspension system

$$
\begin{aligned}
M\ddot{z}_c &= f_f + f_r, \\
J\ddot{\theta} &= a f_f - b f_r, \\
m_f \ddot{z}_f &= -k_{f2}(\eta_f - \mu_f) - f_f, \\
m_r \ddot{z}_r &= -k_{r2}(\eta_r - \mu_r) - f_r,
\end{aligned}
$$

where M and J denote the sprung mass and the mass moment of inertia, m_f and m_r stand for the front and rear unsprung masses, respectively, u_f and u_r indicate the control forces of the active strategy, k_{f1} and k_{r1} represent the stiffness coefficients of the passive suspension elements, b_f and b_r are the damping coefficients of the passive suspension elements for the front and rear assembles, k_{f2} and k_{r2} represent the front and rear tire stiffness, and

$$
\begin{aligned}
f_f &= k_{f1}(\eta_f - z_c - a\theta) + b_f(\dot{\eta}_f - \dot{z}_c - a\dot{\theta}) + \mu_f, \\
f_r &= k_{r1}(\eta_r - z_c + b\theta) + b_r(\dot{\eta}_r - \dot{z}_c + b\dot{\theta}) + \mu_r.
\end{aligned}
\tag{1.49}
$$

The displacements at the front and rear wheels can be calculated as follows:

$$
z_f = z_c + a\theta, \; z_r = z_c - b\theta. \tag{1.50}
$$

Similarly, full-car models and quarter-car models can be developed [71, 77], which are omitted here for brevity.

Semi-Active Suspension

For semi-active suspension systems, as shown in Fig. 1.3 for a quarter-car suspension model, there is no component to produce the control force. Variable damper or other

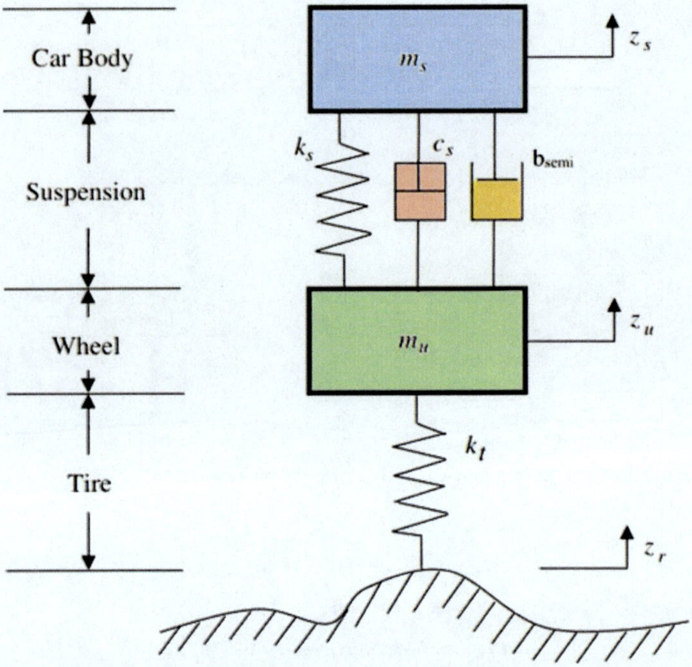

Fig. 1.3 Schematic diagram of a quarter-car semi-active suspension system

variable dissipation components are used to y the damping coefficient. Similar to active suspensions, a semi-active suspension system of a quarter-car can be modeled as

$$
\begin{aligned}
& m_s \ddot{z}_s + b_s(\dot{z}_s - \dot{z}_u) + k_s(z_s - z_u) = -b_{\text{semi}}(z_s - z_u), \\
& m_u \ddot{z}_u + c_s(\dot{z}_u - \dot{z}_s) + k_s(z_u - z_s) + k_t(z_u - z_r) \\
& = b_{\text{semi}}(\dot{z}_s - \dot{z}_u)
\end{aligned} \tag{1.51}
$$

where b_{semi} is the variable damper to be regulated.

1.3.1.2 Vehicle Lateral Dynamics

For the control of vehicle lateral dynamics, the classical bicycle model is mostly used, as shown in Fig. 1.4. A simplified bicycle model consists of two degrees of freedom, i.e., lateral and yaw motions, of which the dynamics can be described as

$$
\begin{aligned}
m V_x(\dot{\beta} + \gamma) &= F_{yf} + F_{yr}, \\
I_z \dot{\gamma} &= l_f F_{yf} - l_r F_{yr} + M_z,
\end{aligned} \tag{1.52}
$$

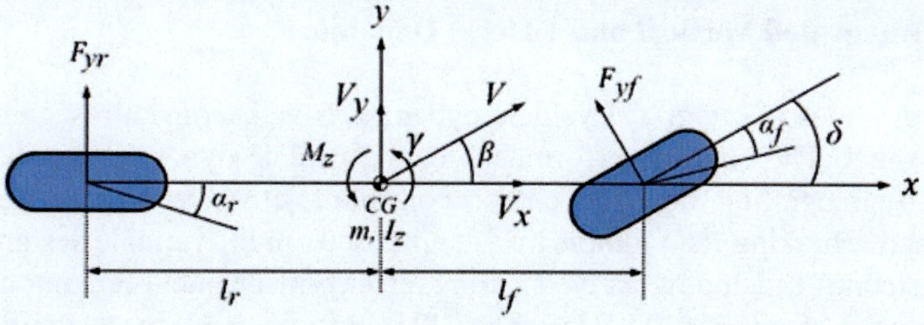

Fig. 1.4 Bicycle model for vehicle lateral dynamics

where m denotes the mass of the vehicle, V_x is the longitudinal vehicle velocity, β stands for the sideslip angle, γ represents the vehicle yaw rate, F_{yf} and F_{yr} denote the lateral tire forces of front and rear wheels, respectively. I_z stands for the moment of inertia around the vertical axis, l_f and l_r denote the distance between the center of gravity and the front and rear axis. In addition, M_z is the external yaw moment.

For normal driving conditions [91], the lateral tire forces in the vehicle model (1.52) can be modeled as

$$F_{\mathrm{yf}} = C_{\mathrm{yf}}\alpha_f, \quad F_{\mathrm{yr}} = C_{\mathrm{yr}}\alpha_r,$$

where C_{yf} and C_{yr} are the cornering stiffness of front and rear tire, respectively. The tire slip angles of the front and rear tires can be respectively represented as

$$\alpha_f = \delta - \frac{l_f\gamma}{V_x} - \beta, \quad \alpha_r = \frac{l_r\gamma}{V_x} - \beta.$$

Then, the vehicle lateral dynamics can be given by the following state-space model as

$$\dot{x} = Ax + B_1u + B_2w, \tag{1.53}$$

where $x = [\beta,\ \gamma]^\top$, $u = M_z$, $w = \delta$, and

$$A = \begin{bmatrix} -\frac{C_{\mathrm{yf}}+C_{\mathrm{yr}}}{mV_x} & \frac{C_{\mathrm{yr}}l_r - C_{\mathrm{yf}}l_f}{mV_x^2} - 1 \\ \frac{C_{\mathrm{yr}}l_r - C_{\mathrm{yf}}l_f}{I_z} & -\frac{C_{\mathrm{yf}}l_f^2 + C_{\mathrm{yr}}l_r^2}{I_zV_x} \end{bmatrix},$$

$$B_1 = \begin{bmatrix} 0 \\ \frac{1}{I_z} \end{bmatrix}, \quad B_2 = \begin{bmatrix} \frac{C_{\mathrm{yf}}}{mV_x} \\ \frac{C_{\mathrm{yf}}l_f}{I_z} \end{bmatrix}.$$

Note that, when the longitudinal-lateral dynamics coupling is taken into account for controller/observer design, it is possible that the lateral velocity V_y and the yaw rate γ are considered together with the longitudinal velocity V_x as the state variables to form the corresponding nonlinear vehicle models [92–94].

1.3.1.3 Integrated Vertical and Lateral Dynamics

The control issues of integrated vehicle dynamics have become a research hotspot in recent years. The control performance of the vehicle dynamics can be greatly improved by integrating the active chassis control of active steering, active suspension, and active braking. The authors in [95] proposed a multi-variable design strategy for chassis control including active steering, electro-mechanical braking actuators, and semi-active suspension. The designed LPV controller leads to an significant performance improvement in critical driving situations. Similarly, suspension systems were considered together with braking systems [86, 87, 90], steering systems [76, 86, 90, 96], and electronic stability program (ESP) systems [72, 87] to enhance the vehicle control performance.

Considering a quarter-car active suspension model together with a vehicle lateral dynamics model, we can obtain the following integrated vehicle model including vertical and lateral dynamics [76]:

$$\dot{x} = A(\rho)x + B_1 u + B_2 w, \tag{1.54}$$

where $x = \left[\beta,\ \gamma,\ \dot{z}_c,\ \dot{\theta},\ \dot{z}_f,\ \dot{z}_r,\ \dot{\eta}_f,\ \dot{\eta}_r,\ z_f,\ z_r,\ \eta_f,\ \eta_r\right]^\top$, $u = \left[\delta_u,\ u_f,\ u_r\right]^\top$, and $w = \left[\delta_c,\ \mu_f,\ \mu_r\right]^\top$. δ_c and δ_u denote the nominal steering angle and the assistant steering angle for the front wheel. The state-space matrices $A(\rho)$, B_1, and B_2 in (1.54) can be found in [76].

Another integrated vehicle control method consists in decomposing the global chassis control into two steps [95]. In the first step, the linear bicycle model is analyzed and a controller is designed to improve the vehicle lateral stability. Then, in the second step, a suspension controller is synthesized to enhance the vehicle vertical performance.

1.3.1.4 Integrated Lateral, Longitudinal, and Braking Dynamics

Taking into account the braking system, the vehicle dynamics system can be given as follows [97]:

$$\begin{bmatrix} \dot{x} \\ z \\ y \end{bmatrix} = \begin{bmatrix} A & B_1 & B_2 \\ C_1 & D_{11} & D_{12} \\ C_2 & 0 & 0 \end{bmatrix} \begin{bmatrix} x \\ w \\ u \end{bmatrix}, \tag{1.55}$$

where $x = \left[\beta,\ \gamma\right]^\top$, $w = \left[\gamma_d,\ \beta_d,\ M_{\mathrm{zd}}\right]^\top$ is the disturbance input, $u = \left[\delta,\ T_{\mathrm{brl}},\ T_{\mathrm{brr}}\right]^\top$ denotes the control input, $y = \left[\gamma,\ \beta\right]^\top$ stands for the measured output, and $z = \left[z_1,\ z_2,\ z_3,\ z_4\right]^\top$ represents the controlled output. Note that z_1 denotes the weighted yaw rate error output signal, z_2 stands for the weighted sideslip angle error output signal, z_3 represents the braking control signal attenuation, z_4 stands for the steering control signal attenuation. The state-space matrices A, B_1, B_2, C_1, D_{11}, D_{12}, and C_2 of the vehicle model (1.55) can be found in [97].

1.3.1.5 Observer-Based Vehicle Models

However, full-state information of vehicle systems are generally not available online due to the issue of sensor costs. For instance, the vehicle sideslip angle cannot be reliably measured with low-cost sensors in practice. Hence, LPV observers have been proposed to estimate important vehicle variables [35, 98, 99]. To design an LPV observer for estimating the vehicle sideslip angle, $\frac{1}{V_x}$ and $\frac{1}{V_x^2}$ were chosen as time-varying parameters of the LPV vehicle system in [98]. For observer design, a triangular polytope was derived to reduce the number of vertices of the polytopic LPV vehicle model. Using Lyapunov stability method, LMI-based design conditions were developed with an augmented vehicle model, including the estimation error dynamics and the vehicle lateral dynamics. Moreover, an energy-to-peak gain specification was taken into account in the observer design to improve the estimation performance under unknown disturbances. Similarly, a robust sideslip angle observer was proposed in [99] for electric ground vehicles based on a lateral vehicle model and the measurement of the yaw rate. The uncertain tire characteristics were taken into account in the observer design. Since the tire cornering stiffness and the vehicle inertial moment cannot be easily measured in practice, a parameter identification procedure was provided with experimental data. The vehicle state and the driver torque are simultaneously estimated in [35] using an LPV observer with an unknown input. To derive the polytopic LPV vehicle model, $\frac{1}{V_x}$ was chosen as the scheduling variable and $\frac{1}{V_x^2}$ is approximated via the first-order Taylor's approximation as

$$\frac{1}{V_x} = \frac{1}{V_0} + \frac{1}{V_1}\theta, \quad \frac{1}{V_x^2} = \frac{1}{V_0^2}\left(1 + 2\frac{V_0}{V_1}\theta\right), \tag{1.56}$$

where the parameter θ is used to describe the variation of V_x between its lower bound $V_{\min}$ and upper bound $V_{\max}$ with $-1 \leq \theta \leq 1$. The terms V_0 and V_1 in (1.56) are given by

$$V_0 = \frac{2V_{\min}V_{\max}}{V_{\min} + V_{\max}}, \quad V_1 = \frac{-2V_{\min}V_{\max}}{V_{\max} - V_{\min}}. \tag{1.57}$$

Based on lateral-longitudinal integrated dynamics, a nonlinear LPV unknown input observer was developed in [94] for a *simultaneous* estimation of the lateral speed, the steering input, and the effective engine torque. The speed-related term $\frac{1}{V_x}$ and the yaw rate γ were used for gain-scheduling purposes.

1.3.2 Choices of Scheduling Parameters for LPV Control

The aforementioned vehicle models can be represented in affine/polytopic LPV forms. Different affine/polytopic LPV models have been established to handle various vehicle control and/or estimation problems. Depending on the considered time-

Table 1.1 LPV approaches for vehicle dynamics control and estimation

Time-varying parameters	Suspension model	Lateral model
Vehicle velocity	[75]	[100–115]
Deflection-related parameters	[70, 71, 77, 84]	–
Braking-related parameters	–	–
Cornering stiffness	–	[120]
Other choices of parameters	[73, 74, 78–81, 83]	[108, 112, 114, 115, 121, 122]
Time-varying parameters	Integrated model	Observer design
Vehicle velocity	[76, 116–118]	[35, 98, 99]
Deflection-related parameters	[95]	–
Braking-related parameters	[86, 87, 90, 95, 119]	–
Cornering stiffness	–	–
Other choices of parameters	[72, 96, 123–125]	[74, 88, 89]

varying parameters, the vehicle models can be reformulated in a general form as follows:

$$\dot{x} = A(\rho)x + B_1(\rho)u + B_2(\rho)w, \tag{1.58}$$

where $A(\rho)$, $B_1(\rho)$, and $B_2(\rho)$ are state-space matrices obtained according to the time-varying variables of vehicle systems. Table 1.1 summarizes the different choices of time-varying variables in vehicles dynamics control and estimation, discussed hereafter.

1.3.2.1 Time-Varying Velocity

Since the vehicle velocity changes during cruising in practice, its related parameters could be regarded as scheduling parameters [75, 76, 108, 116, 117, 126]. In addition, the vehicle lateral acceleration can be also chosen as a scheduling parameter [93].

Since the vehicle velocity is inherently time-varying, the parameters $\rho_1 = V_x$, $\rho_2 = \frac{1}{V_x}$, $\rho_3 = \frac{1}{V_x^2}$ are considered as scheduling parameters in [76]. Then, an integrated vehicle dynamics control strategy is proposed to enhance the vehicle safety and handling performance by combining active front steering and active suspension systems. A polytopic LPV vehicle model with finite vertices, affinely depending on the time-varying longitudinal speed, is built. Then, an $\mathcal{H}_\infty$ gain-scheduled controller is designed using LMI-based techniques and Lyapunov stability arguments. Considering $\rho_1 = V_x$ and $\rho_2 = \sqrt{V_x}$ as scheduling parameters, a velocity-dependent multi-objective LPV control method is presented in [75] to solve the preview control problem with velocity uncertainty. Similarly, $\rho = V_x$ is considered as a scheduling parameter for the LPV control design in [116]. In this work, a variable geometry suspension and a robust suspension control are both considered to enhance the vehicle stability.

The velocity-related terms $\frac{1}{V_x}$ and $\frac{1}{V_x^2}$ are considered as scheduling parameters in [100–107, 127]. With different definitions of vehicle state vectors, $\rho_1 = V_x, \rho_2 = \frac{1}{V_x}$ and $\rho_1 = V_x, \rho_2 = \frac{1}{V_x}, \rho_3 = \frac{1}{V_x^2}$ are adopted as the scheduling parameters in [109] and [118], respectively. In these polytopic LPV settings, the constructed polytope would have 2^n vertexes, where n is the number of scheduling parameters. For example, if the $\rho_1 = \frac{1}{V_x}, \rho_2 = \frac{1}{V_x^2}$, then a quadratic polytope will be obtained as depicted in Fig. 1.5a. However, since the varying parameters are velocity-related, which are not independently variant, some methods are proposed to shrink the polytope in order to reduce the condition conservatism. The trapezoidal polytope in [103, 106, 118], triangle polytope in [100], and two-vertices form in [101, 104, 127] are proposed as shown in Fig. 1.5b–d, where the two vertices are given as

$$\bar{\Omega}_1 = \left(\frac{1}{\overline{V}_x}, \frac{7\underline{V}_x^2 + 2\underline{V}_x\overline{V}_x - \overline{V}_x^2}{8\underline{V}_x^2\overline{V}_x^2} + \frac{(\overline{V}_x - \underline{V}_x)^2}{8\underline{V}_x^2\overline{V}_x^2} N_2(t) \right),$$
$$\bar{\Omega}_2 = \left(\frac{1}{\underline{V}_x}, \frac{7\overline{V}_x^2 + 2\underline{V}_x\overline{V}_x - \overline{V}_x^2}{8\underline{V}_x^2\overline{V}_x^2} + \frac{(\overline{V}_x - \underline{V}_x)^2}{8\underline{V}_x^2\overline{V}_x^2} N_2(t) \right),$$

with $|N_2(t)| \le 1$. Based on the constructed LPV models, $\mathcal{H}_\infty$ control [109], $\mathcal{H}_\infty$ control in the $\mu-$split problem [118], multi-objective energy-to-peak control design with $\mathcal{D}-$stability [101], double layer control strategy considering a tire-force saturation [100, 102], $\mathcal{H}_\infty$ state-feedback with state delay [105], $\mathcal{H}_\infty$ output-feedback control [106], have been investigated.

Apart from the above polytopic reductions, another two-vertices formulation could be found in [49, 109, 126]. To this end, a change of time-varying parameter is performed for $\frac{1}{V_x}$ together with a first-order Taylor approximation as in (1.56).

1.3.2.2 Suspension-Deflection-Related Parameters

Due to the motion of suspension systems, the suspension deflection and its velocity could be regarded as time-varying parameters [70, 77, 95]. Moreover, the authors in [71, 77] also consider the influences of some other varying parameters besides the suspension deflection. For instance, the variations in the suspension deflection and mass are defined as scheduling parameters to form an LPV model in [71]. The resulting LPV controller enables the vehicle suspension systems to prevent from hitting their structural limits. In [77], the suspension deflection and the time-varying parameter representing the road conditions are used for gain-scheduled control purposes. The derived LPV controller allows minimizing either the acceleration or the suspension deflection, directly depending on the magnitude of the suspension deflection.

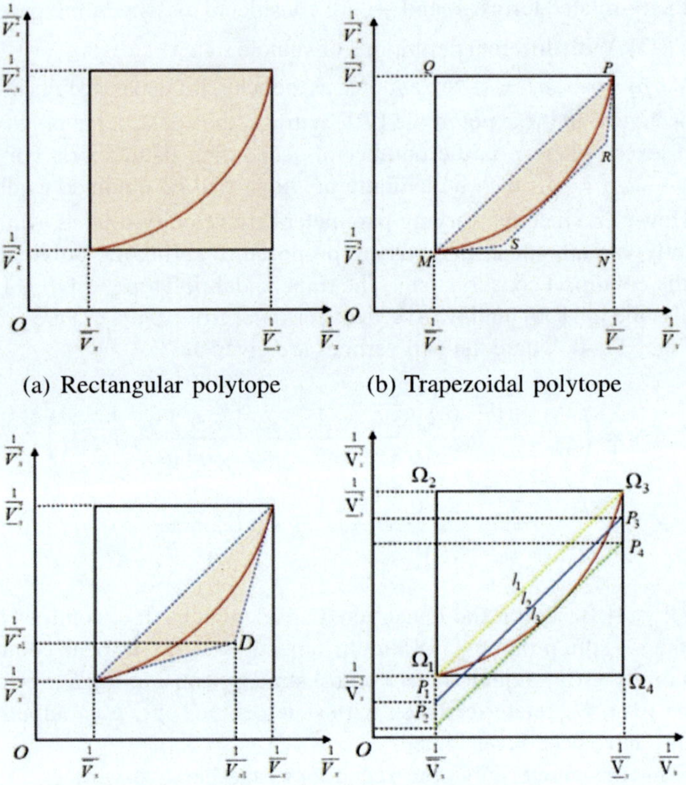

(a) Rectangular polytope (b) Trapezoidal polytope

(c) Triangle polytope (d) Two vertices

Fig. 1.5 Different types of polytopes representing $\left[\frac{1}{V_x}, \frac{1}{V_x^2}\right]$

1.3.2.3 Braking-Monitor-Related Parameters

The braking monitor parameters are considered as time-varying parameters in [86, 87, 95]. In [95], an LPV control strategy is proposed to enhance the vehicle performance in critical driving situations. The scheduling parameters R_b (braking monitor) and R_s (suspension and steering monitor) are used to analyze the system and design three controllers for steering, braking, and semi-active suspension. The parameters R_b and R_s are also chosen as the scheduling parameters in [87]. In the work, the proposed control strategy leads to the situation-dependent objectives in a unified framework. Due to the adaption of R_b and R_s parameters, the control performance is smooth, while satisfying internal stability and minimizing an $\mathcal{L}_2-$gain performance.

Furthermore, a new $\mathcal{H}_\infty$ LPV coordination strategy which aims to improve the vehicle stability using active steering, suspension, and electro-mechanical braking actuators was proposed in [86]. The main idea of this coordination technique is to tune the suspensions in the four corners and to improve the vertical performance by measuring the load transfer distribution of the vehicle while the vehicle is running on irregular road. The results showed that the proposed strategy has a good coordination between braking and steering actuators.

1.3.2.4 Braking-Efficiency-Related Parameters

The braking efficiency and its related weighting function were considered as scheduling parameters in [90, 97, 119, 123–125]. Some of these works have considered the actuator coordination with different parameters. As for [97, 123, 125], to coordinate the steering and braking actuators in different situations, a weighting function-based approach has been used. However, there are some differences among them. The authors in [97, 123] constructed two parameters ρ_1 and ρ_2 to continuously (de)activate the steering and the braking actions. Moreover, the activated rear braking actuator was selected according to the values of ρ_1 and ρ_2 (either 0 or 1). The key difference is that ρ_1 in [97] can only be 0 or 1 while this parameter can be continuously changed between 0 and 1 in [123], allowing for a smoother activation of the steering actuator. The authors in [125] constructed a weighting function of the braking control signal according to a scheduling variable $\rho \in [\underline{\rho}, \overline{\rho}]$. When ρ increases, the braking input is penalized. On the contrary, if ρ decreases, no penalization is performed for the braking control signal.

Different from the three previous papers, the dynamics and the effects of the suspension were also considered in [90, 119, 124]. Similar to the work in [125], the authors in [119] also used a weighting function to achieve the steering–braking coordination. However, the difference is that a function of the braking efficiency according to the scheduling variable $\xi \in [\underline{\xi}, \overline{\xi}]$ was used in [119]. Then, depending on a larger or a smaller value of ξ, the steering input is penalized or not. The authors in [90] proposed two scheduling parameters: $R_b \in [0, 1]$ (braking) and $R_s \in [0, 1]$ (suspension and steering). When the values of these two parameters *gradually* decreases from 1 to 0, the system transits from a normal situation to an intermediate or even a critical situation. The problem of fault tolerant control of a semi-active suspension system was considered in [124]. Using a time-varying parameter $\alpha \in [0, 1]$, estimated by a fault detection and diagnosis strategy, the authors proposed to online adjust the semi-active damper in case of leakage.

Based on the constructed LPV models with different choices of time-varying scheduling parameters, robust $\mathcal{H}_\infty$ gain-scheduled control methods were studied in [97, 123, 125], and $\mathcal{H}_\infty$ LPV global chassis control methods considering the suspension dynamics were investigated in [90, 119, 124].

1.3.2.5 Other Choices of Scheduling Parameters

Apart from the varying longitudinal velocity, the varying lateral velocity, varying yaw rate, and the difference between desired and actual transmitted yaw torque are also considered in [110–113], respectively. The fault-tolerant control is proposed with $\rho_1 = V_x, \rho_2 = V_y, \rho_3 = \frac{1}{V_x}, \rho_4 = \frac{V_y}{V_x^2}$; $\rho_1 = V_x, \rho_2 = \frac{1}{V_x}, \rho_3 = r$, respectively. In [120], the cornering stiffness C_{yf} and C_{yr} are chosen as the varying parameters and the four-vertices rectangular polytope is used to describe them, and a state-feedback multi-objective control is proposed to enhance the vehicle lateral stability. In [114, 121], the tire slip angles α_f and α_r are taken into account as varying parameters, but the difference is [114] also considering the longitudinal velocity V_x as a varying parameter, [121] considering the tire slip angles $\alpha_f(\lambda)$, $\alpha_r(\lambda)$ as a function of adhesion coefficient λ. Moreover, the authors in [115] consider the steering angle δ as a scheduling parameter. However, to avoid the singular point when $\delta \to 0$, a change of variable has been done by shifting the δ interval: $\delta \in [\underline{\delta}, \overline{\delta}] \to \sigma \in [\underline{\delta} + \varepsilon, \overline{\delta} - \varepsilon]$, converting σ into the new scheduling variable and being ε a constant value greater than δ.

1.4 Applications to Autonomous Vehicles

Besides the traditional vehicle dynamics control, the LPV techniques have successfully applied to the path following control problem [128–130] and lateral tracking control [131] for autonomous vehicles. The path following control problem is one of the fundamental challenges for the development of autonomous vehicles. Different from the vehicle dynamics control, there are vehicle-road model, preview model, vehicle dynamics model, and steering system model for the path following problem. Therefore, the design is more challenging.

The authors in [132] studied the path following problem of autonomous ground vehicles via output-feedback control and robust $\mathcal{H}_\infty$ technique. Both the variations of the longitudinal velocity and the cornering stiffness were considered. Based on similar model in [132], the authors in [133] investigated the network-induced delay and data dropouts in path following control problem. The tracking performance can be guaranteed when there is a bounded delay in the steering control input. The research team in [134] and [135] considered different driver steering characteristics for an advanced driver assistant system belonging to a low-level autopilot. The authors in [136] developed a specific LPV lateral motion model of unmanned ground vehicles and proposed a robust gain-scheduled automatic steering controller design method to exploit with the time-varying velocity and external disturbance. With friction force estimation and compensation mechanism, the authors in [137] designed gain-scheduling controller for autonomous vehicles. The model predictive approach was employed to act as a path planner in [138]. Based on the planned trajectory, a tracking controller was designed with the LPV technique. The input saturation constraint was considered for steering control design in [139]. An output feedback gain-scheduled control design was proposed for the path following of autonomous ground vehicles [140], for which the closed-loop transient performance can be improved via the concept of $\mathcal{D}-$stability. In the work [141], the autonomous vehicle has a four-wheel steering system and a four-wheel driving system. A robust $\mathcal{H}_\infty$ controller was designed via the LPV technique for the specific vehicle. The gain-scheduled composite nonlinear feedback control method was utilized to facilitate an impaired driver assistance system to achieve better trajectory tracking performance in [142].

1.5 Applications to Vehicular Powertrain Systems

The powertrain system in which the main function is to convert the power to movement of rotating wheels is one of the most important parts for vehicles. In terms of the power source, there are engine-powered vehicles, motor-powered vehicles, and hybrid vehicles. The powertrain of an engine-powered vehicle generally consists of the engine, clutch, transmission, drive shaft, final drive, and wheels. The powertrain of a motor-powered vehicle is composed of a motor, a motor drive system, a DC–DC converter, a transmission system, and a battery system. In a hybrid vehicle, the vehicle is driven by an electrical motor and an internal combustion (IC) engine. Though the powertrain systems are different, the nonlinearity in the model is a common challenge for powertrain control design. The review of LPV-technique applications on vehicular powertrain systems focuses on the control and estimation of typical powertrain components such as engines, aftertreatment systems, and electric vehicles.

1.5.1 Internal Combustion Engines

The control design for IC engines is known as a challenging problem due to the modeling complexities and the involved nonlinearities [143]. Up to now, many control strategies have been proposed for IC engines. Especially, LPV techniques have been broadly applied in the last decades [25, 144–146]. The simplified diagram of a powertrain system in IC engines is described in Fig. 1.6.

For gasoline engines, LPV technique has been mainly applied to the air–fuel ratio control problem. Feng et al. designed an air–fuel feedback controller by applying LPV control for a spark ignition (SI) engine [144]. This chapter aims to solve the problems of variable time delay in the system and thus maximize the total fuel

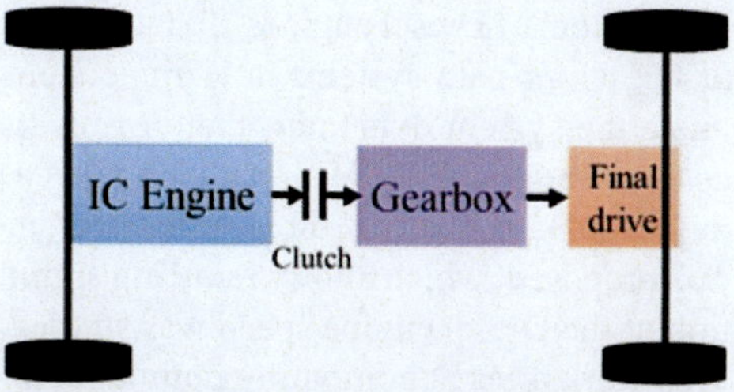

Fig. 1.6 Schematic diagram of a powertrain system in IC engines

economy. Marius et al. constructed an LPV model and developed a gain-scheduled strategy to control the air–fuel ratio in SI engines [147]. In this controller, the engine speed and air flow were regarded as the scheduling parameters to achieve the optimal control. A switching LPV controller was proposed in [148] for air–fuel ratio control of SI engines. The authors represented the system dynamics as a first-order LPV model, which is effective to describe the change of the engine operating points. Then, the control problem was solved through LMI constraints, and the performance of the proposed LPV controller was demonstrated under various simulation tests. Furthermore, LPV control technique was also used for air charge control of gasoline engines in [149]. Abbas et al. first built a quasi-LPV model based on neural state-space model, then developed a discrete-time LPV controller to carry out the air charge control for an SI engine. Similarly, the optimization problem was solved based on LMI constraints and an evolutionary search. A fixed-structure LPV control was also proposed for air charge control of SI engines in [145]. For the air charge control, the authors combined an affine LPV model together with a gain-scheduled PID controller, which allows achieving a good control performance in real-time experiments. An LPV input–output model was proposed in [150] to represent the nonlinear dynamics of air path systems in turbocharged SI engines. Then, LPV control technique was applied to perform the engine charge control. Except for the air–fuel ratio control and the air charge control, the wastegate control was investigated for turbocharged SI engines in [151]. Quasi-LPV modeling was explored to describe the engine nonlinear dynamics, then an internal model control design was proposed to achieve the boost-pressure tracking. LPV control technique was also employed to implement the total engine optimization for SI engines [152]. In this work, a hierarchical control structure, including both upper-level and lower-level controllers, was proposed. Model predictive control framework was adopted as the upper-level controller, and the lower-level controller was designed using LPV control. Both simulations and real track tests were presented to demonstrate the superiority of the developed method.

LPV control methods have been also exploited to address different control problems in Diesel engines. LPV techniques were used to deal with the modeling and control issues of air path systems in Diesel engines [153, 154]. The authors developed a data-driven gray-box model based on quasi-LPV framework to represent the system behaviors, and further designed a gain-scheduled $\mathcal{H}_\infty$ controller to improve the tracking performance. The experimental results showed that the proposed controller is able to control effectively the transient exhaust gas fraction. Based on a physical model, Lihua et al. [155] investigated a quasi-LPV control method to manage the air path system of Diesel engines. To tackle the challenge due to the strong nonlinearities of Diesel air path systems, a Hammerstein quasi-LPV model was used to approximate the system dynamics, then a gain-scheduled law was considered for control design. Besides, an LPV air path model of a turbocharged Diesel engines was studied in [156]. Based on this three-order LPV model, a robust gain-scheduled controller was proposed, which allows reducing significantly the calibration effort. The regulation of the Diesel engine speed was studied through LPV control approach in [157]. This work aims at computing optimal control inputs under fast operating condition changes of Diesel engines and variable transport delays. The performance of the designed controller was validated with hardware-in-the-loop tests. Boost-pressure control is also an important issue in Diesel engines. The authors in [158] explored LPV technique to address the control problem of air path systems. The performance of the proposed LPV method was experimentally validated with an BMW Diesel engine. In [159], the control problem of a common rail injection system was investigated via an $\mathcal{H}_\infty$ LPV control framework. The main contribution of this work was the system modeling through LFT representation under time-varying engine speed, rail pressure, and fuel temperature. Fresh air fraction control was promoted based on LPV hyperbolic systems for Diesel engines in [160]. By stabilizing the LPV hyperbolic system with boundary conditions, the optimal air mass fraction can be obtained, which leads to a good control performance for Diesel engines.

Although most of the works concerning LPV techniques in conventional engines are related to controller design, some investigations on LPV estimation were also carried out in past decades. Zhiyuan et al. proposed an LPV adaptive observer to jointly estimate the system states and parameters in a Diesel engine [161]. This observer is able to compensate the mass air flow sensor error and update an error map online. Finally, the simulation results proved the developed observer is quite qualified for the sensor error compensation. LPV modeling was also applied for fault detection and isolation (FDI) in turbocharged SI engines [162]. The LPV model in this work was mainly used to approximate the nonlinear dynamics of SI engines. Then, $\mathcal{H}_\infty$ Luenberger observers were developed to construct the FDI architecture. Lastly, numerical simulations showed the effectiveness of the proposed LPV model and the designed observers.

1.5.2 Electric Vehicles

Considering the environmental problems, electric vehicles including pure electric vehicles and hybrid electric vehicles (HEVs) would dominate the future worldwide market of ground vehicles, and traditional vehicles equipped with IC engines would be gradually eliminated. For electric vehicles, the powertrain control has been always a hot research topic. In particular, LPV control methods for electric powertrain systems have attracted increasing attention.

The scholars from Hamburg University of Technology focused on the investigation of torque vectoring in electric vehicles using LPV techniques and published several valuable papers [163, 164]. In their work, LPV gain-scheduled controllers were developed based on quadratic Lyapunov functions to track the longitudinal velocity and yaw rate of the vehicle. Besides, torque and slip limiters were considered to address the physical saturation problem in electric motors and wheel slip constraints in the control design. Energy management using LPV control technique is also an appealing idea in electric vehicles. Waleed et al. mainly studied the energy management of electric vehicles with variable power source-fuel cell, battery, and ultra-capacitor [165–167]. An $\mathcal{H}_\infty$ gain-scheduled controller associated to weighting functions was proposed to regulate the fuel cell current and the converter voltage, and to further achieve the optimal power output of the electrical motors. Additionally, the weighting functions in LPV control were selected through a genetic algorithm. Different driving cycles conducted in MATLAB/Simulink showed the good performance of designed controller. LPV strategy was adopted to control the permanent magnet synchronous motor (PMSM) for electric vehicles in [168]. A parameter-dependent Lyapunov function was employed for LPV control design to guarantee a robust stability under time-varying parameters. The simulation results illustrated that the controller could work well in d-q rotating frames.

Four-wheel independently actuated (FWIA) electric vehicle has attracted increasing attention from researchers due to the actuation flexibility. As FWIA electric vehicle is driven through in-wheel motors, the torque control of FWIA electric vehicle is regarded as a type of powertrain control. LPV control is widely used for FWIA electric vehicle, for example, Wang et al. developed an LPV control strategy to ensure the stability and improve the handling of FWIA vehicles [111]. In this article, a novel linear quadratic regulator (LQR)-based controller considering different actuator faults was designed by using LPV techniques. Through the LQR-based LPV control, the negative effect result from actuator faults and external disturbances is minimized. Similarly, robust fault-tolerant control based on LPV techniques was proposed to perform the trajectory tracking control of FWIA vehicles [169]. When faults occurred, the LPV control is used to reallocate the control inputs to stabilize the vehicle. During normal work conditions, the LPV control is able to select energy efficient control actions. LPV-based robust controller was designed to deal with modeling inaccuracies and uncertainties, and further track the desired external yaw rate [170]. Besides, the tire force constraints were also taken into account. A two-degree-of-freedom LPV controller was investigated to implement traction control in independent in-wheel motor electric vehicle [171]. The simulations proved that the proposed LPV control worked properly under limited wheel slip ratio. Reconfigurable control based on LPV methods was developed to achieve velocity and path following for FWIA electric vehicles [172]. The main novelty of reconfigurable control is the accurate torque control and estimation.

HEVs have occupied an important position in the vehicle market these years, and many control strategies comprising LPV control are proposed to deal with the key problems of powertrain in HEVs. The schematic diagram of a typical powertrain system in HEVs is depicted in Fig. 1.7. Firstly, energy management is one of the hottest topics in HEVs. For instance, Wang et al. put forward a discrete-time LPV controller based on LFT to minimize the energy consumption in a parallel HEV [173]. The LPV control scheduled the gain with varying parameters, and split the usage of engine and battery successfully. Nevertheless, a two-layer LPV control method was developed for energy management of an HEV [174]. The authors aimed at determining the optimal control actions under inputs and state constraints and time-varying parameters. Model predictive control with LPV model was designed in [175] to improve the fuel economy of HEVs. The vehicle speed was considered as the varying parameter in LPV model. The effectiveness of the developed controller was verified with a MATLAB/SIMULINK simulation platform. Torque ripple reduction is another control

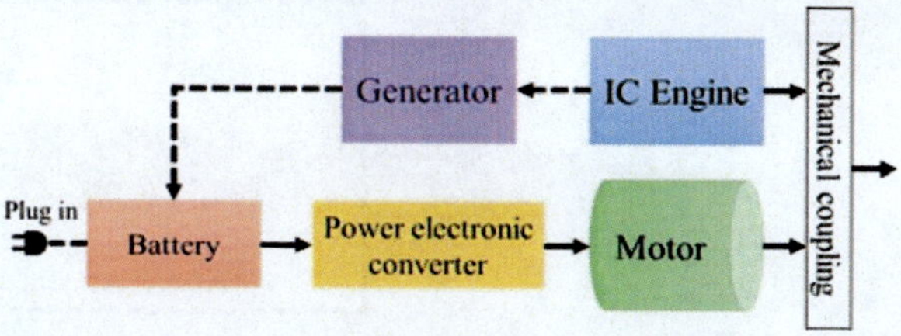

Fig. 1.7 Schematic diagram of a typical powertrain system in HEVs [180]

issue of HEVs. The authors in [176] presented an LPV control framework to reduce the torque ripple in HEVs with Diesel engine and IC engine, respectively. In these works, the rotation speed was selected as a time-varying parameter, and internal model principle concerning multi-sinusoidal persistent disturbances was employed to enhance the LPV control performance. Some researches about the control of series HEVs have been also carried out through LPV control framework. The authors in [177] studied the control of a Diesel auxiliary power unit (APU) in an HEV. The nonlinear system, including a Diesel engine, a synchronous generator, and a three-phase diode rectifier, was represented by a quasi-LPV model. Then, a robust LPV control framework was used to solve the control problem. A reduced-order robust LPV control was introduced in [178] for APU control in a series HEV. In that paper, the nonlinear system was formed by a simple LPV model with parametric uncertainties, while the reduced-order LPV controller was developed based on LMI-based optimization. Finally, the effectiveness of the proposed gain-scheduled controller was illustrated with simulation results. In [179], a two-degree-of-freedom LPV controller was put forward to implement the torque vectoring in an HEV. The designed controller associated to an anti-windup control scheme could guaranteed the system stability and trajectory tracking performance.

In summary, LPV techniques are mainly applied to the control issues of electric vehicles powertrain, and only few works related to estimation. A robust LPV observer was proposed to estimate the thermally derated torque for HEVs in [181]. As the temperature variations would deteriorate control effect, it's necessary to estimate the thermally derated torque performance. Finally, the shortened FUDS test cycle was conducted to validate the designed observer. On the basis of the work in [181], the authors continued to develop an LPV controller to manage the thermally derated torque in HEV [182]. Besides, feedback field-oriented control was utilized for torque control as well. The federal urban driving test cycle was carried out, and the results demonstrated the superiority of proposed observer and controller.

1.5.3 *Aftertreatment Systems*

Diesel engine aftertreatment system is an important component for vehicle powertrain systems, which is usually used to eliminate the NO_x and the particulate matter (PM) in Diesel engine emissions. The aftertreatment system comprised of three parts:

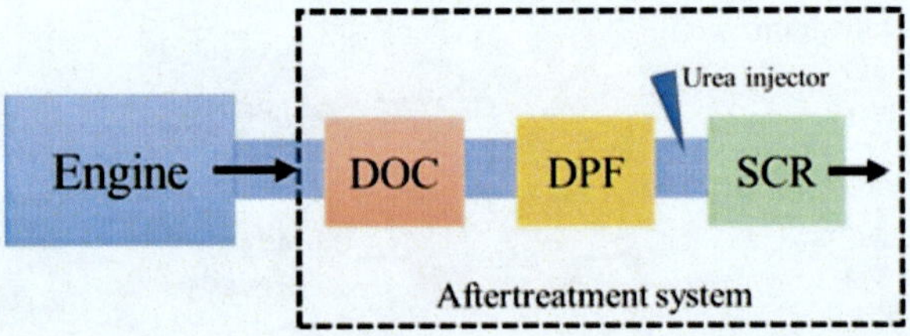

Fig. 1.8 Schematic diagram of a typical aftertreatment system for Diesel engines. DOC refers to Diesel oxidation catalyst, DPF is the abbreviation of Diesel particulate filter, SCR is selective catalytic reduction (SCR) system

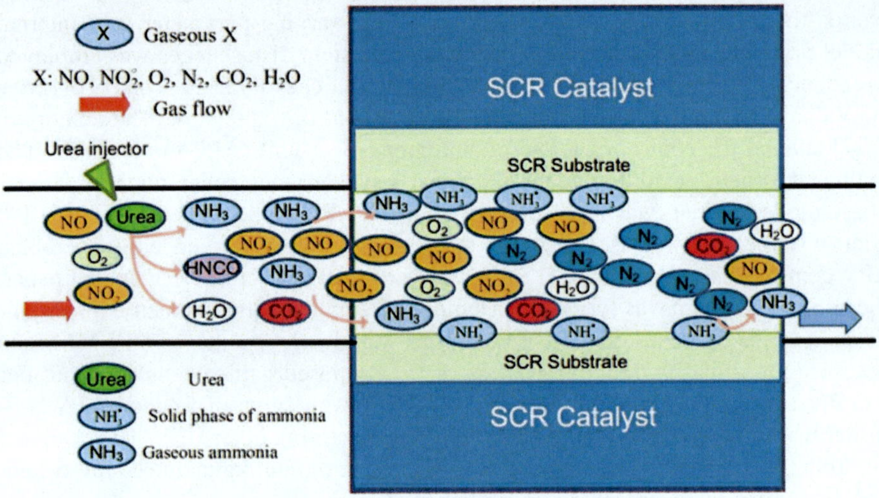

Fig. 1.9 Principle of a urea-based SCR system for diesel engines [183]

Diesel oxidation catalyst (DOC), Diesel particulate filter (DPF), and selective catalytic reduction (SCR) system. The schematic diagram of a typical aftertreatment system for current Diesel engines is shown in Fig. 1.8. The main function of DOC component is to convert CO to CO_2 and convert hydrocarbons to H_2O and CO_2. The DPF component is used to capture the particulate matters while the SCR system is employed to reduce the NO_x emissions.

Numerous model-based control strategies have been proposed for SCR systems to reduce the NO_x emissions and to constrain the ammonia slips, simultaneously. The principle of an SCR system is illustrated in Fig. 1.9. As reported in [183], considering the kinetic dynamics, the three-state SCR model can be established. Note that there are strong nonlinearities involved in this SCR model. Moan et al. proposed an adaptive LPV control strategy to simultaneously minimize the emission and the ammonia slip in urea-SCR systems [184]. In this work, a three-state nonlinear model with time-varying parameters was utilized. To improve the performance of the LPV controller, the authors in [184] proposed a quasi-LPV model to represent the complex nonlinear dynamics of urea-SCR systems [185, 186]. Moreover, the technique of principal component analysis (SPCA) was employed to reduce the complexity of the LPV model and the computational load for micro-control implementation. At last, performance comparisons between controllers based on a low-order model and a high-order model were conducted on GET-Power and MATLAB/SIMULINK platforms. Temperature management through LPV control in SCR systems was put forward as well [187–189]. To this end, LPV state-space model was adopted to describe the temperature propagation in an SCR system, then a linear-quadratic-Gaussian (LQG) controller was developed for the temperature regulation. Furthermore, to achieve a better control performance, a Kalman filter-based observer was developed to estimate the system states. Finally, the LPV model and controller were validated under a simulator environment. A robust LPV controller was designed for the thermal management of a Diesel particulate filter [190]. For LPV control design, a simplified physical model was established. The performance of the designed LPV controller was compared with that of a baseline PID controller.

LPV-based observers have been also extensively applied for the estimation issues of aftertreatment systems. A Luenberger-like observer was developed to estimate the air fraction for Diesel engine and aftertreatment system [191]. The dynamic model for the Diesel engine and coupled aftertreatment were built in form of LPV model, which is efficient to represent the parametric uncertainties and disturbances. The designed observer was analyzed through Lyapunov functions and validated under real Diesel experiments. A nonlinear observer was promoted to estimate the immeasurable ammonia coverage ratio for an SCR system [192]. The nonlinear behaviors of the SCR system were first modeled in a quasi-LPV form. Then, an LPV observer was developed by stabilizing the estimation error system. Finally, an experimental test was conducted to validate the effectiveness of the designed LPV observer. Besides, an LPV proportional-multiple-integral (PMI) observer was investigated to estimate the NO_x sensor ammonia-cross-sensitivity factor, which is essential to compensate the sensor errors in SCR systems [193]. Applying Lyapunov stability theorem, the gain-scheduled PMI observer was developed while guaranteeing an $\mathcal{H}_\infty$ performance for the nonlinear error system. The simulation results proved that the LPV-based observer can provide a good estimation performance under disturbances. In addition, the authors in [194] studied a gain-scheduled Luenberger observer to estimate the NO and NO_2 concentrations in aftertreatment systems, including DOC and DPF, of Diesel engines. In this work, the nonlinear dynamics was modeled using LPV representations, and the observer was designed by analyzing the stability of the estimated error system. Similarly, ammonia and NO_x concentrations in a two-cell SCR system were also estimated through LPV-based techniques in [195]. Two observers were developed for the upstream cell of the SCR system, and one observer was designed for the downstream cell. The first observer in the upstream cell was used for unknown input and state estimation, while the second LPV observer was used to estimate the ammonia coverage ratio. The observer in the downstream cell aimed at estimating the NO_x emission and the ammonia coverage ratio. Finally, the excellent estimation performance of the designed observers was experimentally demonstrated.

1.6 Future Research Trends and Challenges

Polytopic LPV techniques for IASs go toward maturity step by step, while some challenges still remain to be exploited. Moreover, with the emerging of new theories and hardware, more opportunities would follow constantly. Therefore, we focus hereafter on challenges and future trends related to polytopic LPV research topics from both theoretical and application viewpoints.

1.6.1 LPV Complexity Reduction

Polytopic LPV paradigms have become a standard formalism in stability analysis, estimation, and control design of nonlinear systems [2, 27]. From the theoretical viewpoint, it is possible to derive *necessary and sufficient* stability conditions for quasi-LPV systems [65]. However, in practice these stability conditions are conceptual rather than implementable since the computational burden swiftly increases in a way that most numerical solvers crash [66]. Hence, for stability analysis and control design of complex nonlinear systems such as IASs, it is crucial to study the reductions, either of LPV models or LMI constraints. At the same time, such numerical reductions must theoretically guarantee all specifications predefined for the initial model. A promising solution consists in exploiting data-based approaches such as SPCA-based technique [26] or deep neural network-based technique [196] to reduce the number of polytope vertices. As a result, the amount of LMI constraints, decision variables, online computational load, and hardware resources requirements can be significantly reduced. The resulting integrated method of LPV model-based and data-based approaches would extend the application ranges of polytopic LPV techniques.

1.6.2 Fault Detection and Fault-Tolerant Control

With a constant increase in complexity, the demand on reliability and safety of IASs becomes more and more stringent. As a result, the issues of fault detection and fault-tolerant control have played a key role to minimize the performance degradation and to avoid dangerous situations [197]. Within this context, the ability of control reconfiguration is decisive to have a robust and resilient system operation, i.e., actuator and/or sensor faults can be effectively dealt with while still guaranteeing an acceptable closed-loop performance [198]. A promising solution to this problem is based on a two-step design procedure for which a fault detection algorithm and a fault-tolerant control scheme are separately designed. To this end, polytopic LPV technique can be used to model virtual actuators and/or sensors for fault detection algorithm. Then, an LPV observer-based control scheme can be formulated with suitable closed-loop specifications to achieve the reconfiguration goal.

1.6.3 *Limited Capacities of Perception and Motion Planning*

Autonomous vehicles have been regarded as the future of vehicle industry. Quasi-LPV techniques have been successfully applied to path following control of autonomous vehicles [126, 131, 132, 140], which is one of the most challenging problems in automated driving technology. For path following control, it is generally assumed that the perception is ideal and the vehicle can accurately obtain the positions of itself and of the surrounding obstacles. Then, the vehicle motion can be planned in function of the predicted obstacle trajectories. However, it always takes a certain time duration for the detection of the lane and the surrounding environment. Moreover, a smooth planned path requires an optimal iterative search which is also time-consuming [15]. Therefore, when an autonomous vehicle is with limited capacities of perception and motion planning, the planned path trajectory and the measured signals would be delayed. Finding an effective solution to obtain a robust path following control performance in presence of delayed signals would be the key to the widespread acceptance of autonomous vehicles.

1.6.4 *Driver-Automation Shared Driving Control*

Recent advances in actuation, perception technologies, and artificial intelligence have prompted the intensive investigations driving assistance and highly automated driving in both academic and industry settings. However, fully automated driving is still prone to errors in the human presence in the control loop [199, 200]. Driver-automation shared control has been shown as an effective scheme permitting to better meet the design guidelines of automation [201–204]. To this end, several novel shared control architectures, allowing for the driver-automation cooperation at the tactical level (decision-making level) and at the operational level (guidance level), have been proposed and experimentally validated under various driving scenarios [91, 199]. The obtained results have shown a strong interest of the shared driving control concept in reducing both the driver workload and the driver-automation conflict. However, many challenges still remain which offer great opportunities for the research on shared driving control in the future, for instance

- how to integrate the decision-making information into a robust vehicle control scheme to effectively handle hazardous situations (driver failure, undetected obstacles, sudden driving transition phases, etc.);
- how to integrate a *self-learning* ability into the shared control architecture such that the automation could *analyze* and *understand* the driver's actions during the shared control mode and the manual control mode.

Remark 1.8 Several survey articles related to this chapter are available in the open literature, e.g., the surveys on LPV theory [11, 12], and on optimal control theory for vehicle dynamics applications [13]. Compared to the works in [11, 12], which were done in 2000, we have introduced more recent techniques for the control design of LPV systems. In addition, the successful applications on IASs are special highlights. The survey in [13] mostly focused on linear control methods such as MPC control, LQR control for automotive systems. However, nonlinearities and uncertainties are unavoidable in IASs. In these cases, the surveyed approaches may not be suitable to deal with efficiently the related control problems. In our work, we have demonstrated the strengths of polytopic LPV approaches not only from the theory side, but also from the application side.

1.7 Concluding Remarks

A review on recent advances in polytopic LPV approaches for IASs has been carried out. First, fundamental theories on polytopic LPV control were discussed. Some key techniques to derive less conservative results for stability analysis and robust control design were reviewed. Second, applications of polytopic LPV techniques to vehicle dynamics modeling, vehicle vertical dynamics control, vehicle lateral dynamics control, path-following control, and vehicle powertrain control were summarized. Finally, some challenges and future trends were given.

Acknowledgements This chapter is mainly from the previous work in [205], and some typos are corrected here.

References

1. J. Mohammadpour, C. Scherer, Control of linear parameter varying systems with applications. Springer Science and Business Media (2012)
2. C. Briat, Filtering and Control, in *Linear Parameter-Varying and Time-Delay Systems: Analysis, Observation*. (Springer, Berlin, Heidelberg, 2014)
3. D. Liberzon, Switching in Systems and Control. Springer Science and Business Media (2003)
4. R. Goebel, R. Sanfelice, A. Teel, Hybrid dynamical systems. IEEE Control Syst. Mag. **29**(2), 28–93 (2009)
5. V. Yakubovich, A. Fradkov, D. Hill, A. Proskurnikov, Dissipativity of T-periodic linear systems. IEEE Trans. Autom. Control **52**(6), 1039–1047 (2007)
6. C. Hoffmann, H. Werner, A survey of LPV control applications validated by experiments or high-fidelity simulations. IEEE Trans. Control Syst. Technol. **23**(2), 416–433 (2014)
7. O. Sename, P. Gaspar, J. Bokor, Robust Control and Linear Parameter Varying Approaches: Application to Vehicle Dynamics, vol 437 (Springer, Berlin Heidelberg, 2013)
8. K. Bengler, K. Dietmayer, B. Farber, M. Maurer, C. Stiller, H. Winner, Three decades of driver assistance systems: review and future perspectives. IEEE Intell. Transp. Syst. Mag. **6**(4), 6–22 (2014)
9. A. White, G. Zhu, J. Choi, Linear Parameter-Varying Control for Engineering Applications. (Springer, 2013)
10. R. Tóth, *Modeling and Identification of Linear Parameter-Varying Systems*, vol. 403 (Springer, Berlin, Heidelberg, 2010)

11. D. Leith, W. Leithead, Survey of gain-scheduling analysis and design. Int. J. Control **73**(11), 1001–1025 (2000)
12. W. Rugh, J. Shamma, Research on gain scheduling. Automatica **36**(10), 1401–1425 (2000)
13. R. Sharp, H. Peng, Vehicle dynamics applications of optimal control theory. Veh. Syst. Dyn. **49**(7), 1073–1111 (2011)
14. A. Eskandarian, *Handbook of Intelligent Vehicles*, vol. 2 (Springer, London, 2012)
15. B. Paden, M. Čáp, S.-Z. Yong, D. Yershov, E. Frazzoli, A survey of motion planning and control techniques for self-driving urban vehicles. IEEE Trans. Intell. Veh. **1**(1), 33–55 (2016)
16. H. Amer, H. Zamzuri, K. Hudha, A. Kadir, Modelling and control strategies in path tracking control for autonomous ground vehicles: a review of state of the art and challenges. J. Intell. Robot. Syst. **86**(2), 225–254 (2017)
17. A. Vahidi, A. Eskandarian, Research advances in intelligent collision avoidance and adaptive cruise control. IEEE Trans. Intell. Transp. Syst. **4**(3), 143–153 (2003)
18. MathWorks. (2020) Adaptive cruise control system. Matlab R2020b, Library: MPC Toolbox / Automated Driving. [Online]. Available: https://fr.mathworks.com/help/mpc/ref/adaptivecruisecontrolsystem.html
19. B. Németh, P. Gáspár, R. Orjuela, M. Basset, LPV-based control design of an adaptive cruise control system for road vehicles. IFAC-PapersOnLine **48**(14), 62–67 (2015)
20. R. Attia, R. Orjuela, M. Basset, Combined longitudinal and lateral control for automated vehicle guidance. Vehicle Syst. Dyn. **52**(2), 261–279 (2014)
21. D. Hrovat, S. Di Cairano, H. Tseng, I. Kolmanovsky, *"The development of Model Predictive Control in automotive industry: a survey," in IEEE Int* (Dubrovnik, Croatia, Conference Control Applications, 2012), pp. 295–302
22. A. Marcos, G.J. Balas, Development of linear-parameter-varying models for aircraft. J Guidance Control Dyn **27**(2), 218–228 (2004)
23. W. Tan, A. Packard, G. Balas, Quasi-LPV modeling and LPV control of a generic missile. Am. Control Conf., **5**, 3692–3696 (2000) IEEE
24. J. Shamma, J. Cloutier, Gain-scheduled missile autopilot design using linear parameter varying transformations. J. Guidance Control Dyn. **16**(2), 256–263 (1993)
25. M. Jung, K. Glover, Calibratable linear parameter-varying control of a turbocharged diesel engine. IEEE Trans. Control Syst. Technol. **14**(1), 45–62 (2005)
26. A. Kwiatkowski, H. Werner, PCA-based parameter set mappings for LPV models with fewer parameters and less overbounding. IEEE Trans. Control Syst. Technol. **16**(4), 781–788 (2008)
27. K. Tanaka, H. Wang, Fuzzy Control Systems Design and Analysis: A Linear Matrix Inequality Approach (Wiley, 2004)
28. S. Boyd, L. El Ghaoui, E. Feron, V. Balakrishnan, Linear Matrix Inequalities in System and Control Theory, ser. Studies in Applied Mathematics, vol 15 (Philadelphia, PA, SIAM, 1994)
29. E. Feron, P. Apkarian, P. Gahinet, Analysis and synthesis of robust control systems via parameter-dependent Lyapunov functions. IEEE Trans. Autom. Control **41**(7), 1041–1046 (1996)
30. J. Daafouz, J. Bernussou, Parameter dependent lyapunov functions for discrete time systems with time varying parametric uncertainties. Syst. Control Lett. **43**(5), 355–359 (2001)
31. B. Lu, F. Wu, Switching LPV control designs using multiple parameter-dependent Lyapunov functions. Automatica **40**(11), 1973–1980 (2004)
32. P. Cox, S. Weiland, R. Tóth, Affine parameter-dependent Lyapunov functions for LPV systems with affine dependence. IEEE Trans. Autom. Control **63**(11), 3865–3872 (2018)
33. Y. Ebihara, D. Peaucelle, D. Arzelier, *S-Variable Approach to LMI-Based Robust Control* (Springer, London, 2015)
34. C. Scherer, LMI relaxations in robust control. Eur. J. Control **12**(1), 3–29 (2006)
35. A.-T. Nguyen, T.-M. Guerra, C. Sentouh, H. Zhang, Unknown input observers for simultaneous estimation of vehicle dynamics and driver torque: theoretical design and hardware experiments. IEEE/ASME Trans. Mechatron. **24**(6), 2508–2518 (2019)
36. L. Mozelli, R. Palhares, G. Avellar, A systematic approach to improve multiple Lyapunov function stability and stabilization conditions for fuzzy systems. Inf. Sci. **179**(8), 1149–1162 (2009)

37. F. Blanchini, S. Miani, A new class of universal Lyapunov functions for the control of uncertain linear systems. IEEE Trans. Autom. Control **44**(3), 641–647 (1999)
38. R. Ambrosino, M. Ariola, F. Amato, A convex condition for robust stability analysis via polyhedral Lyapunov functions. SIAM J. Control Opt. **50**(1), 490–506 (2012)
39. T. Hu, F. Blanchini, Non-conservative matrix inequality conditions for stability/stabilizability of linear differential inclusions. Automatica **46**(1), 190–196 (2010)
40. L. Xie, S. Shishkin, M. Fu, Piecewise Lyapunov functions for robust stability of linear time-varying systems. Syst. Control Lett. **31**(3), 165–171 (1997)
41. M. Johansson, A. Rantzer, K.-E. Årzén, Piecewise quadratic stability of fuzzy systems. IEEE Trans. Fuzzy Syst. **7**(6), 713–722 (1999)
42. G. Chesi, A. Garulli, A. Tesi, A. Vicino, Homogeneous Lyapunov functions for systems with structured uncertainties. Automatica **39**(6), 1027–1035 (2003)
43. G. Chesi, A. Garulli, A. Tesi, A. Vicino, Polynomially parameter-dependent Lyapunov functions for robust stability of polytopic systems: an LMI approach. IEEE Trans. Autom. Control **50**(3), 365–370 (2005)
44. G. Chesi, LMI techniques for optimization over polynomials in control: A survey. IEEE Trans. Autom. Control **55**(11), 2500–2510 (2010)
45. J. Shamma, Analysis and design of gain scheduled control systems, Ph.D. dissertation, Massachusetts Institute of Technology (1988)
46. J. Shamma, M. Athans, Gain scheduling: potential hazards and possible remedies. IEEE Control Syst. Mag. **12**(3), 101–107 (1992)
47. V. Syrmos, C. Abdallah, P. Dorato, K. Grigoriadis, Static output feedback-a survey. Automatica **33**(2), 125–137 (1997)
48. A.-T. Nguyen, P. Chevrel, F. Claveau, Gain-scheduled static output feedback control for saturated LPV systems with bounded parameter variations. Automatica **89**, 420–424 (2018)
49. A.-T. Nguyen, P. Chevrel, F. Claveau, LPV static output feedback for constrained direct tilt control of narrow tilting vehicles. IEEE Trans. Control Syst. Technol. **28**(2), 661–670 (2020)
50. B. Sereni, E. Assuncao, M. Carvalho Minhoto Teixeira, New gain-scheduled static output feedback controller design strategy for stability and transient performance of LPV systems. IET Contr. Theory Appl., **14**(5), 717–725 (2020)
51. A. White, Z. Ren, G. Zhu, J. Choi, Mixed $\mathcal{H}_2/\mathcal{H}_\infty$ observer-based LPV control of a hydraulic engine cam phasing actuator. IEEE Trans. Control Syst. Technol. **21**(1), 229–238 (2013)
52. C.-C. Ku, G.-W. Chen, New observer-based controller design for LPV stochastic systems with multiplicative noise. Int. J. Robust Nonlinear Control **29**(13), 4315–4327 (2019)
53. C.-C. Ku, G.-W. Chen, Relaxed observer-based controller design method of discrete-time multiplicative noised LPV systems via an extended projective lemma. Int. J. Control **93**(3), 462–472 (2020)
54. P. Apkarian, P. Gahinet, G. Becker, Self-scheduled $\mathcal{H}_\infty$ control of linear parameter-varying systems: a design example. Automatica **31**(9), 1251–1261 (1995)
55. F. Wu, X.H. Yang, A. Packard, G. Becker, Induced $\mathcal{L}_2-$norm control for LPV systems with bounded parameter variation rates. Int. J. Robust Nonlinear Control **6**(9–10), 983–998 (1996)
56. P. Apkarian, R. Adams, Advanced gain-scheduling techniques for uncertain systems. IEEE Trans. Control Syst. Technol. **6**(1), 21–32 (1998)
57. C. Scherer, LPV control and full block multipliers. Automatica **37**(3), 361–375 (2001)
58. J. De Caigny, J. Camino, R. Oliveira, P. Peres, J. Swevers, Gain-scheduled dynamic output feedback control for discrete-time LPV systems. Int. J. Robust Nonlinear Control **22**(5), 535–558 (2012)
59. C. Scherer, P. Gahinet, M. Chilali, Multiobjective output-feedback control via LMI optimization. IEEE Trans. Autom. Control **42**(7), 896–911 (1997)
60. Y. Shi, H. Tuan, P. Apkarian, Nonconvex spectral optimization algorithms for reduced-order LPV-LFT controllers. Int. J. Robust Nonlinear Control **27**(18), 4421–4442 (2017)
61. T.-M. Guerra, L. Vermeiren, LMI-based relaxed non-quadratic stabilization conditions for nonlinear systems in the Takagi-Sugeno's form. Automatica **40**(5), 823–829 (2004)

62. H. Wang, K. Tanaka, M. Griffin, An approach to fuzzy control of nonlinear systems: stability and design issues. IEEE Trans. Fuzzy Syst. **4**(1), 14–23 (1996)
63. H. Tuan, P. Apkarian, T. Narikiyo, Y. Yamamoto, Parameterized linear matrix inequality techniques in fuzzy control system design. IEEE Trans. Fuzzy Syst. **9**(2), 324–332 (2001)
64. X. Liu, Q. Zhang, New approaches to $\mathcal{H}_\infty$ controller designs based on fuzzy observers for T-S fuzzy systems via LMI. Automatica **39**(9), 1571–1582 (2003)
65. A. Sala, Ariño, Asymptotically necessary and sufficient conditions for stability and performance in fuzzy control: applications of Polya's theorem. Fuzzy Sets Syst, **158**(24), 2671–2686 (2007)
66. A.-T. Nguyen, T. Taniguchi, L. Eciolaza, V. Campos, R. Palhares, M. Sugeno, Fuzzy control systems: past, present and future. IEEE Comput. Intell. Mag. **14**(1), 56–68 (2019)
67. T. Takagi, M. Sugeno, Fuzzy identification of systems and its applications to modeling and control. IEEE Trans. Syst. Man, Cybern. B, Cybern., **SMC-15**(1), 116–132 (1985)
68. D. Rotondo, V. Puig, F. Nejjari, M. Witczak, Automated generation and comparison of Takagi-Sugeno and polytopic quasi-LPV models. Fuzzy Sets Syst. **277**, 44–64 (2015)
69. K. Tanaka, T. Hori, H.O. Wang, A multiple Lyapunov function approach to stabilization of fuzzy control systems. IEEE Trans. Fuzzy Syst. **11**(4), 582–589 (2003)
70. M. Morato, M. Nguyen, O. Sename, L. Dugard, Design of a fast real-time LPV model predictive control system for semi-active suspension control of a full vehicle. J. Franklin Inst. **356**(3), 1196–1224 (2019)
71. C. Onat, I.B. Kucukdemiral, S. Sivrioglu, I. Yuksek, LPV model based gain-scheduling controller for a full vehicle active suspension system. J. Vib. Control **13**(11), 1629–1666 (2007)
72. J. Sun, J. Cong, L. Gu, M. Dong, Fault-tolerant control for vehicle with vertical and lateral dynamics. Proc. Inst. Mech. Engineers, Part D: J. Automobile Eng., **233**(12), 3165–3184 (2019)
73. M. Dezasse, F. Svaricek, J. Brembeck, Damper fault-tolerant linear parameter-varying semi-active suspension control. IFAC-PapersOnLine **50**(1), 8592–8599 (2017)
74. M. Flepsdezasse, F. Svaricek, J. Brembeck, Design and experimental assessment of an active fault-tolerant LPV vertical dynamics controller. IEEE Trans. Control Syst. Technol. **27**(3), 1267–1274 (2019)
75. P. Li, J. Lam, K. Cheung, Velocity-dependent multi-objective control of vehicle suspension with preview measurements. Mechatronics **24**(5), 464–475 (2014)
76. X. Jin, G. Yin, C. Bian, J. Chen, P. Li, N. Chen, Robust gain-scheduled vehicle handling stability control via integration of active front steering and suspension systems. J. Dyn. Syst. Meas. Control, **138**(1), 014 501–12 (2016)
77. I. Fialho, G. Balas, Road adaptive active suspension design using linear parameter-varying gain-scheduling. IEEE Trans. Control Syst. Technol. **10**(1), 43–54 (2002)
78. J. Esmaeili, A. Akbari, H. Karimi, Load-dependent LPV/$\mathcal{H}_2$ output-feedback control of semi-active suspension systems equipped with MR damper. Int. J. Veh. Design **68**, 119–140 (2015)
79. C. Onat, I. Kucukdemiral, S. Sivrioglu, I. Yuksek, G. Cansever, LPV gain-scheduling controller design for a non-linear quarter-vehicle active suspension system. Trans. Inst. Meas. Control **31**(1), 71–95 (2009)
80. C. Poussot-Vassal, O. Sename, L. Dugard, P. Gaspar, Z. Szabo, J. Bokor, A new semi-active suspension control strategy through LPV technique. Control Eng. Pract. **16**, 1519–1534 (2008)
81. P. Li, Y. Wu, X. Sun, Z. Lang, Gain-scheduled control of linear differential inclusions subject to actuator saturation. IEEE Trans. Ind. Electron. **66**(10), 8051–8059 (2019)
82. H. Gao, J. Lam, C. Wang, Multi-objective control of vehicle active suspension systems via load-dependent controllers. J. Sound Vib. **290**(3), 654–675 (2006)
83. J. Wu, H. Zhou, Z. Liu, M. Gu, A load-dependent PWA-$\mathcal{H}_\infty$ controller for semi-active suspensions to exploit the performance of MR dampers. Mech. Syst. Signal Process. **127**, 41–62 (2019)
84. H. Zebiri, B. Mourllion, M. Basset, Frequency-limited $\mathcal{H}_\infty$-controller order reduction for linear parameter-varying systems. Int. J. Control **90**(9), 2031–2046 (2017)

85. Z. Zhang, H. Liang, H. Ma, Y. Pan, Reliable fuzzy control for uncertain vehicle suspension systems with random incomplete transmission signals and sensor failure. Mech. Syst. Signal Process. **130**, 776–789 (2019)
86. S. Fergani, O. Sename, L. Dugard, An LPV suspension control with performance adaptation to roll behavior, embedded in a global vehicle dynamic control strategy, European Control Conf., pp. 487–492 (2013)
87. C. Poussot-Vassal, O. Sename, L. Dugard, P. Gaspar, Z. Szaba, J. Bokor, Attitude and handling improvements through gain-scheduled suspensions and brakes control. Control Eng. Pract. **19**(3), 252–263 (2011)
88. A. Karimi, Z. Emedi, $\mathcal{H}_\infty$ gain-scheduled controller design for rejection of time-varying disturbances with application to an active suspension system. Conf. Decision Control, pp. 7540–7545 (2013)
89. M. Morato, O. Sename, L. Dugard, M. Nguyen, Fault estimation for automotive electro-rheological dampers: LPV-based observer approach. Control Eng. Pract. **85**, 11–22 (2019)
90. S. Fergani, O. Sename, L. Dugard, An LPV/$\mathcal{H}_\infty$ integrated vehicle dynamic controller. IEEE Trans. Veh. Technol. **65**(4), 1880–1889 (2016)
91. A.-T. Nguyen, C. Sentouh, J.-C. Popieul, Driver-automation cooperative approach for shared steering control under multiple system constraints: design and experiments. IEEE Trans. Ind. Electron. **64**(5), 3819–3830 (2017)
92. R. Rajamani, *Vehicle Dynamics and Control* (Springer, US, 2012)
93. S. Fergani, L. Menhour, O. Sename, L. Dugard, B. D'Andrea-Novel, Integrated vehicle control through the coordination of longitudinal/lateral and vertical dynamics controllers: flatness and LPV/$\mathcal{H}_\infty$-based design. Int. J. Robust Nonlinear Control **27**(18), 4992–5007 (2017)
94. A.-T. Nguyen, T.-Q. Dinh, T.-M. Guerra, J. Pan, Takagi-Sugeno fuzzy unknown input observers to estimate nonlinear dynamics of autonomous ground vehicles: theory and real-time verification. IEEE/ASME Trans. Mechatron., pp. 1–1 (2021), https://doi.org/10.1109/TMECH.2020.3049070
95. S. Fergani, O. Sename, L. Dugard, Performances improvement through an LPV/$\mathcal{H}_\infty$ control coordination strategy involving braking, semi-active suspension and steering systems, in 51st Conf. Decision Control (2012), pp. 4384–4389
96. B. Nemeth, D. Fenyes, P. Gaspar, J. Bokor, Coordination of independent steering and torque vectoring in a variable-geometry suspension system. IEEE Trans. Control Syst. Technol. **27**(5), 2209–2220 (2019)
97. M. Doumiati, O. Sename, J. Martinez, L. Dugard, C. Vassal, Gain-scheduled LPV/$\mathcal{H}_\infty$ controller based on direct yaw moment and active steering for vehicle handling improvements, in 49th IEEE Conf. Decision Control (2010), pp. 6427–6432
98. H. Zhang, X. Huang, J. Wang, H.R. Karimi, Robust energy-to-peak sideslip angle estimation with applications to ground vehicles. Mechatronics **30**, 338–347 (2014)
99. H. Zhang, G. Zhang, J. Wang, Sideslip angle estimation of an electric ground vehicle via finite-frequency $\mathcal{H}_\infty$ approach. IEEE Trans. Transp. Elect. **2**(2), 200–209 (2016)
100. N. Wada, A. Takahashi, M. Saeki, M. Nishimura, Vehicle yaw control using an AFS system with measurements of lateral tire forces. J. Robot. Mechatron. **23**(1), 83–93 (2011)
101. H. Zhang, X. Zhang, J. Wang, Robust gain-scheduling energy-to-peak control of vehicle lateral dynamics stabilization. Vehicle Syst. Dyn. **52**(3), 309–340 (2014)
102. R. Wang, H. Zhang, J. Wang, F. Yan, N. Chen, Robust lateral motion control of four-wheel independently actuated electric vehicles with tire force saturation consideration. J. Franklin Inst. **352**(2), 645–668 (2015)
103. X. Jin, G. Yin, N. Chen, Gain-scheduled robust control for lateral stability of four-wheel-independent-drive electric vehicles via linear parameter-varying technique. Mechatronics **30**, 286–296 (2015)
104. H. Zhang, J. Wang, Vehicle lateral dynamics control through AFS/DYC and robust gain-scheduling approach. IEEE Trans. Veh. Technol. **65**(1), 489–494 (2016)
105. X. Jin, G. Yin, Y. Li, J. Li, Stabilizing vehicle lateral dynamics with consideration of state delay of AFS for electric vehicles via gain-scheduling control. Asian J. Control **18**(1), 89–97 (2016)

106. X. Jin, G. Yin, X. Zeng, J. Chen, Robust gain-scheduled output feedback yaw stability control for in-wheel-motor-driven electric vehicles with external yaw-moment. J. Franklin Inst. **355**(18), 9271–9297 (2017)
107. Q. Liu, G. Kaiser, S. Boonto, H. Werner, F. Holzmann, B. Chretien, M. Korte, Two-degree-of-freedom LPV control for a through-the-road hybrid electric vehicle via torque vectoring, in 44th IEEE Conf. Decision Control (2011), pp. 1274–1279
108. T. Besselmann, M. Morari, Autonomous vehicle steering using explicit LPV-MPC, in European Control Conf., (2009), pp. 2628–2633
109. T. Raharijaona, F. Duc, S. Mammar, Linear parameter-varying control and $\mathcal{H}_\infty$ synthesis dedicated to lateral driving assistance, in IEEE Intell. Vehi. Symp., (2004), pp. 407–412
110. R. Wang, H. Zhang, J. Wang, Linear parameter-varying-based fault-tolerant controller design for a class of over-actuated non-linear systems with applications to electric vehicles. IET Control Theory Appl. **8**(9), 705–717 (2014)
111. R. Wang, H. Zhang, J. Wang, LPV controller design for four-wheel independently actuated electric ground vehicles with active steering systems. IEEE Trans. Control Syst. Technol. **1281–1296**(4), 22 (2014)
112. G. Kaiser, Q. Liu, C. Hoffmann, M. Korte, H. Werner, Torque vectoring for an electric vehicle using an LPV drive controller and a torque and slip limiter, in 51st IEEE Conference on Decision and Control, (2012), pp. 5016–5021
113. A. Mihaly, P. Gaspar, Reconfigurable control of an in-wheel electric vehicle based on LPV methods, in International Symposium on Computational Intelligence and Informatics, (2014), pp. 97–102
114. B. Nemeth, P. Gaspar, J. Bokor, Improvement of the LPV-based vehicle control design considering the polynomial invariant set analysis, in American Control Conference (2015)
115. E. Alcala, V. Puig, J. Quevedo, T. Escobet, Gain-scheduling LPV control for autonomous vehicles including friction force estimation and compensation mechanism. IET Control Theory Appl. **12**(12), 1683–1693 (2018)
116. B. Nemeth, P. Gaspar, Control design of variable-geometry suspension considering the construction system. IEEE Trans. Veh. Technol. **62**(8), 4104–4109 (2013)
117. B. Nemeth, P. Gaspar, Nonlinear analysis and control of a variable-geometry suspension system. Control Eng. Pract. **61**, 279–291 (2017)
118. L. Palladino, G. Duc, R. Pothin, LPV control for μ-split braking assistance of a road vehicle," in 44th IEEE Conference on Decision and Control (2005)
119. C. Vassal, O. Sename, L. Dugard, A LPV/$\mathcal{H}_\infty$ global chassis controller for handling improvements involving braking and steering systems, in 47th IEEE Conference on Decision and Control (2008), pp. 5366–5371
120. H. Du, N. Zhang, Robust yaw moment control for vehicle handling and stability improvement, in 24th Chinese Control Decision Conference (2012), pp. 4221–4226
121. S. Baslamisli, I. Kose, G. Anlas, Gain-scheduled integrated active steering and differential control for vehicle handling improvement. Vehicle Syst. Dyn., pp. 99–119 (2009)
122. M. Li, Y. Jia, J. Du, LPV control with decoupling performance of 4WS vehicles under velocity-varying motion. IEEE Trans. Control Syst. Technol. **22**(5), 1708–1724 (2014)
123. C. Vassal, O. Sename, L. Dugard, S. Savaresi, Vehicle dynamic stability improvements through gain-scheduled steering and braking control. Vehicle Syst. Dyn. **49**(10), 1597–1621 (2011)
124. O. Sename, J. Martinez, S. Fergani, LPV methods for fault-tolerant vehicle dynamic control, in Conference on Control and Fault-Tolerant Systems (IEEE, 2013), pp. 116–130
125. M. Doumiati, O. Sename, L. Dugard, J.-J. Martinez-Molina, P. Gaspar, Z. Szabo, Integrated vehicle dynamics control via coordination of active front steering and rear braking. Eur. J. Control **19**(2), 121–143 (2013)
126. A.-T. Nguyen, J. Rath, T.-M. Guerra, R. Palhares, H. Zhang, Robust set-invariance based fuzzy output tracking control for vehicle autonomous driving under uncertain lateral forces and steering constraints. IEEE Trans. Intell. Transp. Syst., pp. 1–12 (2020) https://doi.org/10.1109/TITS.2020.3021292

127. H. Zhang, J. Wang, Modeling, Dynamics and Control of Electrified Vehicles (Elsevier Inc, 2018), ch. Chapter 10-Robust gain-scheduling control of vehicle lateral dynamics through AFS/DYC, pp. 339–368
128. W. Li, Z. Xie, J. Zhao, P.K. Wong, Velocity-based robust fault tolerant automatic steering control of autonomous ground vehicles via adaptive event triggered network communication. Mech. Syst. Signal Process. **143**, 106798 (2020)
129. J. Guo, J. Wang, Y. Luo, K. Li, Robust lateral control of autonomous four-wheel independent drive electric vehicles considering the roll effects and actuator faults. Mech. Syst. Signal Process. **143**, 106773 (2020)
130. T. Chen, L. Chen, X. Xu, Y. Cai, H. Jiang, X. Sun, Passive fault-tolerant path following control of autonomous distributed drive electric vehicle considering steering system fault. Mech. Syst. Signal Process. **123**, 298–315 (2019)
131. W. Zhang, A robust lateral tracking control strategy for autonomous driving vehicles. Mech. Syst. Signal Process. **150**, 107238 (2021)
132. C. Hu, H. Jing, R. Wang, F. Yan, M. Chadli, Robust $\mathcal{H}_\infty$ output-feedback control for path following of autonomous ground vehicles. Mech. Syst. Signal Process. **70–71**, 414–427 (2016)
133. R. Wang, H. Jing, C. Hu, F. Yan, N. Chen, Robust $\mathcal{H}_\infty$ path following control for autonomous ground vehicles with delay and data dropout. IEEE Trans. Intell. Transp. Syst. **17**(7), 2042–2050 (2016)
134. J. Wang, G. Zhang, R. Wang, S.C. Schnelle, J. Wang, A gain-scheduling driver assistance trajectory-following algorithm considering different driver steering characteristics. IEEE Trans. Intell. Transp. Syst. **18**(5), 1097–1108 (2017)
135. J. Wang, M. Dai, G. Yin, N. Chen, Output-feedback robust control for vehicle path tracking considering different human drivers' characteristics. Mechatronics **50**, 402–412 (2018)
136. J. Guo, Y. Luo, K. Li, Robust gain-scheduling automatic steering control of unmanned ground vehicles under velocity-varying motion. Vehicle Syst. Dyn. **57**(4), 595–616 (2019)
137. E. Alcala, V. Puig, J. Quevedo, T. Escobet, Gain-scheduling LPV control for autonomous vehicles including friction force estimation and compensation mechanism. IET Contr. Theory Appl. **12**(12), 1683–1693 (2018)
138. E. Alcala, V. Puig, J. Quevedo, LPV-MP planning for autonomous racing vehicles considering obstacles. Robot. Autonomous Syst. **124**, 103392 (2020)
139. A.-T. Nguyen, C. Sentouh, J.-C. Popieul, Fuzzy steering control for autonomous vehicles under actuator saturation: design and experiments. J. Franklin Inst. **355**(18), 9374–9395 (2018)
140. A.-T. Nguyen, C. Sentouh, H. Zhang, J.-C. Popieul, Fuzzy static output feedback control for path following of autonomous vehicles with transient performance improvements. IEEE Trans. Intell. Transp. Syst. **21**(7), 3069–3079 (2020)
141. L.F. Hang Peng, Chen Xinbo, LPV/$\mathcal{H}_\infty$ controller design for path tracking of autonomous ground vehicles through four-wheel steering and direct yaw-moment control. Int. J. Automotive Technol., **20**
142. Y. Chen, C. Hu, J. Wang, Impaired driver assistance control with gain-scheduling composite nonlinear feedback for vehicle trajectory tracking. ASME J. Dyn. Syst., Meas., Control, **142**(7) (2020)
143. A.-T. Nguyen, M. Dambrine, J. Lauber, Lyapunov-based robust control design for a class of switching non-linear systems subject to input saturation: application to engine control. IET Control Theory Appl. **8**(17), 1789–1802 (2014)
144. F. Zhang, K. Grigoriadis, M. Franchek, I. Makki, Linear parameter-varying lean burn air-fuel ratio control for a spark ignition engine. J. Dyn. Sys., Meas. **129**, 404–414 (2007)
145. A. Kwiatkowski, H. Werner, J. Blath, A. Ali, M. Schultalbers, Linear parameter varying PID controller design for charge control of a spark-ignited engine. Control Eng. Pract. **17**(11), 1307–17 (2009)
146. M. Postma, R. Nagamune, Air-fuel ratio control of spark ignition engines using a switching LPV controller. IEEE Trans. Control Syst. Technol. **20**(5), 1175–1187 (2011)
147. M. Postma, R. Nagamune, LPV-based air-fuel ratio control of spark ignition engines using two gain scheduling parameters. ASME Dyn. Syst. Control Conf., pp. 665–672 (2010)

148. M. Postma, R. Nagamune, Air-fuel ratio control of spark ignition engines using a switching LPV controller. IEEE Trans. Control Syst. Technol. **20**(5), 1175–87 (2011)
149. H. Abbas, H. Werner, Polytopic quasi-LPV models based on neural state-space models and application to air charge control of a SI engine. IFAC Proc. **41**(2), 6466–71 (2008)
150. A. Kominek, S. Remolina, S. Boonto, H. Werner, M. Garwon, M. Schultalbers, Low-complexity LPV input-output identification and control of a turbocharged combustion engine, in 51st IEEE Conference on Decision and Control (2012), pp. 4492–4497
151. Z. Qiu, J. Sun, M. Jankovic, M. Santillo, Nonlinear internal model controller design for wastegate control of a turbocharged gasoline engine. Control Eng. Pract. **46**, 105–14 (2016)
152. P. Majecki, M. Grimble, I. Haskara, Y. Hu, C. Chang, Total engine optimization and control for SI engines using linear parameter-varying models, in American Control Conference (2017), pp. 3631–3637
153. X. Wei, R. Del, Gain scheduled $\mathcal{H}_\infty$ control for air path systems of Diesel engines using LPV techniques. IEEE Trans. Control Syst. Technol. **15**(3), 406–15 (2007)
154. X. Wei, R.L. Del, L. Liu, Air path identification of diesel engines by LPV techniques for gain scheduled control. Math. Comput. Model. Dyn. Syst. **14**(6), 495–513 (2008)
155. L. Liu, X. Wei, T. Zhu, Quasi-LPV gain scheduling control for the air path system of diesel engines, in Chinese Control Decision Conference (2008), pp. 4893–4898
156. M. Jung, K. Glover, Calibratable linear parameter-varying control of a turbocharged diesel engine. IEEE Trans. Control Syst. Technol. **14**(1), 45–62 (2005)
157. Q. Song, K. Grigoriadis, Diesel engine speed regulation using linear parameter varying control, in American Control Conference (2003), pp. 779–784
158. X. Wei, R.L. Del, Modeling and control of the boost pressure for a diesel engine based on LPV techniques, in American Control Conference (2006), pp. 1892–1897
159. G. Christophe, O. Sename, L. Dugard, G. Meissonnier, An $\mathcal{H}_\infty$ LPV controller for a diesel engine common rail injection system, in European Control Conference (2007), pp. 1932–1939
160. F. Castillo, E. Witrant, C. Prieur, V. Talon, L. Dugard, Fresh air fraction control in engines using dynamic boundary stabilization of LPV hyperbolic systems. IEEE Trans. Control Syst. Technol. **23**(3), 963–74 (2014)
161. Z. Liu, C. Wang, An LPV adaptive observer for updating a map applied to an MAF sensor in a diesel engine. Sensors **15**(10), 27 142–59 (2015)
162. G. Gagliardi, F. Tedesco, A. Casavola, A LPV modeling of turbocharged SI automotive engine oriented to fault detection and isolation purposes. J. Franklin Inst. **355**(14), 10–45 (2018)
163. M. Bartels, Q. Liu, G. Kaiser, H. Werner, LPV torque vectoring for an electric vehicle using parameter-dependent lyapunov functions, in American Control Conference (2013), pp. 2153–2158
164. G. Kaiser, Q. Liu, C. Hoffmann, M. Korte, H. Werner, LPV torque vectoring for an electric vehicle with experimental validation, IFAC Proc. Vol., **47**(3), 12 010–5 (2014)
165. W. Nwesaty, A. Bratcu, O. Sename, LPV control for power source coordination-application to electric vehicles energy management systems, in European Control Conference (2014), pp. 2649–2654
166. W. Nwesaty, A.I. Bratcu, O. Sename, Optimal frequency separation of power sources by multivariable LPV/$\mathcal{H}_\infty$ control: application to on-board energy management systems of electric vehicles, in 53rd IEEE Conference on Decision and Control (2014), pp. 5636–5641
167. W. Nwesaty, A.I. Bratcu, O. Sename, Power sources coordination through multivariable LPV $\mathcal{H}_\infty$ control with application to multi-source electric vehicles. IET Contr. Theory Appl. **10**(16), 2049–59 (2016)
168. Y. Altun, K. Gulez, Linear parameter varying control of permanent magnet synchronous motor via parameter-dependent Lyapunov function for electrical vehicles, in IEEE International Conference on Vehicular Electronics and Safety (2012), pp. 340–345
169. A. Mihaly, P. Gaspar, B. Nemeth, Robust fault-tolerant control of in-wheel driven bus with cornering energy minimization. J. Mech. Eng. **63**(1), 35–44 (2017)
170. R. Wang, H. Zhang, J. Wang, F. Yan, N. Chen, Robust lateral motion control of four-wheel independently actuated electric vehicles with tire force saturation consideration. J. Franklin Inst. **352**(2), 645–68 (2015)

171. F. Jia, Z. Liu, A LPV traction control approach for independent in-wheel electric motor vehicle, in 11th World Congress on Intelligent Control and Automation (2014), pp. 1992–1997
172. A. Mihaly, P. Gaspar, Reconfigurable control of an in-wheel electric vehicle based on LPV methods, in 15th International Symposium on Computational Intelligence and Informatics (2014) pp. 97–102
173. T. Wang, O. Sename, J. Martinez-Molina, A LPV/$\mathcal{H}_\infty$ approach for fuel consumption minimization of the PHEV with battery life prolongation. IFAC Proc. **46**(21), 378–83 (2013)
174. C. Fauvel, F. Claveau, P. Chevrel, A two-layer LPV based control strategy for input and state constrained problem: application to energy management, in European Control Conf (2015), pp. 1127–1133
175. Y. Takahashi, K. Hidaka, Model predictive control for hybrid electric vehicles with linear parameter-varying model, in 18th International conference on control, automation and systems (2018), pp. 1501–1506
176. S. Cauet, P. Coirault, M. Njeh, Diesel engine torque ripple reduction through LPV control in hybrid electric vehicle powertrain: experimental results. Control Eng. Pract. **21**(12), 1830–40 (2013)
177. B. He, M. Yang, Robust LPV control of diesel auxiliary power unit for series hybrid electric vehicles. IEEE Trans. Power Electron. **21**(3), 791–8 (2006)
178. B. He, M. Ouyang, J. Li, Reduced order robust gain scheduling control of the diesel APU for series hybrid vehicles. Asian J. Control **8**(3), 227–36 (2006)
179. Q. Liu, G. Kaiser, S. Boonto, H. Werner, F. Holzmann, B. Chretien, M. Korte, Two-degree-of-freedom LPV control for a through-the-road hybrid electric vehicle via torque vectoring, in 50th IEEE Conference on Decision and Control Conference (2011), pp. 1274–1279
180. Hybrid-electric system truth test: Energy analysis of Toyota Prius IV in real urban drive conditions. Sustain. Energy Technol. Assess., **37**, 100573 (2020)
181. A. Hanif, A.I. Bhatti, Q. Ahmed, Estimation of thermally de-rated torque of an HEV drive using robust LPV observer, in American Control Conference (2016), pp. 1530–1535
182. A. Hanif, A.I. Bhatti, Q. Ahmed, Managing thermally derated torque of an electrified powertrain through LPV control. IEEE/ASME Trans. Mechatron. **23**(1), 364–76 (2017)
183. H. Zhang, J. Wang, Adaptive sliding-mode observer design for a selective catalytic reduction system of ground-vehicle diesel engines. IEEE/ASME Trans. Mechatron. **21**(4), 2027–2038 (2016)
184. M. Meisami-Azad, J. Mohammadpour, K. Grigoriadis, M. Harold, An adaptive control strategy for urea-SCR aftertreatment system, in American Control Conference (2010), pp. 3027–3032
185. M. Meisami-Azad, J. Mohammadpour, K. Grigoriadis, M. Harold, F.M., LPV gain-scheduled control of SCR aftertreatment systems. Int. J. Control, **85**(1), 114–33 (2012)
186. M. Meisami-Azad, J. Mohammadpour, K. Grigoriadis, M. Harold, F. M., PCA-based linear parameter varying control of SCR aftertreatment systems, in American Control Conference (2011), pp. 1543–1548
187. S. Tayamon, A. Larsson, B. Westerberg, Model-based temperature control of a selective catalytic reduction system. IET Contr. Theory Appl. **9**(2), 211–21 (2014)
188. S. Tayamon, J. Sjoberg, Modelling of selective catalytic reduction systems using discrete-time linear parameter varying models. IFAC Proc. **47**(3), 5685–90 (2014)
189. S. Tayamon, Nonlinear system identification and control applied to selective catalytic reduction systems, Ph.D. dissertation, Uppsala University (2014)
190. K. Bencherif, F. Benaicha, S. Sadai, M. Sorine, Diesel particulate filter thermal management using model-based design. SAE Technical Paper (2009)
191. P. Chen, J. Wang, Observer-based estimation of air-fractions for a diesel engine coupled with aftertreatment systems. IEEE Trans. Control Syst. Technol. **21**(6), 2239–50 (2012)
192. H. Zhang, J. Wang, Y. Wang, Nonlinear observer design of diesel engine selective catalytic reduction systems with NO_x sensor measurements. IEEE/ASME Trans. Mechatron., **20**(4), 1585–94 (2014)

193. H. Zhang, J. Wang, NO_x sensor ammonia-cross-sensitivity factor estimation in diesel engine selective catalytic reduction systems. J. Dyn. Syst., Meas., Control, **137**(6), 061 015–9 (2015)
194. H. Zhang, J. Wang, Y. Wang, Sensor reduction in diesel engine two-cell selective catalytic reduction systems for automotive applications. IEEE/ASME Trans. Mechatron. **20**(5), 22–33 (2014)
195. H. Zhang, J. Wang, Improved NO and NO_2 concentration estimation for a diesel-engine-aftertreatment system. IEEE/ASME Trans. Mechatron. **23**(1), 190–9 (2017)
196. P. Koelewijn, R. Tth, Scheduling dimension reduction of lpv models—a deep neural network approach, in American Control Conference (Denver, CO, USA, 2020), pp. 1111–1117
197. Z. Gao, C. Cecati, S.X. Ding, A survey of fault diagnosis and fault-tolerant techniques-Part i: fault diagnosis with model-based and signal-based approaches. IEEE Trans. Ind. Electron. **62**(6), 3757–3767 (2015)
198. I. Bessa, V. Puig, R.M. Palhares, TS fuzzy reconfiguration blocks for fault tolerant control of nonlinear systems. J. Franklin Inst. **357**(8), 4592–4623 (2020)
199. C. Sentouh, A.-T. Nguyen, M. Benloucif, J.-C. Popieul, Driver-automation cooperation oriented approach for shared control of lane keeping assist systems. IEEE Trans. Control Syst. Technol. **27**(5), 1962–1978 (2018)
200. A. Benloucif, A.-T. Nguyen, C. Sentouh, J.-C. Popieul, Cooperative trajectory planning for haptic shared control between driver and automation in highway driving. IEEE Trans. Ind. Electron. **66**(12), 9846–9857 (2019)
201. A.-T. Nguyen, C. Sentouh, J.-C. Popieul, Sensor reduction for driver-automation shared steering control via an adaptive authority allocation strategy. IEEE/ASME Trans. Mechatron. **23**(1), 5–16 (2017)
202. Z. Wang, R. Zheng, T. Kaizuka, K. Nakano, The effect of a haptic guidance steering system on fatigue-related driver behavior. IEEE Trans. Human-Machine Syst. **47**(5), 741–748 (2017)
203. M. Flad, L. Fröhlich, S. Hohmann, Cooperative shared control driver assistance based on motion primitives and differential games. IEEE Trans. Human-Machine Syst. **47**(5), 711–722 (2017)
204. M. Li, H. Cao, X. Song, Y. Huang, Z. Huang, Shared control driver assistance based on driving intention and situation assessment. IEEE Trans. Indus. Inform. **14**(11), 82–94 (2018)
205. P. Li, A.-T. Nguyen, H. Du, Y. Wang, H. Zhang, Polytopic LPV approaches for intelligent automotive systems: state of the art and future challenges. Mech. Syst. Signal Process. **161**(12), 107931 (2021)

Chapter 2
$\mathcal{H}_\infty$ Observer Design for LPV Systems with Uncertain Measurements on Scheduling Variables: Application to an Electric Ground Vehicle

Abstract In this chapter, we aim to study the observer design problem for polytopic linear-parameter-varying (LPV) systems with uncertain measurements on scheduling variables. Due to the uncertain measurements, the uncertainties are considered in the weighting factors. It is assumed that the vertices of polytope are the same when the measurements on scheduling variables are uncertain and perfect. Then, an LPV system with the uncertain weighting factors can be transferred to an LPV system with uncertainties. To deal with the uncertainties and unknown disturbance in the observer design problem, we propose a gain-scheduled sliding mode observer. Defining the estimation error as the state vector minus the estimated state vector, the estimation error dynamics is established. The sliding mode observer design method is developed based on analysis results of the established estimation error system. The proposed observer design method is then applied to an electric ground vehicle (EGV) in which the measurement of longitudinal velocity is assumed to be uncertain. Experimental tests and comparisons are given to show the advantages of the proposed design method and the designed observer.

2.1 Introduction

The last two decades have witnessed the increasing attention to linear-parameter-varying (LPV) systems; see [1–6] and the references therein. The main reason arises from the fact that the LPV technique is very useful to deal with smooth nonlinearities. In LPV systems, the state-space models depend on exogenous nonstationary parameters. When the bounds of exogenous parameters are available, an LPV system can be reformulated into a convex combination of linear-time-invariant (LTI) systems and the well-developed results on LTI systems can be extended to the LPV systems. If the nonlinearities are determined with exogenous nonstationary parameters, the nonlinear system can be transferred into an LPV system and the analysis and synthesis of the nonlinear system are available to be carried out under the framework of linear systems. Due to the significant advantages, the research on LPV systems has gained numerous attention from both theoretical development and application sides. The authors in [7] investigated the model predictive control for LPV systems with

H. Zhang et al., *Robust Gain-Scheduled Estimation and Control of Electrified Vehicles via LPV Technique*, Key Technologies on New Energy Vehicles,
https://doi.org/10.1007/978-981-19-8509-6_2

bounded rates of parameter changes. The theory of MPC control was extended to LPV systems. The dynamic output-feedback controller design for discrete-time LPV systems was presented in [8]. The uncertain issue of LPV systems was discussed in [9], and the observer-based controller was designed for discrete-time LPV system. The finite-time stability for LPV systems was discussed in [10]. At the application side, the LPV techniques have been successfully applied to a number of practical setups, to name a few, such as PEM fuel cell systems in [11], Diesel engines in [12], inverted pendulums in [13], and wind turbine systems in [14].

When the exogenous nonstationary parameters are measurable online, it is desired that the controllers and observers for LPV systems are also dependent on the measurable parameters. In this way, the conservativeness of the design can be reduced. This is the idea of gain-scheduled approach. In most of gain-scheduled controller or observer design work, it is generally assumed that the measurements on the scheduling variables are perfect and the controller or observer directly employs the measured scheduling variables. However, in many practical applications, the measurements on the scheduling variables cannot be so ideal. For example, the vehicle longitudinal velocity is quite important for vehicle control and diagnosis. For engine-powered vehicles, the longitudinal velocity is estimated with the engine speed. Without the vehicle slip, this approach may be accurate. However, the vehicle should maintain the slip ratio at a certain level to fulfill a good road cohesive condition. Therefore, the estimated vehicle longitudinal velocity is inaccurate. If the inaccurate or uncertain measurements of the scheduling variables are directly taken as the ideal measurements, the performance of the designed controllers and observers cannot be guaranteed. Recently, there are some research works on the uncertain measurements of LPV scheduling variables. In [9], the uncertain-measurement-induced issue was regarded as a disturbance. However, when the scheduling variables lie in the system matrix and the uncertainties are large, definitely, the stability will be degraded and the disturbance method cannot be applied. In [15, 16], the difference between the weighting factors calculated with ideal measurements and the estimated values satisfies with a Lipschitz condition.

The main principle is to design an observer to estimate the unavailable states [17–22]. In the literature, there are a lot of well-known types of observer such as the Kalman filter [23, 24], the Luenberger observer [25, 26], the $\mathcal{H}_\infty$ filter [27, 28], the $\mathcal{H}_2$ filter [29, 30], the mixed $\mathcal{H}_2/\mathcal{H}_\infty$ filter [31, 32], and the sliding mode observer [33, 34]. Each type of observer has its own advantages and application fields. The $\mathcal{H}_\infty$ filters are robust to system uncertainties. However, the uncertainties should be well studied or modeled before the filters design. The sliding model observers or controllers are also capable to deal with system uncertainties and only the bounds of uncertainties are necessary [35, 36].

In this work, we aim to study the observer design problem for LPV systems with uncertain measurements on scheduling variables. The contributions of this work can be summarized as: (1) We employ a different method to describe the LPV systems with uncertain measurements on scheduling variables; (2) we propose robust $\mathcal{H}_\infty$ sliding mode observer for LPV systems subject to the exogenous disturbances, unknown inputs, and uncertainties on scheduling variables; (3) slack matrices are introduced to reduce the conservativeness of obtained condition; and (4) the proposed design method is successfully applied to an EGV to estimate the sideslip angle.

Notation: The notations used in this chapter are standard. Superscript 'T' and '−1' indicate matrix transposition and matrix inverse; in symmetric block matrices or long matrix expressions, we use $*$ as an ellipsis for the terms that are introduced by symmetry; and $\|X\|$ represents the Euclidean norm for vectors and the spectral norm for matrices. Matrices, if their dimensions are not explicitly stated, are assumed to be compatible for algebraic operations. For a real symmetric matrix P, the notations $P > 0$, and $P < 0$, mean that P is positive definite and negative definite, respectively.

2.2 Problem Formulation and Preliminary

Consider a class of LPV system with the following expression:

$$\begin{aligned} \dot{x} &= A(\alpha)x + B_1(\alpha)u + B_2(\alpha)\omega \\ y &= Cx + Dv \end{aligned} \tag{2.1}$$

where x is the system state vector, u is the input vector, ω is the disturbance, y is the measured output vector, and v denotes the measurement noise. $A(\alpha)$, $B_1(\alpha)$, $B_2(\alpha)$, C, and D are matrices with compatible dimensions. Suppose that the number of scheduling variables $\rho_l (l = 1, \cdots, r)$ is r and the scheduling variable is independent with each other. Then, if the bounds of scheduling variables are known and the scheduling variables are measurable, the number of polytope vertices to describe the system is 2^r and the system matrix set $\mathcal{S} = (A(\alpha), B_1(\alpha), B_2(\alpha))$ expressed as:

$$\mathcal{S} = \left\{ \Omega \middle| \Omega = \sum_{i=1}^{2^r} \alpha_i \Omega_i; 0 \leq \alpha_i \leq 1; \sum_{i=1}^{2^r} \alpha_i = 1 \right\}, \tag{2.2}$$

where $\Omega_i = (A_i, B_{1,i}, B_{2,i})$ and the value of matrix set for each vertex is known. Since the scheduling variables are measurable online, the value of weighting factor α_i for each vertex can be determined online. To formulate the problem, we make the following assumptions:

Assumption 1: The matrix A_i is a Hurwitz matrix and each sub-system is observable.

Assumption 2: The symbol u denotes the control input which is known and the symbol ω is the disturbance which is unknown but bounded as

$$\|\omega\| \leq \delta_1, \tag{2.3}$$

where δ_1 is a known positive scalar.

Assumption 3: Though the measurements on the scheduling variables may not be accurate, the vertices of the describing polytope do not change, that is, we use one polytope to describe the system.

In this work, we dedicate to develop a gain-scheduled sliding mode observer for the LPV system. If the scheduling variables are perfectly measurable, the determined weighting factors can be directly used in the gain-scheduled observer design. However, each physical sensor has its own resolution and the measurement cannot be totally perfect. If the weighting factors of the original system in (2.1) are directly used in the gain-scheduled observer, the mismatched weighting factors may deteriorate the designed observer. Therefore, it is necessary to consider the asynchronous weighting factors. Consider a generalized case for the scheduling variables with uncertain measurements as:

$$\underline{k}_i\hat{\alpha}_i \leq \alpha_i = k_i\hat{\alpha}_i \leq \bar{k}_i\hat{\alpha}_i, \forall\, i = 1, \cdots, 2^r, \tag{2.4}$$

where $\hat{\alpha}_i$ denotes the weighting factor for the ith vertex calculated with the imprecisely measured scheduling variables, k_i is the uncertain factor, and $\underline{k}_i$ and $\bar{k}_i$ are known bounds of k_i. With the expression in (2.4), one gets:

$$\begin{aligned} A(\alpha) &= \sum_{i=1}^{2^r} \alpha_i A_i = \sum_{i=1}^{2^r} \hat{\alpha}_i \left(A_i + (k_i - 1)A_i\right) \\ &= \sum_{i=1}^{2^r} \hat{\alpha}_i \left(A_i + \Delta A_i\right), \sum_{i=1}^{2^r} \hat{\alpha} = 1 \\ B_1(\alpha) &= \sum_{i=1}^{2^r} \alpha_i B_{1,i} = \sum_{i=1}^{2^r} \hat{\alpha}_i \left(B_{1,i} + (k_i - 1)B_{1,i}\right) \\ &= \sum_{i=1}^{2^r} \hat{\alpha}_i \left(B_{1,i} + \Delta B_{1,i}\right), \\ B_2(\alpha)\omega &= \sum_{i=1}^{2^r} \hat{\alpha}_i k_i B_{2,i}\omega = \sum_{i=1}^{2^r} \hat{\alpha}_i B_{2,i}\omega_1, \\ \Delta A_i &= (k_i - 1)A_i,\ \Delta B_{1,i} = (k_i - 1)B_{1,i},\ \omega_1 = k_i\omega. \end{aligned} \tag{2.5}$$

Here, $X(\hat{\alpha})$ represents a matrix determined with the measured scheduling variables. To facilitate the observer design, we make the following practical assumptions:

$$\|\Delta A_i\| \leq \zeta_i, \text{ and } \|\omega_1\| \leq \delta_2, \tag{2.6}$$

with positive scalars ζ_i and δ_2. The assumptions are used to constrain the induced uncertainty and the external noise, which would benefit the sliding mode observer design. Since k_i has a lower bound and a upper bound, the matrix $\Delta B_{1,i}$ can be rewritten as:

$$\Delta B_{1,i} = \delta_{3,i} M B_{1,i} \tag{2.7}$$

where $\delta_{3,i}$ is a positive scalar with the value of $\delta_{3,i} = \max\{|\bar{k}_i - 1|, |\underline{k}_i - 1|\}$, and $\|M\| \leq 1$.

For the LPV system in (2.1), we propose the following sliding mode observer:

$$\begin{aligned} \dot{\hat{x}} &= A(\hat{\alpha})\hat{x} + B_1(\hat{\alpha})u + L(\hat{\alpha})(y - \hat{y}) + \varphi_1(\hat{\alpha}) + \varphi_2(\hat{\alpha}), \\ \hat{y} &= C\hat{x}, \end{aligned} \tag{2.8}$$

where $\hat{x}$ is the estimated state vector, $L(\hat{\alpha})$ is the observer gain to be tuned, and $\varphi_1(\hat{\alpha})$ and $\varphi_2(\hat{\alpha})$ are two discontinuous functions to compensate for the induced uncertainty $\Delta A(\hat{\alpha})$ and the disturbance ω_1, respectively. The main objective of observer design is to make the estimated state vector $\hat{x}$ to follow the state vector x quickly. Defining the estimation error as $e = x - \hat{x}$, the estimation error dynamics is expressed as:

$$\begin{aligned} \dot{e} = \Big[&(A(\hat{\alpha}) - L(\hat{\alpha})C)e + \Delta A(\hat{\alpha})x + \hat{B}_1(\hat{\alpha})\nu \\ &+ B_2(\hat{\alpha})\omega_1 - \varphi_1(\hat{\alpha}) - \varphi_2(\hat{\alpha})\Big], \end{aligned} \tag{2.9}$$

where

$$\hat{B}_1(\hat{\alpha}) = \bar{B}_1(\hat{\alpha}) + \Delta_3(\hat{\alpha})M\tilde{B}_1(\hat{\alpha}),\ \bar{B}_1(\hat{\alpha}) = \begin{bmatrix} 0 & -L(\hat{\alpha})D \end{bmatrix},$$

$$\Delta_3(\hat{\alpha}) = \delta_3(\hat{\alpha})I,\ \tilde{B}_1(\hat{\alpha}) = \begin{bmatrix} B_1(\hat{\alpha}) & 0 \end{bmatrix},\ \nu = \begin{bmatrix} u \\ v \end{bmatrix}.$$

We can see from (2.9) that the gains to be tuned, and state vector x, the unknown external input, $\varphi_1(\hat{\alpha})$ and $\varphi_2(\hat{\alpha})$ are also involved in the estimation error system. Since the main criterion of selecting the gain $L(\hat{\alpha})$ is to make the estimation error system stable such that the estimation error will converge to zero, we define a new variable as:

$$z = Ge, \tag{2.10}$$

with a constant matrix G. Then, the challenges and objectives are to tune the observer gain $L(\hat{\alpha})$ and discontinuous functions $\varphi_1(\hat{\alpha})$ and $\varphi_2(\hat{\alpha})$ such that

O1: The estimation error system in (2.9) is asymptotically stable when the external input is zero.

O2: The effect from the external input ν to the signal z is constrained as:

$$\|z\|_2 < \gamma\|\nu\|_2, \tag{2.11}$$

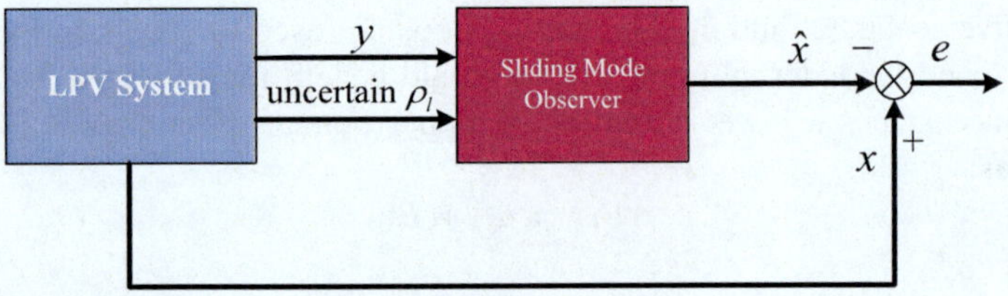

Fig. 2.1 Schematic diagram of the observer design problem

where $\|X\|_2$ denotes the 2 norm of $\mathcal{L}_2$-bounded signal X and γ is the $\mathcal{H}_\infty$ performance index.
O3: The discontinuous functions $\varphi_1(\hat{\alpha})$ and $\varphi_2(\hat{\alpha})$ can compensate for the effects of $\Delta A(\hat{\alpha})x$ and $B_2(\hat{\alpha})\omega_1$, respectively.

Remark 2.1 The schematic diagram of the observer design problem is illustrated in Fig. 2.1. The primary objective is to design the sliding model observer to estimate the signal z for the LPV system. The signal z can be the states or the combination of the states. The available signals for the sliding mode observer include the output of the LPV system and the scheduling variables. Different from most of the existing works on LPV systems, we consider the uncertain measurements on the scheduling variables, that is, the scheduling variables are not precisely known. Though we discuss and formulate the problem for polytopic LPV systems with uncertain measurements on scheduling variables, the case with precise measurements on scheduling variables is a special one of the uncertain case. If $k_i = 1, \forall\, i = 1, \cdots, 2^r$, ΔA_i and $\Delta B_{1,i}$ become zero matrices. The formulated problem is reduced to the case with precise measurements. For the matrices C and D in the output y, they are not assumed to depend on the scheduling variables. The main reason is that, for most of practical applications, the outputs are only the states or the combination of states besides the measurement noises. The weighting factors are not involved in the output matrices.

Remark 2.2 The proposed observer in (2.8) is different from the Luenberger-like observer since there are two discontinuous terms $\varphi_1(\hat{\alpha})$ and $\varphi_2(\hat{\alpha})$. The discontinuous terms are used to compensate for the uncertainties and the unknown inputs. This type of sliding mode observers was original developed in [37]. Due to the good performance, the idea was then extended to various setups and applications; see [38–40]. Motivated by the above works, we develop the sliding mode observer for LPV systems. It is not a straightforward extension of the existing works. The significant differences include that: (1) The studied setups are LPV systems with uncertain measurements; (2) not only the stability but also the $\mathcal{H}_\infty$ performance [41, 42] will be discussed in this work; and (3) the proposed design method will be applied to a practical application.

2.3 Observer Design

In this section, we propose the observer design method based on the analysis conditions for the estimation error system. Since the observer gain is unknown and to be determined, we assume that the gain $L(\hat{\alpha})$ is given during the stability and $\mathcal{H}_\infty$ performance analysis. Before proceeding, we introduce the following useful lemmas.

Lemma 2.1 *[43]: If there exist real matrices $\Xi = \Xi^{\mathrm{T}}$, $\hat{E}$ and $\hat{F}$ with compatible dimensions, and M satisfying $\|M\| \leq 1$, then, the following condition*

$$\Xi + \hat{E}M\hat{F} + (\hat{E}M\hat{F})^{\mathrm{T}} < 0, \tag{2.12}$$

is satisfied if and only if there exists a positive scalar $\psi > 0$ such that the following condition

$$\begin{bmatrix} \Xi & \hat{E} & \psi\hat{F}^{\mathrm{T}} \\ * & -\psi I & 0 \\ * & * & -\psi I \end{bmatrix} < 0, \tag{2.13}$$

is satisfied.

Lemma 2.2 *For any matrices $\hat{E}$ and $\hat{F}$ with compatible dimensions and a positive scalar ψ, the following condition holds:*

$$\hat{E}^{\mathrm{T}}\hat{F} + (\hat{E}^{\mathrm{T}}\hat{F})^{\mathrm{T}} \leq \psi\hat{E}^{\mathrm{T}}\hat{E} + \psi^{-1}\hat{F}^{\mathrm{T}}\hat{F}. \tag{2.14}$$

Observe that the uncertain term M in the condition of (2.12) is eliminated in the condition of (2.13). Therefore, Lemma 2.1 can be used to deal with the bounded uncertainties. The following theorem gives the conditions which guarantee the asymptotically stable and the $\mathcal{H}_\infty$ performance of the estimation error system in (2.9).

Theorem 2.1 *Given a positive scalar γ, the estimation error system in (2.9) is asymptotically stable and the $\mathcal{H}_\infty$ performance is guaranteed if there exists a positive-definite matrix P and positive scalars ψ_1, ψ_2, and ψ_3 such that the following condition is satisfied:*

$$\begin{bmatrix} \Lambda_1 & P\hat{B}_1(\hat{\alpha}) & G^{\mathrm{T}} & P \\ * & -\gamma^2 I & 0 & 0 \\ * & * & -I & 0 \\ * & * & * & -\psi_1 I \end{bmatrix} < 0, \tag{2.15}$$

where

$$
\begin{aligned}
\Lambda_1 =& P(A(\hat{\alpha}) - L(\hat{\alpha})C) + (P(A(\hat{\alpha}) - L(\hat{\alpha})C))^{\mathrm{T}} \\
&+ \psi_2 \zeta^2(\hat{\alpha}) I + \psi_3 I, \\
e_y =& y - \hat{y}, \\
\varphi_1(\hat{\alpha}) =& \sum_{i=1}^{2^r} \hat{\alpha}_i \varphi_{1,i} \\
=& \sum_{i=1}^{2^r} \hat{\alpha}_i \delta_2^2 \psi_3^{-1} \frac{\| P B_{2,i} \|^2}{2 e_y^{\mathrm{T}} e_y} P^{-1} C^{\mathrm{T}} e_y, \text{ if } e_y \neq 0, \\
\varphi_2(\hat{\alpha}) =& \sum_{i=1}^{2^r} \hat{\alpha}_i \varphi_{2,i} \\
=& \sum_{i=1}^{2^r} \hat{\alpha}_i \psi_1 (1 + \psi_4) \zeta_i^2 \frac{\hat{x}^{\mathrm{T}} \hat{x}}{2 e_y^{\mathrm{T}} e_y} P^{-1} C^{\mathrm{T}} e_y, \text{ if } e_y \neq 0, \\
\varphi_1(\hat{\alpha}) =& \sum_{i=1}^{2^r} \hat{\alpha}_i \varphi_{1,i} = 0, \text{ if } e_y = 0, \\
\varphi_2(\hat{\alpha}) =& \sum_{i=1}^{2^r} \hat{\alpha}_i \varphi_{2,i} = 0, \text{ if } e_y = 0, \\
\psi_4 =& \frac{\psi_1}{\psi_2 - \psi_1}.
\end{aligned}
\tag{2.16}
$$

Proof It infers from [44] that the asymptotically stable and the $\mathcal{H}_\infty$ performance of the estimation error system are both achieved if the following statement is satisfied:

$$
\mathcal{J} = \dot{V} + z^{\mathrm{T}} z - \gamma^2 \nu^{\mathrm{T}} \nu < 0, \tag{2.17}
$$

where V is the Lyapunov function and selected as $V = e^{\mathrm{T}} P e$ with a positive-definite matrix P. Considering the trajectories of the estimation error system, the expression $\mathcal{J}$ can be evaluated as:

$$
\begin{aligned}
\mathcal{J} &= \dot{e}^{\mathrm{T}} P e + e^{\mathrm{T}} P \dot{e} + z^{\mathrm{T}} z - \gamma^2 \nu^{\mathrm{T}} \nu \\
&= e^{\mathrm{T}}((A(\hat{\alpha}) - L(\hat{\alpha})C)^{\mathrm{T}} P + P(A(\hat{\alpha}) - L(\hat{\alpha})C))e. \\
&+ x^{\mathrm{T}} \Delta A^{\mathrm{T}}(\hat{\alpha}) P e + (x^{\mathrm{T}} \Delta A^{\mathrm{T}}(\hat{\alpha}) P e)^{\mathrm{T}} \\
&+ 2 e^{\mathrm{T}} P \hat{B}_1(\hat{\alpha}) \nu + 2 e^{\mathrm{T}} P B_2(\hat{\alpha}) \omega_1 \\
&- 2 \varphi_1^{\mathrm{T}}(\hat{\alpha}) P e - 2 e^{\mathrm{T}} P \varphi_2(\hat{\alpha}) + e^{\mathrm{T}} G^{\mathrm{T}}(\hat{\alpha}) G(\hat{\alpha}) e - \gamma^2 \nu^{\mathrm{T}} \nu.
\end{aligned}
\tag{2.18}
$$

In terms of Lemma 2.2, one gets:

$$\begin{aligned}
&x^{\mathrm{T}}\Delta A^{\mathrm{T}}(\hat{\alpha})Pe+(x^{\mathrm{T}}\Delta A^{\mathrm{T}}(\hat{\alpha})Pe)^{\mathrm{T}}\\
&\leq \psi_1^{-1}e^{\mathrm{T}}P^2e+\psi_1x^{\mathrm{T}}\Delta A^{\mathrm{T}}(\hat{\alpha})\Delta A(\hat{\alpha})x\\
&=\psi_1^{-1}e^{\mathrm{T}}P^2e+\psi_1(\hat{x}+e)^{\mathrm{T}}\Delta A^{\mathrm{T}}(\hat{\alpha})\Delta A(\hat{\alpha})(\hat{x}+e)\\
&\leq \psi_1^{-1}e^{\mathrm{T}}P^2e+\psi_1\zeta^2(\hat{\alpha})(\hat{x}+e)^{\mathrm{T}}(\hat{x}+e)\\
&=\psi_1^{-1}e^{\mathrm{T}}P^2e+\psi_1\zeta^2(\hat{\alpha})(\hat{x}^{\mathrm{T}}\hat{x}+e^{\mathrm{T}}e+\hat{x}^{\mathrm{T}}e+e^{\mathrm{T}}\hat{x})\\
&=\psi_1^{-1}e^{\mathrm{T}}P^2e+\psi_1\zeta^2(\hat{\alpha})(\hat{x}^{\mathrm{T}}\hat{x}+e^{\mathrm{T}}e+\hat{x}^{\mathrm{T}}e+e^{\mathrm{T}}\hat{x})\\
&\leq \psi_1^{-1}e^{\mathrm{T}}P^2e+\psi_2\zeta^2(\hat{\alpha})e^{\mathrm{T}}e+\psi_1(1+\psi_4)\zeta^2(\hat{\alpha})\hat{x}^{\mathrm{T}}\hat{x}
\end{aligned} \tag{2.19}$$

where ψ_2 is defined as $\psi_2=\frac{\psi_1(\psi_4+1)}{\psi_4}$. If e_y is zero, since each sub-system is observable, the estimation error is zero. If e_y is nonzero, we have the following expressions:

$$-2\varphi_1^{\mathrm{T}}(\hat{\alpha})Pe=-\delta_2^2\psi_3^{-1}\|PB_2(\hat{\alpha})\|^2. \tag{2.20}$$

$$2e^{\mathrm{T}}PB_2(\hat{\alpha})w_1\leq \psi_3e^{\mathrm{T}}e+\delta_2^2\psi_3^{-1}\|PB_2(\hat{\alpha})\|^2, \tag{2.21}$$

$$-2\varphi_2^{\mathrm{T}}(\hat{\alpha})Pe=-\psi_1(1+\psi_4)\zeta^2(\hat{\alpha})\hat{x}^{\mathrm{T}}\hat{x}. \tag{2.22}$$

Then, the expression $\mathcal{J}$ equals to

$$\begin{aligned}
\mathcal{J}&=e^{\mathrm{T}}\Gamma e+2e^{\mathrm{T}}P\hat{B}_1(\hat{\alpha})\nu-\gamma^2\nu^{\mathrm{T}}\nu\\
&=\begin{bmatrix} e\\ \nu\end{bmatrix}^{\mathrm{T}}\begin{bmatrix}\Gamma & P\hat{B}_1(\hat{\alpha})\\ * & -\gamma^2 I\end{bmatrix}\begin{bmatrix} e\\ \nu\end{bmatrix},
\end{aligned} \tag{2.23}$$

where $\Gamma=P(A(\hat{\alpha})-L(\hat{\alpha})C)+(P(A(\hat{\alpha})-L(\hat{\alpha})C))^{\mathrm{T}}+\psi_1^{-1}P^2+\psi_2\zeta^2(\hat{\alpha})I+\psi_3I+G^{\mathrm{T}}G$. Performing Schur compliment to (2.15), we can get

$$\begin{bmatrix}\Gamma & P\hat{B}_1(\hat{\alpha})\\ * & -\gamma^2 I\end{bmatrix}<0. \tag{2.24}$$

Obviously, the condition in (2.17) is satisfied when the condition (2.15) holds. Therefore, both the stability and the $\mathcal{H}_\infty$ performance are achieved when the condition (2.15) holds. The proof is completed. □

Though Theorem 2.1 provides the condition which can guarantee the stability and the $\mathcal{H}_\infty$ performance, the varying matrix M is involved. Since the value of matrix M is uncertain, the condition in (2.15) has infinite dimensions. In the following theorem, Lemma 2.1 will be used to eliminate the matrix M.

Theorem 2.2 *Given a positive scalar γ, the estimation error system in (2.9) is asymptotically stable and the $\mathcal{H}_\infty$ performance is guaranteed if there exists a positive-definite matrix P and positive scalars ψ_1, ψ_2, ψ_3, and $\psi_5(\hat{\alpha})$ such that the following condition is satisfied:*

$$\begin{bmatrix} \Lambda_1 & P\bar{B}_1(\hat{\alpha}) & G^{\mathrm{T}} & P & P\Delta_3(\hat{\alpha}) & 0 \\ * & -\gamma^2 I & 0 & 0 & 0 & \psi_5(\hat{\alpha})\tilde{B}_1^{\mathrm{T}}(\hat{\alpha}) \\ * & * & -I & 0 & 0 & 0 \\ * & * & * & -\psi_1 I & 0 & 0 \\ * & * & * & * & -\psi_5(\hat{\alpha})I & 0 \\ * & * & * & * & * & -\psi_5(\hat{\alpha})I \end{bmatrix} < 0. \tag{2.25}$$

Proof The condition in (2.15) can be rewritten as:

$$\begin{aligned} &\begin{bmatrix} \Lambda_1 & P\bar{B}_1(\hat{\alpha}) & G^{\mathrm{T}} & P \\ * & -\gamma^2 I & 0 & 0 \\ * & * & -I & 0 \\ * & * & * & -\psi_1 \end{bmatrix} + \begin{bmatrix} P\Delta_3(\hat{\alpha}) \\ 0 \\ 0 \\ 0 \end{bmatrix} M \begin{bmatrix} 0 & \tilde{B}_1(\hat{\alpha}) & 0 & 0 \end{bmatrix} \\ &+ \left(\begin{bmatrix} P\Delta_3(\hat{\alpha}) \\ 0 \\ 0 \\ 0 \end{bmatrix} M \begin{bmatrix} 0 & \tilde{B}_1(\hat{\alpha}) & 0 & 0 \end{bmatrix} \right)^{\mathrm{T}} < 0. \end{aligned} \tag{2.26}$$

In terms of Lemma 2.1, the conditions (2.15) and (2.25) are equivalent with each other. □

The main disadvantage of polytopic LPV systems is that the number of vertices would exponentially increase with respect to the number of scheduling variables. When the number is large, the computational load becomes heavy and it is difficult to find the feasible solution. Thus, it is necessary to reduce the conservativeness of obtained condition. The following theorem offers less conservative results.

Theorem 2.3 *Given a positive scalar γ, the estimation error system in (2.9) is asymptotically stable and the $\mathcal{H}_\infty$ performance is guaranteed if there exists a positive-definite matrix P, random matrices $E(\hat{\alpha})$ and $F(\hat{\alpha})$, and positive scalars ψ_1, ψ_2, ψ_3, and $\psi_5(\hat{\alpha})$ such that the following conditions are satisfied:*

$$\Psi = \begin{bmatrix} \Lambda_2 & \Lambda_3 & F(\hat{\alpha})\bar{B}_1(\hat{\alpha}) & G^{\mathrm{T}} & P & F(\hat{\alpha})\Delta_3(\hat{\alpha}) & 0 \\ * & -E(\hat{\alpha}) - E^{\mathrm{T}}(\hat{\alpha}) & E^{\mathrm{T}}(\hat{\alpha})\bar{B}_1(\hat{\alpha}) & 0 & 0 & E^{\mathrm{T}}(\hat{\alpha})\Delta_3(\hat{\alpha}) & 0 \\ * & * & -\gamma^2 I & 0 & 0 & 0 & \psi_5(\hat{\alpha})\tilde{B}_1^{\mathrm{T}}(\hat{\alpha}) \\ * & * & * & -I & 0 & 0 & 0 \\ * & * & * & * & -\psi_1 I & 0 & 0 \\ * & * & * & * & * & -\psi_5(\hat{\alpha})I & 0 \\ * & * & * & * & * & * & -\psi_5(\hat{\alpha})I \end{bmatrix} < 0, \tag{2.27}$$

where $\Lambda_2 = F(\hat{\alpha})(A(\hat{\alpha}) - L(\hat{\alpha})C) + (F(\hat{\alpha})(A(\hat{\alpha}) - L(\hat{\alpha})C))^{\mathrm{T}} + \psi_2\zeta^2(\hat{\alpha}) + \psi_3 I$ *and* $\Lambda_3 = P - F(\hat{\alpha}) + (A(\hat{\alpha}) - L(\hat{\alpha})C)^{\mathrm{T}} E(\hat{\alpha})$.

Proof The gap between the condition (2.27) and the condition (2.25) can be bridged by multiplying Φ on the left of Ψ and the transpose on the right of Ψ with

$$\Phi = \begin{bmatrix} I & (A(\hat{\alpha}) - L(\hat{\alpha})C)^{\mathrm{T}} & 0 & 0 & 0 & 0 & 0 \\ 0 & \bar{B}_1^{\mathrm{T}}(\hat{\alpha}) & I & 0 & 0 & 0 & 0 \\ 0 & 0 & 0 & I & 0 & 0 & 0 \\ 0 & 0 & 0 & 0 & I & 0 & 0 \\ 0 & \Delta_3^{\mathrm{T}}(\hat{\alpha}) & 0 & 0 & 0 & I & 0 \\ 0 & 0 & 0 & 0 & 0 & 0 & I \end{bmatrix}. \tag{2.28}$$

□

Since the weighting factor vector $\hat{\alpha}$ is involved in the condition of Theorem 2.3, the condition has infinite dimensions. The following theorem is going to project the condition to the vertices of the polytope such that the conditions can be used to calculate the observer gain.

Theorem 2.4 *Given a positive scalar γ and scalars λ_i, the estimation error system in (2.9) is asymptotically stable and the $\mathcal{H}_\infty$ performance is guaranteed if there exists a positive-definite matrix P, random matrices F_i and $\hat{L}_i$, and positive scalars ψ_1, ψ_2, ψ_3, and $\psi_{5,i}$ such that the following conditions are satisfied:*

$$\Psi_{ij} + \Psi_{ji} < 0 \tag{2.29}$$

where

$$\Psi_{ij} = \begin{bmatrix} \Lambda_{2,ij} & \Lambda_{3,ij} & \begin{bmatrix} 0 & -\hat{L}_i D \end{bmatrix} & G^{\mathrm{T}} & P & F_i\Delta_{3,j} & 0 \\ * & -\lambda_i F_i - \lambda_i F_i^{\mathrm{T}} & \begin{bmatrix} 0 & -\lambda_i \hat{L}_i D \end{bmatrix} & 0 & 0 & \lambda_i F_i \Delta_{3,j} & 0 \\ * & * & -\gamma^2 I & 0 & 0 & 0 & \psi_{5,i}\tilde{B}_{1,j}^{\mathrm{T}} \\ * & * & * & -I & 0 & 0 & 0 \\ * & * & * & * & -\psi_1 I & 0 & 0 \\ * & * & * & * & * & -\psi_{5,i} I & 0 \\ * & * & * & * & * & * & -\psi_{5,i} I \end{bmatrix},$$

$\Lambda_{2,ij} = F_i A_j - \hat{L}_i C + (F_i A_j - \hat{L}_i C)^{\mathrm{T}} + \psi_2 \zeta_i^2 I + \psi_3 I$, $\Lambda_{3,ij} = P - F_i + \lambda_i A_j^{\mathrm{T}} F_i - \lambda_i C^{\mathrm{T}} \hat{L}_i^{\mathrm{T}}$, *and* $1 \leq i \leq j \leq 2^r$. *In addition, the observer gain $L(\hat{\alpha})$ can be calculated via the following expression:*

$$L(\hat{\alpha}) = F^{-1}(\hat{\alpha})\hat{L}(\hat{\alpha}). \tag{2.30}$$

Proof Note that, in Theorem 2.3, both $E(\hat{\alpha})$ and $F(\hat{\alpha})$ are coupled with the observer gain $L(\hat{\alpha})$. Moreover, $E(\hat{\alpha})$, $F(\hat{\alpha})$, and $L(\hat{\alpha})$ are all unknown. In order to eliminate the coupled terms, we make the following assumptions

$$\begin{aligned} \psi_5(\hat{\alpha}) &= \sum_{i=1}^{2^r} \hat{\alpha}_i \psi_i, \\ F(\hat{\alpha}) &= \sum_{i=1}^{2^r} \hat{\alpha}_i F_i, \\ E^{\mathrm{T}}(\hat{\alpha}) &= \sum_{i=1}^{2^r} \lambda_i \hat{\alpha}_i F_i. \end{aligned} \tag{2.31}$$

Define a new variable as $\hat{L}(\hat{\alpha}) = F(\hat{\alpha})L(\hat{\alpha})$ and assume that

$$\hat{L}(\hat{\alpha}) = \sum_{i=1}^{2^r} \hat{\alpha}_i \hat{L}_i. \tag{2.32}$$

Then, it is obvious that

$$\Psi = \sum_{i=1}^{2^r} \hat{\alpha}_i^2 \Psi_{ii} + \sum_{i=1}^{2^r-1} \sum_{j=i+1}^{2^r} \hat{\alpha}_i \hat{\alpha}_j (\Psi_{ij} + \Psi_{ji}). \tag{2.33}$$

Since $\hat{\alpha}_i \hat{\alpha}_j$ is non-negative, Ψ is negative-definite if the conditions in (2.29) are satisfied. In addition, it infers from (2.29) that F_i is nonsingular. Therefore, $F(\hat{\alpha})$ is nonsingular and the inverse exists. With the definition of $\hat{L}(\hat{\alpha})$, the observer gain can be calculated in terms of (2.30). □

It is necessary to mention that the $\mathcal{H}_\infty$ performance index γ indicates the effect of the disturbance to the signal z. It is desired that the performance index γ is as small as possible. The optimal observer is defined as the one with the gain which is obtained by minimizing the value for γ. The minimization can be done in terms of the following corollary.

Corollary 2.1 *The minimum $\mathcal{H}_\infty$ performance index γ^* in Theorem 2.4 can be found by solving the following convex optimization problem:*

$$\begin{aligned} \gamma^* &= \min \gamma. \\ &s.\ t.\ (2.29) \end{aligned}$$

Remark 2.3 It was reported in [37] that the sliding mode observers are different from Luenberger observers since there are nonlinear discontinuous terms incorporated in the observer expressions. The discontinuous terms are dependent on the output estimation error. It can be seen from (2.8) and (2.16) that there are two output-estimation-error-dependent discontinuous terms. Therefore, the designed observer in (2.8) is a sliding mode observer. The sliding surface is the estimation error space. Moreover, we can see from Theorem 2.1 that the discontinuous functions $\varphi_1(\hat{\alpha})$ and $\varphi_2(\hat{\alpha})$ may have too large values when the residual e_y is quite small. These large values of discontinuous functions are harmful for the design observer. To avoid this phenomenon, the discontinuous functions in Theorem 2.1 can be reformulated as

$$
\begin{aligned}
\varphi_1(\hat{\alpha}) &= \sum_{i=1}^{2^r} \hat{\alpha}_i \varphi_{1,i} \\
&= \sum_{i=1}^{2^r} \hat{\alpha}_i \delta_2^2 \psi_3^{-1} \frac{\|PB_{2,i}\|^2}{2e_y^{\mathrm{T}} e_y} P^{-1} C^{\mathrm{T}} e_y, \text{ if } \|e_y\| > \phi, \\
\varphi_2(\hat{\alpha}) &= \sum_{i=1}^{2^r} \hat{\alpha}_i \varphi_{2,i} \\
&= \sum_{i=1}^{2^r} \hat{\alpha}_i \psi_1 (1+\psi_4) \zeta_i^2 \frac{\hat{x}^{\mathrm{T}} \hat{x}}{2e_y^{\mathrm{T}} e_y} P^{-1} C^{\mathrm{T}} e_y, \text{ if } \|e_y\| > \phi, \\
\varphi_1(\hat{\alpha}) &= \sum_{i=1}^{2^r} \hat{\alpha}_i \varphi_{1,i} = 0, \text{ if } \|e_y\| \le \phi, \\
\varphi_2(\hat{\alpha}) &= \sum_{i=1}^{2^r} \hat{\alpha}_i \varphi_{2,i} = 0, \text{ if } \|e_y\| \le \phi,
\end{aligned}
$$

where ϕ is a positive scalar which can be obtained by using the trial-and-error method. Then, the residual e_y would be bounded to ϕ.

Remark 2.4 In robust control and filtering problems, conservativeness is one of the most frequently discussed issues since conservative results may lead to infeasible solutions. Generally, there are two main approaches to derive less conservative results. One is to introduce slack matrices, and the other one is to develop parameter-dependent conditions. In this work, we introduce two slack matrices $F(\hat{\alpha})$ and $E(\hat{\alpha})$ in Theorem 2.3. In Theorem 2.1 and Theorem 2.2, the positive-definite P is coupled with the system matrix $A(\hat{\alpha}) - L(\hat{\alpha})C$. However, in Theorem 2.3, a nonsingular matrix $F(\hat{\alpha})$ is coupled with the system matrix instead of the positive-definite matrix. Therefore, the derived condition in Theorem 2.3 is relaxed. Moreover, both $F(\hat{\alpha})$ and $E(\hat{\alpha})$ are dependent on the scheduling variables.

Remark 2.5 Observe from (2.30) that the calculation of observer gain consists of online computation and offline computation. The offline computation is to solve the conditions in (2.29) and derive the values for F_i and $\hat{L}_i$, $\forall i = 1, \cdots, 2^r$. The online calculation includes the computation of weighting factors $\hat{\alpha}_i$ and the matrix operation in (2.30). Moreover, there are some prescribed scalars λ_i, $\forall i = 1, \cdots, 2^r$ in Theorem 2.4. In order to simply the observer design, the scalars can be selected

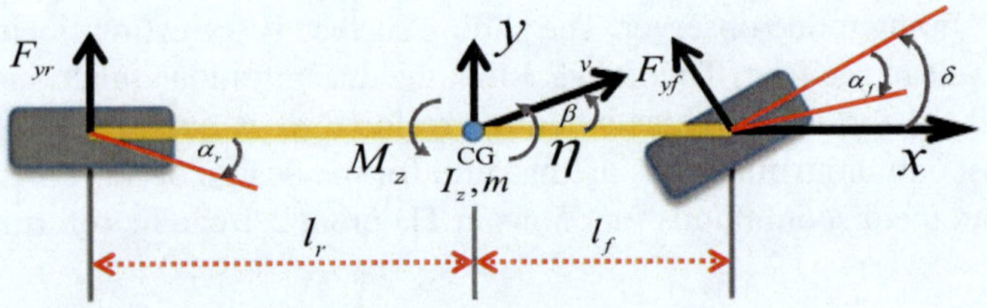

Fig. 2.2 Schematic diagram of a simplified lateral dynamics model

as one. However, in order to derive smaller $\mathcal{H}_\infty$ performance index, the nonlinear optimization algorithms such as the Matlab functions 'fmincon' and 'fminsearch' can be employed.

2.4 Application to an EGV

The lateral stability is of importance for ground vehicles, and the vehicle sideslip angle is a critical index for the lateral stability. It is well known that the vehicle sideslip angle is necessary for electronic stability program (ESP) systems. Unfortunately, currently, the sideslip angle is not measurable by using affordable physical sensors [45–48]. In order to deal with the conflict, an alternative approach is to estimate the sideslip angle with relatively cheap physical sensors. We can see from the existing work that the yaw rate measurement is a good choice for the sideslip angle estimation since the commercialized gyroscope can get the accurate yaw rate.

To analyze the lateral dynamics and establish the model, we employ a two-degree-of-freedom bicycle model as shown in Fig. 2.2. The importation notations are given as follows: The total ground vehicle mass is represented by m; the inertia moment about the yaw axis through its center of gravity (CG) is represented by I_z; l_f and l_r are used to denote distances of the front axis and the rear axis from the CG, respectively; the front-wheel steering angle is δ which is controlled by the driver and can be measured online via the encoder mounted on the steering column; α_f and α_r stand for the wheel slip angles of the front and rear tires, which are decided the front and rear lateral tire forces F_{yf} and F_{yr}, respectively. In addition, M_z denotes the direct yaw moment about the yaw axis, which is induced by the un-equal tracking/braking forces on the force tires. Due to the fact that tracking/braking forces are difficult to measure or estimate as the forces are related to the variable road condition, it is reasonable to assume that the direct yaw moment M_z is unknown.

In this application, the EGV which is developed in The Ohio State University [49] is driven on good road condition and the tire sideslip angles are not too large. Then, we can assume that the tires are working in the linear areas and the corresponding state-space model of the lateral dynamics is governed as:

$$\dot{x} = Ax + B_1\delta + B_2M_z, \tag{2.34}$$

where

$$x = \begin{bmatrix} \beta \\ \eta \end{bmatrix},\ B_1 = \begin{bmatrix} \frac{\bar{c}_f}{mv_x} \\ \frac{l_f \bar{c}_f}{I_z} \end{bmatrix},\ B_2 = \begin{bmatrix} 0 \\ \frac{1}{I_z} \end{bmatrix},$$

$$A = \begin{bmatrix} \frac{-\bar{c}_f - \bar{c}_r}{mv_x} & \frac{l_r \bar{c}_r - l_f \bar{c}_f}{mv_x^2} - 1 \\ \frac{l_r \bar{c}_r - l_f \bar{c}_f}{I_z} & \frac{-l_f^2 \bar{c}_f - l_r^2 \bar{c}_r}{I_z v_x} \end{bmatrix},$$

β is the vehicle sideslip angle, v_x is the longitudinal velocity, and the approximated values of corning stiffness are denoted by $\bar{c}_f$ and $\bar{c}_r$, respectively.

Since the longitudinal velocity v_x is not always constant, the model in (2.34) is not an LTI system but a nonlinear system [50]. It can be converted to an LPV representative. Suppose that the longitudinal velocity v_x is within the range $\begin{bmatrix} \underline{v}_x & \bar{v}_x \end{bmatrix}$ in which $\underline{v}_x$ is the lower bound of the velocity and $\bar{v}_x$ is the upper bound of the velocity. The scheduling variables are chosen as $\rho_1 = \frac{1}{v_x}$ and $\rho_2 = \frac{1}{v_x^2}$. Then, the nonlinear model in (2.34) can be represented by:

$$\begin{aligned} \dot{x} &= A(\alpha)x + B_1(\alpha)\delta + B_2 M_z \\ &= \sum_{i=1}^{4} \alpha_i (A_i x + B_{1,i}\delta + B_2 M_z) \end{aligned} \tag{2.35}$$

where

$$A_1 = \begin{bmatrix} \frac{-\bar{c}_f - \bar{c}_r}{m\underline{v}_x} & \frac{l_r \bar{c}_r - l_f \bar{c}_f}{m\underline{v}_x^2} - 1 \\ \frac{l_r \bar{c}_r - l_f \bar{c}_f}{I_z} & \frac{-l_f^2 \bar{c}_f - l_r^2 \bar{c}_r}{I_z \underline{v}_x} \end{bmatrix},\ B_{1,1} = \begin{bmatrix} \frac{\bar{c}_f}{m\underline{v}_x} \\ \frac{l_f \bar{c}_f}{I_z} \end{bmatrix},$$

$$A_2 = \begin{bmatrix} \frac{-\bar{c}_f - \bar{c}_r}{m\bar{v}_x} & \frac{l_r \bar{c}_r - l_f \bar{c}_f}{m\underline{v}_x^2} - 1 \\ \frac{l_r \bar{c}_r - l_f \bar{c}_f}{I_z} & \frac{-l_f^2 \bar{c}_f - l_r^2 \bar{c}_r}{I_z \bar{v}_x} \end{bmatrix},\ B_{1,2} = \begin{bmatrix} \frac{\bar{c}_f}{m\bar{v}_x} \\ \frac{l_f \bar{c}_f}{I_z} \end{bmatrix},$$

$$A_3 = \begin{bmatrix} \frac{-\bar{c}_f - \bar{c}_r}{m\underline{v}_x} & \frac{l_r \bar{c}_r - l_f \bar{c}_f}{m\bar{v}_x^2} - 1 \\ \frac{l_r \bar{c}_r - l_f \bar{c}_f}{I_z} & \frac{-l_f^2 \bar{c}_f - l_r^2 \bar{c}_r}{I_z \underline{v}_x} \end{bmatrix},\ B_{1,3} = \begin{bmatrix} \frac{\bar{c}_f}{m\underline{v}_x} \\ \frac{l_f \bar{c}_f}{I_z} \end{bmatrix},$$

$$A_4 = \begin{bmatrix} \frac{-\bar{c}_f - \bar{c}_r}{m\bar{v}_x} & \frac{l_r \bar{c}_r - l_f \bar{c}_f}{m\bar{v}_x^2} - 1 \\ \frac{l_r \bar{c}_r - l_f \bar{c}_f}{I_z} & \frac{-l_f^2 \bar{c}_f - l_r^2 \bar{c}_r}{I_z \bar{v}_x} \end{bmatrix},\ B_{1,4} = \begin{bmatrix} \frac{\bar{c}_f}{m\bar{v}_x} \\ \frac{l_f \bar{c}_f}{I_z} \end{bmatrix},$$

$$\alpha_1 = \frac{|\frac{1}{v_x} - \frac{1}{\bar{v}_x}||\frac{1}{v_x^2} - \frac{1}{\bar{v}_x^2}|}{|\frac{1}{\underline{v}_x} - \frac{1}{\bar{v}_x}||\frac{1}{\underline{v}_x^2} - \frac{1}{\bar{v}_x^2}|},\ \alpha_2 = \frac{|\frac{1}{v_x} - \frac{1}{\underline{v}_x}||\frac{1}{v_x^2} - \frac{1}{\bar{v}_x^2}|}{|\frac{1}{\underline{v}_x} - \frac{1}{\bar{v}_x}||\frac{1}{\underline{v}_x^2} - \frac{1}{\bar{v}_x^2}|},$$

$$\alpha_3 = \frac{|\frac{1}{v_x} - \frac{1}{\bar{v}_x}||\frac{1}{v_x^2} - \frac{1}{\underline{v}_x^2}|}{|\frac{1}{\underline{v}_x} - \frac{1}{\bar{v}_x}||\frac{1}{\underline{v}_x^2} - \frac{1}{\bar{v}_x^2}|},\ \alpha_4 = \frac{|\frac{1}{v_x} - \frac{1}{\underline{v}_x}||\frac{1}{v_x^2} - \frac{1}{\underline{v}_x^2}|}{|\frac{1}{\underline{v}_x} - \frac{1}{\bar{v}_x}||\frac{1}{\underline{v}_x^2} - \frac{1}{\bar{v}_x^2}|}.$$

It is necessary to mention that in traditional vehicles the longitudinal velocity v_x is generally estimated with the engine speed and in electric vehicles the longitudinal velocity v_x is estimated with the motor speed. Considering the variable vehicle slip in the longitudinal direction, the estimated longitudinal velocity cannot be accurate, that is, the calculated weighting factors α_1, α_2, α_3, and α_4 with the estimated longitudinal

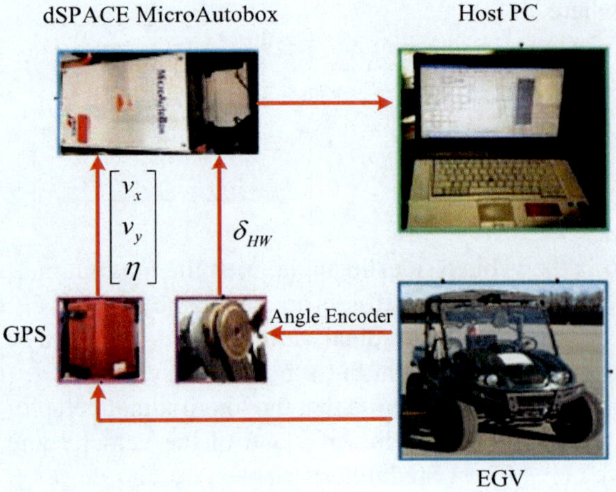

Fig. 2.3 Acquisition system and corresponding sensors

velocity are inaccurate. It is reasonable to assume that

$$k_i \in \begin{bmatrix} 0.9 \ 1.1 \end{bmatrix}, \forall\, i = 1, 2, 3, 4. \tag{2.36}$$

Then, $\|\Delta A_i\|$ can be calculated. In addition, the upper bound of M_z is the maximal longitudinal force multiplied by one half of vehicle width. The proposed algorithm of observer design is applied to estimate the sideslip angle with the yaw rate measurements, that is, the output is

$$y = Cx + v, \tag{2.37}$$

with $C = \begin{bmatrix} 0 \ 1 \end{bmatrix}$. In this application, we design two observers for comparisons:

Observer 1: $k_i \in \begin{bmatrix} 0.9 \ 1.1 \end{bmatrix}, \forall\, i = 1, 2, 3, 4.$
Observer 2: $k_i = 1, \forall\, i = 1, 2, 3, 4.$

Since the main objective is to estimate the sideslip angle, the value of matrix G is selected as $G = \begin{bmatrix} 1 \ 0 \end{bmatrix}$. For a fair comparison, all the conditions are identical when designing the two observers.

In order to evaluate the performance of the designed observers, the EGV is equipped with an acquisition system. The system is consisted of a steering angle encoder and a high-end RT3003 navigation system from Oxford Technical Solutions. It is necessary to mention that the navigation system can measure the longitudinal, the lateral, the yaw, and the pitch motions with a high accuracy. The encoder placed on the steering column can measure the hand-wheel steering angle. The front-wheel steering angle is available since the hand-wheel steering angle and the front-wheel steering angle have a fixed relationship. The acquisition system is described in Fig. 2.3 . The utilized outputs of the GPS include the longitudinal speed (v_x), the lateral speed (v_y), and the yaw rate (η). During the experiments, the measured signals are collected by a dSPACE MicroAutoBox and the data is then transmitted and stored in a Host

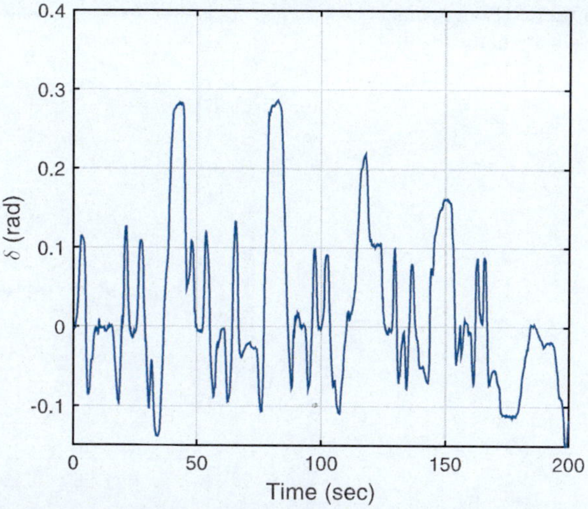

Fig. 2.4 Front-wheel steering angle in the first test

PC. It is necessary to mention that the longitudinal velocity (v_x) can be precisely measured in this application since the equipped GPS has a high performance. However, the expensive GPS cannot equip to vehicles in the market. In order to simulate the inaccurate longitudinal velocity measurements, the GPS-measured longitudinal speed is multiplied by a factor within the range $\left[\,0.95\ 1.05\,\right]$ and the factor randomly changes with respect to time.

We carry out the experimental tests at Transportation Research Center (TRC) in East Liberty, Ohio, USA. Figure 2.4 depicts the front-wheel steering angle during the first test. The maximal steering angle is around 16°, and the front wheels are turned from left to right frequently. Figure 2.5 shows the longitudinal velocity v_x in the test. For the safety issue, the EGV is not driven too fast. The red-dash curve is the actual velocity which is measured with the accurate GPS. Since the revolution is high, the red curve can be taken as the actual value. The blue-solid curve illustrates the simulated uncertain measurements, and the uncertain factor is realized with the Simulink block Uniform Random Number (minimum 0.95 and maximal 1.05). It is obvious that the weighting factors calculated with the measured velocity are not the same with the ones calculated by using the actual velocity. Therefore, the uncertain weighting factors should be considered from the application perspective.

With the experimental data, both observers are used to estimate the sideslip angle during the test. Since the longitudinal speed v_x and the lateral speed v_y are available from the accurate GPS, the measured sideslip angle is the ratio of v_y and v_x when v_x is nonzero. When v_x is close to zero, the sideslip angle is defined as zero. Figure 2.6 shows the yaw rate comparison between the measured signal and the estimated values with the designed two observers. We can see from the zoom-in figure that both observers can estimate the yaw rate well and there is almost no estimation error. Figure 2.7 illustrates the sideslip angle comparison between the measured value and the estimated values. It infers from the figure that the observer 1 can achieve better

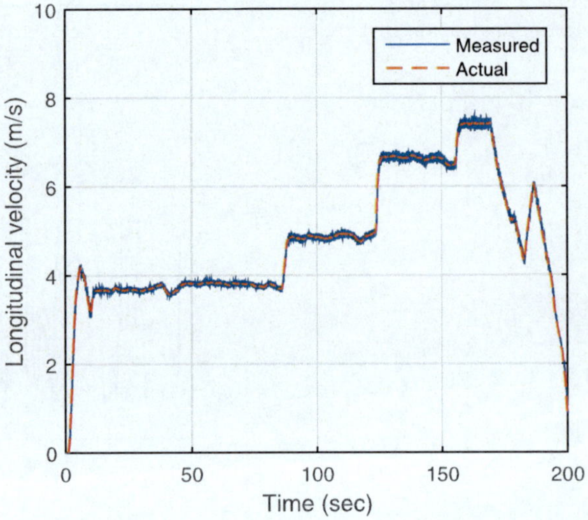

Fig. 2.5 Longitudinal velocity in the first test

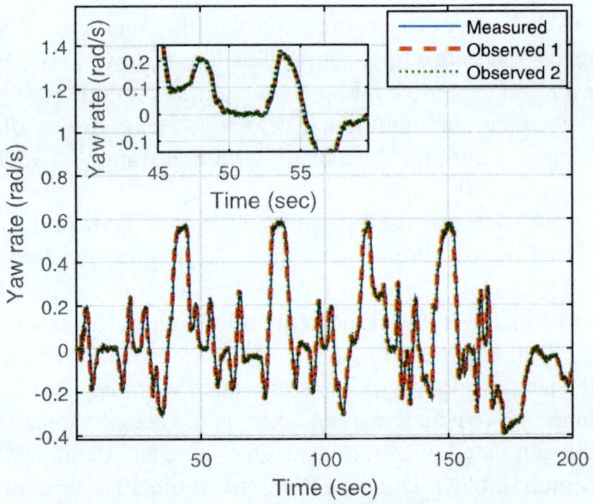

Fig. 2.6 Yaw rate comparison for the designed observers in the first test

performance than observer 2. The sideslip angle estimation error e_β is defined as the measured value minus the estimate value. The errors are sampled with the sampling period of 0.1s. Table 2.1 lists the comparison of sideslip angle estimation error e_β for two observers. Observe that, in terms of 2 norm and ∞ norm of e_β, observer 1 can lead to better performance than observer 2 and the relative improvements are 21.94% and 13.48%, respectively.

In the first test, the data is recorded when the vehicle starts up. Therefore, the measured initial sideslip angle is the same with the estimated initial sideslip angle. In the second test, the measured initial conditions are nonzero and the initial conditions

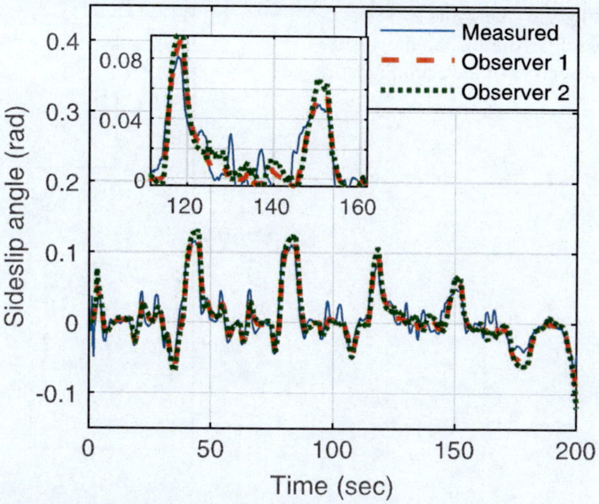

Fig. 2.7 Sideslip angle comparison for the designed observers in the first test

Table 2.1 Comparison of sideslip angle estimation error e_β for two observers in the first test

Method	2 norm of e_β	∞ norm of e_β
Observer 2	0.7673	0.0742
Observer 1	0.6121	0.0642
Relative improvement (%)	21.94	13.48

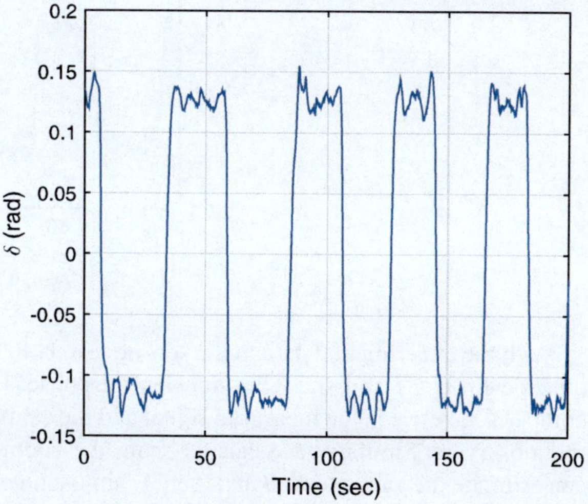

Fig. 2.8 Front-wheel steering angle in the second test

of the designed observers are zeros. Figure 2.8 depicts the front-wheel steering angle in the second test. The measured longitudinal velocity is also multiplied by a random factor similar to the first test for the design observers.

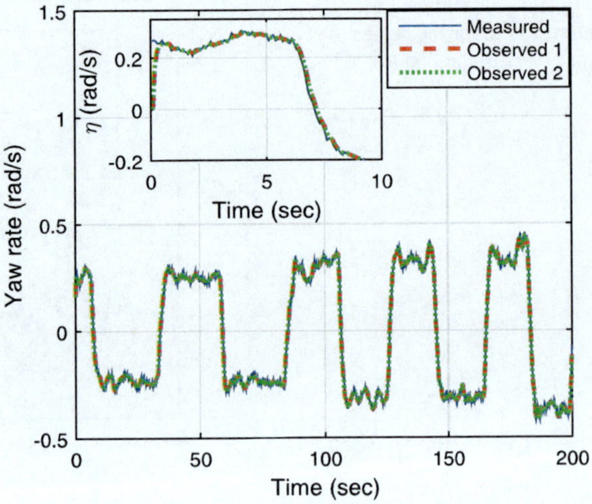

Fig. 2.9 Yaw rate comparison for the designed observers in the second test

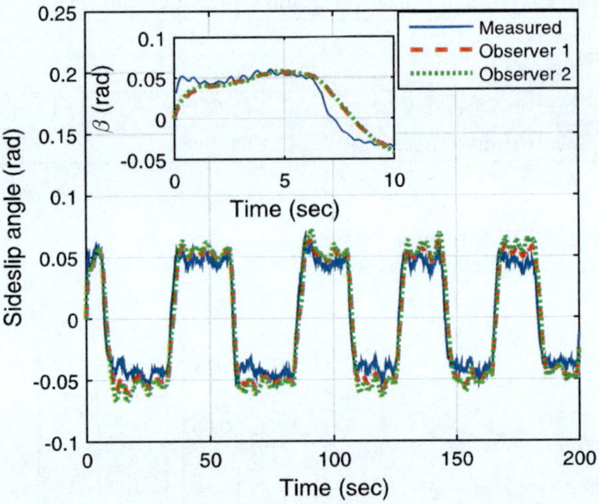

Fig. 2.10 Sideslip angle comparison for the designed observers in the second test

With the experimental data in the second test, both observers are used to estimate the sideslip angle and show the transient response. Figure 2.9 shows the yaw rate comparison between the measured signal and the estimated values with the designed two observers. Similarly, we can see from the zoom-in figure that both observers can estimate the yaw rate well and there is almost no estimation error. Moreover, the estimated yaw rate can increase from zero to the measured value quickly. Figure 2.10 illustrates the sideslip angle comparison between the measured value and the estimated values. Again, observe that the observer 1 can achieve better performance than observer 2. The estimated value can follow the actual value in the first two seconds, that is, the transient response is excellent. Table 2.2 lists the comparison of sideslip angle estimation error e_β for two observers. We can seen that, in terms of 2 norm

Table 2.2 Comparison of sideslip angle estimation error e_β for two observers in the second test

Method	2 norm of e_β	∞ norm of e_β
Observer 2	0.6533	0.0456
Observer 1	0.5295	0.0418
Relative improvement (%)	18.95	8.33

and ∞ norm of e_β, observer 1 can lead to better performance than observer 2 and the relative improvements are 18.95% and 8.33%, respectively. We can conclude that, with the consideration of uncertain scheduling parameters during the design, the designed observer is more reasonable and can achieve better performance.

2.5 Conclusion

In this work, we have investigated the gain-scheduled observer design for LPV systems with uncertain scheduling variables. Due to the uncertain measurements on scheduling variables, the corresponding weighting factors were not the same with the actual ones and assumed to be uncertain times of the actual values. Then, the system was converted to an LPV system with uncertainties. A gain-scheduled sliding mode observer was proposed to estimate the system state vector. Both the stability and the $\mathcal{H}_\infty$ performance were studied for the estimation error system. Based on the analytic results, the observer design method was developed. The developed method was applied to an EGV. Experimental tests were carried out, and the comparison results validated the advantages of the designed observer. In the future research, we will focus on the sliding mode controller design [51, 52] and model predictive control [53, 54].

Acknowledgements This chapter is mainly from the previous work in [55], and some typos are corrected here.

References

1. M. Butcher, A. Karimi, Linear parameter-varying iterative learning control with application to a linear motor system. IEEE/ASME Trans. Mechatron. **15**(3), 412–420 (2010)
2. S. Zhang, J.-J. Yang, G. Zhu, LPV modeling and mixed constrained $\mathcal{H}_2/\mathcal{H}_\infty$ control of an electronic throttle. IEEE/ASME Trans. Mechatron., accepted and in press. https://doi.org/10.1109/TMECH.2014.2364538 (2014)
3. P. Apkarian, P. Gahinet, G. Becker, Self-scheduled $\mathcal{H}_\infty$ control of linear parameter-varying systems: a design example. Automatica **31**(9), 1251–1261 (1995)
4. J.S. Shamma, M. Athans, Guaranteed properties of gain scheduled control for linear parameter-varying plants. Automatica **27**(3), 559–564 (1991)

5. G. Becker, A. Packard, Robust performance of linear parametrically varying systems using parametrically-dependent linear feedback. Syst. Contr. Lett. **23**(3), 205–215 (1994)
6. L. Wu, X. Yang, F. Li, Nonfragile output tracking control of hypersonic air-breathing vehicles with an LPV model. IEEE/ASME Trans. Mechatron. **18**(4), 1280–1288 (2013)
7. D. Li, Y. Xi, The feedback robust MPC for LPV systems with bounded rates of parameter changes. IEEE Trans. Autom. Contr. **55**(2), 503–507 (2010)
8. J. De Caigny, J.F. Camino, R.C.L.F. Oliveira, P.L.D. Peres, J. Swevers, Gain-scheduled dynamic output feedback control for discrete-time LPV systems. Int. J. Robust Nonlinear Control **22**(5), 535–558 (2012)
9. W. Heemels, J. Daafouz, G. Millerioux, Observer-based control of discrete-time LPV systems with uncertain parameters. IEEE Trans. Autom. Contr. **55**(9), 2130–2135 (2010)
10. F. Amato, M. Ariola, C. Cosentino, Finite-time stability of linear time-varying systems: Analysis and controller design. IEEE Trans. Autom. Contr. **55**(4), 1003–1008 (2010)
11. F. Bianchi, C. Kunusch, C. Ocampo-Martinez, R. Sanchez-Pena, A gain-scheduled LPV control for oxygen stoichiometry regulation in PEM fuel cell systems. IEEE Trans. Control Syst. Technol. **22**(5), 1837–1844 (2014)
12. X. Wei, L. del Re, Gain scheduled $\mathcal{H}_\infty$ control for air path systems of diesel engines using LPV techniques. IEEE Trans. Control Syst. Technol. **15**(3), 406–415 (2007)
13. D. Robert, O. Sename, D. Simon, An $\mathcal{H}_\infty$ LPV design for sampling varying controllers: Experimentation with a T-inverted pendulum. IEEE Trans. Control Syst. Technol. **18**(3), 741–749 (2010)
14. P. Gebraad, J. van Wingerden, P. Fleming, A. Wright, LPV identification of wind turbine rotor vibrational dynamics using periodic disturbance basis functions. IEEE Trans. Control Syst. Technol. **21**(4), 1183–1190 (2013)
15. D. Ichalal, B. Marx, J. Ragot, D. Maquin, State estimation of Takagi-Sugeno systems with unmeasurable premise variables. IET Control Theory Appl. **4**(5), 897–908 (2010)
16. K.-Y. Lian, C.-H. Chiang, H.-W. Tu, LMI-based sensorless control of permanent-magnet synchronous motors. IEEE Trans. Ind. Electron. **54**(5), 2769–2778 (2007)
17. K. Cho, J. Kim, S.B. Choi, S. Oh, A high-precision motion control based on a periodic adaptive disturbance observer in a PMLSM. IEEE/ASME Trans. Mechatron. **20**(5), 2158–2171 (2015)
18. G. Phanomchoeng, R. Rajamani, Real-time estimation of rollover index for tripped rollovers with a novel unknown input nonlinear observer. IEEE/ASME Trans. Mechatron. **19**(2), 743–754 (2014)
19. W. He, S.S. Ge, Vibration control of a nonuniform wind turbine tower via disturbance observer. IEEE/ASME Trans. Mechatron. **20**(1), 237–244 (2015)
20. H. Zhang, J. Wang, Y.-Y. Wang, Nonlinear observer design of diesel engine selective catalytic reduction systems with $\mathcal{H}_\infty$ sensor measurements. IEEE/ASME Trans. Mechatron. **20**(4), 1585–1594 (2015)
21. H. Hur, H.-S. Ahn, Unknown input $\mathcal{H}_\infty$ observer-based localization of a mobile robot with sensor failure. IEEE/ASME Trans. Mechatron. **19**(6), 1830–1838 (2014)
22. A. Mujumdar, B. Tamhane, S. Kurode, Observer-based sliding mode control for a class of noncommensurate fractional-order systems. IEEE/ASME Trans. Mechatron. **20**(5), 2504–2512 (2015)
23. R. E. Kalman, A new approach to linear filtering and prediction problems. Trans. ASME, J. Basic Eng. **82**(Series D), 35–45 (1960)
24. S. Helm, M. Kozek, S. Jakubek, Combustion torque estimation and misfire detection for calibration of combustion engines by parametric Kalman filtering. IEEE Trans. Ind. Electron. **59**(11), 4326–4337 (2012)
25. D. Luenberger, Observing the state of a linear system. IEEE Trans. Mil. Electron. **8**(2), 74–80 (1964)
26. K. Erazo, E.M. Hernandez, A model-based observer for state and stress estimation in structural and mechanical systems: Experimental validation. Mech. Syst. Signal. Pr. **43**(1–2), 141–152 (2014)

27. H. Gao, T. Chen, $\mathcal{H}_\infty$ estimation for uncertain systems with limited communication capacity. IEEE Trans. Autom. Contr. **52**(11), 2070–2084 (2007)
28. M. Sahebsara, T. Chen, S.L. Shah, Optimal $\mathcal{H}_\infty$ filtering in networked control systems with multiple packet dropouts. Syst. Contr. Lett. **57**(9), 696–702 (2008)
29. V. Dragan, A.-M. Stoica, Optimal $\mathcal{H}_2$ filtering for a class of linear stochastic systems with sampling. Automatica **48**(10), 2494–2501 (2012)
30. L. Wu, W.X. Zheng, Reduced-order $\mathcal{H}_2$ filtering for discrete linear repetitive processes. Signal Process. **91**(7), 1636–1644 (2011)
31. W.M. Haddad, D.S. Bernstein, D. Mustafa, Mixed-norm $\mathcal{H}_2/\mathcal{H}_\infty$ regulation and estimation: The discrete-time case. Syst. Contr. Lett. **16**(4), 235–247 (1991)
32. D.J.N. Limebeer, B.D.O. Anderson, B. Hendel, A Nash game approach to the mixed $\mathcal{H}_2/\mathcal{H}_\infty$ control problem. IEEE Trans. Autom. Contr. **39**(1), 69–82 (1994)
33. M.-F. Hsieh, J. Wang, Sliding-mode observer for urea-selective catalytic reduction (SCR) mid-catalyst ammonia concentration estimation. Int. J. Auto. Technol. **12**(3), 321–329 (2011)
34. L. Zhao, J. Huang, H. Liu, B. Li, W. Kong, Second-order sliding-mode observer with online parameter identification for sensorless induction motor drives. IEEE Trans. Ind. Electron. **61**(10), 5280–5289 (2014)
35. Y. Lin, Y. Shi, R. Burton, Modeling and robust discrete-time sliding mode contril design for a fluid power electrohydraulic actuator (EHA) system. IEEE/ASME Trans. Mechatron. **18**(1), 1–10 (2013)
36. F. Li, L. Wu, P. Shi, C.C. Lim, State estimation and sliding mode control for semi-Markovian jump systems with mismatched uncertainties. Automatica **51**(1), 385–393 (2015)
37. C.P. Tan, C. Edwards, An LMI approach for designing sliding mode observers. Int. J. Control **74**(6), 1559–1568 (2001)
38. P. Bergsten, R. Palm, D. Driankov, Observers for Takagi-Sugeno fuzzy systems. IEEE Trans. Syst. Man Cybern. B, **32**(1), 114–121 (2002)
39. C.P. Tan, C. Edwards, Sliding mode observers for detection and reconstruction of sensor faults. Automatica **38**(10), 1815–1821 (2002)
40. A. Akhenak, M. Chadli, J. Ragot, D. Maquin, Design of sliding mode unknown input observer for uncertain Takagi-Sugeno model. in *Control Automation, 2007. MED '07. Mediterranean Conference on* (Athens, Greece, June 2007), pp. 1–6
41. M. Fallah, R. Bhat, W.F. Xie, Optimized control of semiactive suspension systems using $\mathcal{H}_\infty$ robust control theory and current signal estimation. IEEE/ASME Trans. Mechatron. **17**(4), 767–778 (2012)
42. S. Ibaraki, S. Suryanarayanan, M. Tomizuka, Design of Luenberger state observers using fixed-structure $\mathcal{H}_\infty$ optimization and its application to fault detection in lane-keeping control of automated vehicles. IEEE/ASME Trans. Mechatron. **10**(1), 34–42 (2005)
43. L. Xie, Y.C. Soh, Robust control of linear systems with generalized positive real uncertainty. Automatica **33**(5), 963–967 (1997)
44. H. Zhang, Y. Shi, A. Saadat Mehr, Robust weighted $\mathcal{H}_\infty$ filtering for networked systems with intermitted measurements of multiple sensors. Int. J. Adapt. Control Signal Process. **25**(4), 313–330 (2011)
45. R. Wang, H. Zhang, J. Wang, Linear parameter-varying controller design for four wheel independently-actuated electric ground vehicles with active steering systems. IEEE Trans. Control Syst. Technol. **22**(4), 1281–1296 (2014)
46. D. Piyabongkarn, R. Rajamani, J.A. Grogg, J.Y. Lew, Development and experimental evaluation of a slip angle estimator for vehicle stability control. IEEE Trans. Control Syst. Technol. **17**(1), 78–88 (2009)
47. Y.H. Hsu, S.M. Laws, J.C. Gerdes, Estimation of tire slip angle and friction limits using steering torque. IEEE Trans. Control Syst. Technol. **18**(4), 896–907 (2010)
48. S.H. You, J.O. Hahn, H. Lee, New adaptive approaches to real-time estimation of vehicle sideslip angle. Control Eng. Pract. **17**(12), 1367–1379 (2009)
49. R. Wang, Y. Chen, D. Feng, X. Huang, J. Wang, Development and performance characterization of an electric ground vehicle with independently actuated in-wheel motors. J. Power Sources **196**(8), 3962–3971 (2011)

50. H. Zhang, X. Zhang, J. Wang, Robust gain-scheduled energy-to-peak control of vehicle lateral dynamics stabilisation. Veh. Syst. Dyn. **52**(3), 309–340 (2014)
51. H. Li, P. Shi, D. Yao, L. Wu, Observer-based adaptive sliding mode control of nonlinear markovian jump systems. Automatica **64**(1), 133–142 (2016)
52. H. Li, J. Wang, P. Shi, Output-feedback based sliding mode control for fuzzy systems with actuator saturation. IEEE Trans. Fuzzy Syst., Accepted and in press. https://doi.org/10.1109/TFUZZ.2015.2513085 (2016)
53. H. Li, Y. Shi, Robust distributed model predictive control of constrained continuous-time nonlinear systems: A robustness constraint approach. IEEE Trans. Autom. Contr. **59**(6), 1673–1678 (2014)
54. H. Li, Y. Shi, Robust distributed receding horizon control of continuous-time nonlinear systems: Handling communication delays and disturbances. Automatica **50**(41), 1264–1271 (2014)
55. H. Zhang, G. Zhang, J. Wang, $\mathcal{H}_\infty$ observer design for LPV systems with uncertain measurements on scheduling variables: Application to an electric ground vehicle. IEEE/ASME Trans. Mechatron. **21**(3), 1659–1670 (2016)

Chapter 3
Sideslip Angle Estimation of An Electric Ground Vehicle Via Finite-Frequency $\mathcal{H}_\infty$ Approach

Abstract The lateral stability is critically important especially when the vehicle is steered at a high longitudinal speed. Loss of lateral stability would lead to severe accidents. Therefore, the study of lateral stability has been a hot topic for decades. In order to monitor the lateral stability or improve the stability by using the feedback control, the sideslip angle is an important index. However, the sideslip angle is not measurable by using an affordable physical sensor. An alternative approach is to estimate the sideslip angle with the measurements of relatively cheap sensors. In this chapter, we investigate the problem of sideslip angle observer design for an electric ground vehicle (EGV). The EGV is equipped with an advanced navigation system. The lateral velocity, the longitudinal velocity, and the yaw rate are available. Thus, the sideslip angle which is defined as the ratio of lateral velocity and longitudinal velocity is also available. Meanwhile, the hand-wheel steering angle is also measurable in this application. The main work is to estimate the sideslip angle with the measurements of yaw rate based on the vehicle lateral dynamics. The dynamic model is first established, and the parameters are identified with experimental data. Since the dynamic model is nonlinear, in order to facilitate the system analysis and observer design, the nonlinear model is transformed to a linear-parameter-varying (LPV) system. An observer is proposed based on the LPV form. By defining the estimation error, a compact system which contains the estimation error and the original dynamics is obtained. Considering the frequency of the front-wheel steering angle, the finite-frequency $\mathcal{H}_\infty$ performance of the compact system is exploited. An optimal observer design method is then developed. For the EGV, the observer is designed according to the developed method. The performance of the designed observer is illustrated with experimental test data.

3.1 Introduction

The last decade has witnessed the increasing attention to electric vehicles (EVs) and hybrid electric vehicles (HEVs) not only due to environmental and energy concerns but also due to the new emerging technologies; see [1–5] and the references therein. The research on the EVs and HEVs focuses on various areas such as active suspension

H. Zhang et al., *Robust Gain-Scheduled Estimation and Control of Electrified Vehicles via LPV Technique*, Key Technologies on New Energy Vehicles,
https://doi.org/10.1007/978-981-19-8509-6_3

energy regeneration in [6], power management in [7], component design in [8, 9], smart grid in [10], and control and estimation in [11–15]. For the traditional lateral stability side, there are some new challenges for EVs and HEVs including the control allocation [16], fault-tolerant control [17], etc. In the stability analysis and synthesis, the sideslip angle is of importance. However, currently, the sideslip angle is not measurable by an affordable physical sensor. To deal with the conflict, a promising strategy is to develop observers to estimate the sideslip angle with measurements obtained by affordable physical sensor. The authors in [18] employed the GPS-based velocity measurements to design the observer. In the work of [19], the authors utilized a combination of model-based estimation and kinematics-based estimation to estimate the angle. The presence of road bank angle and variations in tire-road characteristics were compensated. Different from [19], the steering torque was used in [20] to achieve the similar task. The recursive least square and Kalman filter approaches can be seen in [21] for the estimation of an EV.

For the model-based observer design, typically, there are Kalman and extended Kalman filters [22], $\mathcal{H}_\infty$ filters [23, 24], and energy-to-peak filters [25]. Different strategies have their own advantages. In the $\mathcal{H}_\infty$ filter design, the knowledge on the external noise is not required. Moreover, there is a good relationship between the time-domain understanding and the frequency-domain explanation. Thus, the $\mathcal{H}_\infty$ filtering has been paid relatively more efforts. However, most of the $\mathcal{H}_\infty$ filter design was done for the entire frequency. For many practical applications, the external disturbance may not cover the entire frequency. In this case, the finite-frequency method may be more useful. In the literature, there are some finite-frequency filter design works such as [26–28]. But these works only focused on the theory studies.

Regarding the sideslip angle estimation, in the literature, there are many interesting works. The authors in [29] studied the sideslip angle estimation using lateral tire force sensors. With the estimated angle, the lateral stability problem was investigated. The estimation work was extended in [30] in which both the sideslip and roll angles of an electric vehicles were estimated through RLS and Kalman filter approaches. In [31], the problem of robust vehicle sideslip angle estimation was explored. The estimation work was done through a disturbance rejection filter, and a Magnetometer with GPS was equipped. The neural networks were employed to estimate the sideslip angle in [32]. The authors in [33] used a variable structure extended Kalman filter to estimate the vehicle sideslip angle for vehicles on a low friction road. Though there are a lot of existing works on the estimation of sideslip angle, few researchers focused on the road condition with relatively large variation.

In this chapter, we aim to design the robust sideslip angle observer for an EGV based on the lateral model and the measurements of yaw rate. To achieve the target, the lateral model of the EGV is studied with consideration of uncertain tire characteristics. Since the cornering stiffness and the inertial moment are not easy to measure, a parameter identification procedure is provided with experimental data. Then, the model validation is also offered. When the lateral model is determined, a model-based state observer is proposed. With the defined estimation error, an uncertain compact model is obtained. Both the robust stability and the finite-frequency $\mathcal{H}_\infty$ performance are investigated for the compact system. Then, an optimal $\mathcal{H}_\infty$

observer design method is proposed. The designed procedure and method are illustrated via the EGV. It infers from the experimental results that the designed observer is effective. The contributions of the work can be summarized as: (1) We have done a comprehensive study from the system modeling and identification to the observer design; (2) we extend the finite-frequency $\mathcal{H}_\infty$ performance for uncertain LPV systems and investigate the conditions which can guarantee the robust finite-frequency $\mathcal{H}_\infty$ performance; and (3) experimental tests are carried out to evaluate the designed filter.

3.2 Problem Formulation and Preliminary

3.2.1 Introduction of the Electric Ground Vehicle

In this work, we aim to study the dynamics of an EGV shown in Fig. 3.1 and design a state estimator based on the dynamics. The EGV is actuated by four permanent-magnet brushless direct-current electric motors which are mounted in four wheels and each in-wheel motor has the power of 7.5 kW. The power supplier of the EGV is a 72 V lithium-ion power battery pack, and the pack is composed of 22 battery cells which are connected in series. The pack can provide the peak power of 40 kW. Therefore, the battery pack matches the motors well. In addition, the vehicle is steered by front wheels. More details on the development of the EGV and the specifications can be referred to the work in [34].

Fig. 3.1 Picture view of the EGV and navigation system

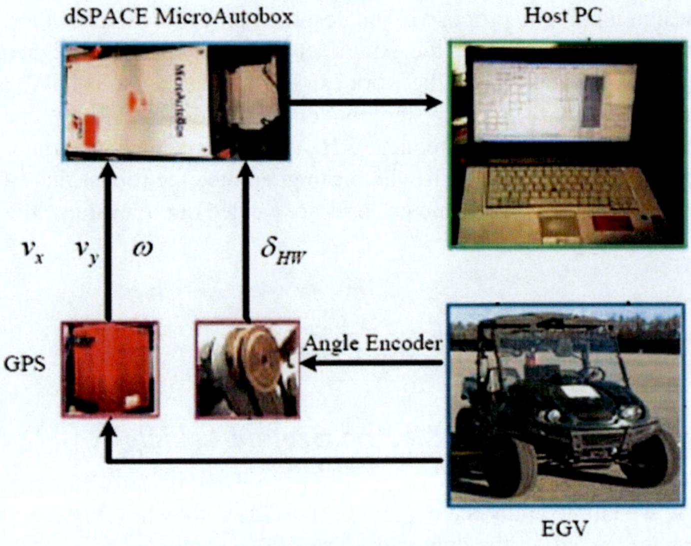

Fig. 3.2 Acquisition system and corresponding sensors

To study the dynamics of the EGV, an acquisition system is equipped, which is consisted of a high-end RT3003 navigation system from Oxford Technical Solutions and a steering angle encoder. The navigation system includes a dual-antenna differential global positioning system (GPS) and an integrated inertial measurement system. Thus, the navigation system can measure the longitudinal, the lateral, the yaw, and the pitch motions with a high accuracy. The encoder is placed on the steering column, and the hand-wheel steering angle can be measured. Since the hand-wheel steering angle and the front-wheel steering angle have a fixed relationship, the front-wheel steering angle is available in this application. Specifically, the acquisition system is described in Fig. 3.2. The front-wheel steering angle of the EGV is measured by the encoder. Because the GPS is placed on the EGV, the longitudinal speed (v_x), the lateral speed (v_y), and the yaw rate (ω) are available. The measured signals are collected by a dSPACE MicroAutoBox, and the data is transmitted and stored in a Host PC. It is necessary to mention that the yaw rate is measured by the GPS in this application. However, in the market, much more cheaper sensors are available for the yaw rate measurement. In addition, the sideslip angle is defined as the ratio of v_y and v_x. Since both velocities v_y and v_x are available, the sideslip angle is also available.

3.2.2 System Modeling and Identification

Though there are four independent-driven wheels, we can analyze the dynamics by employing a two-degree-of-freedom bicycle model. For the lateral dynamics, Fig. 3.3

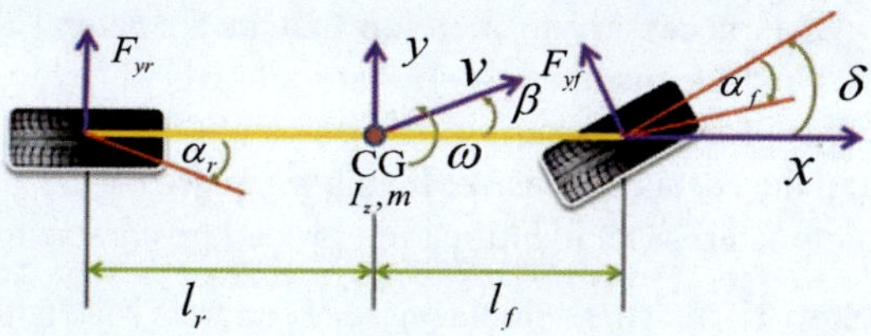

Fig. 3.3 Schematic diagram of a simplified lateral dynamics model

depicts the notations in the study. The total vehicle mass is represented by m. The inertia moment about the yaw axis through its center of gravity (CG) is I_z. The front axis is located at the distance l_f from the CG, and the rear axis is located at the distance l_r from the CG. The steering angle δ is controlled by the driver, and it can be measured online via the encoder mounted on the steering column. The wheel slip angles α_f and α_r of the front and rear tires are decided the front and rear lateral tire forces F_{yf} and F_{yr}, respectively. As there are two tires on each axis, the lateral tire force is the lumped lateral force on each axis. For the longitudinal side, the longitudinal forces are distributed equally. Thus, there is no yaw moment induced by the longitudinal forces on four tires.

Due to the nonlinearity in the lateral tire model, the relationship between the lateral force and the wheel slip angle is complicated. However, the nonlinear lateral tire model describes the loss of lateral force when the force reaches the peak value and the wheel slip angle is relatively large. Therefore, we can adopt the uncertain tire model developed in [35]

$$\begin{aligned} F_{yf} &\approx (c_f + \Delta c_f N(t))\alpha_f = \bar{c}_f \alpha_f, \\ F_{yr} &\approx (c_r + \Delta c_r N(t))\alpha_r = \bar{c}_r \alpha_r. \end{aligned} \tag{3.1}$$

Here, the approximated values of corning stiffness are denoted by $\bar{c}_f$ and $\bar{c}_r$. The equivalent nominal coefficients are c_f and c_r. Meanwhile, the absolute value of the scalar $N(t)$ is no more than one and the bounds of the uncertainties Δc_f and Δc_r can be estimated before hand. Similarly, the values of $\bar{c}_f$ and $\bar{c}_r$ are the lumped ones of two tires.

By defining the vehicle sideslip angle β at the CG and studying the yaw motion, a state-space model of the lateral dynamics can be obtained as:

$$\dot{x}(t) = Ax(t) + B\delta, \tag{3.2}$$

where

$$x(t) = \begin{bmatrix} \beta \\ \omega \end{bmatrix}, \quad B = \begin{bmatrix} \frac{\bar{c}_f}{mv_x} \\ \frac{l_f \bar{c}_f}{I_z} \end{bmatrix},$$

$$A = \begin{bmatrix} \frac{-\bar{c}_f - \bar{c}_r}{mv_x} & \frac{l_r \bar{c}_r - l_f \bar{c}_f}{mv_x^2} - 1 \\ \frac{l_r \bar{c}_r - l_f \bar{c}_f}{I_z} & \frac{-l_f^2 \bar{c}_f - l_r^2 \bar{c}_r}{I_z v_x} \end{bmatrix}.$$

It is necessary to mention that in the lateral dynamics (3.2), the vehicle mass (m) and the distances (l_f, l_r) are relatively easy to be determined. However, precise values for the cornering stiffness and the inertia moment require specific facilities and the cost is expensive. In this work, we use an alternative method to identify these parameters. The identification procedures are as follows:

Step 1: Test the vehicle on a dry concrete road and collect the corresponding signals including the steering angle, the yaw rate, the longitudinal velocity, and the lateral velocity. In order to guarantee the identification accuracy, the maneuver should cover various longitudinal velocity v_x and various steering angle δ.

Step 2: Establish the lateral model in Matlab/Simulink without uncertainties and set the parameters $\bar{c}_f$, $\bar{c}_f$, and I_z as the inputs. Employ the nonlinear optimization method to minimize the following cost function:

$$J = \int_0^{t_{\text{test}}} (|\beta_s - \beta_m| + \lambda|\omega_s - \omega_m|)d\tau, \tag{3.3}$$

where t_{test} is the time duration of the test, β_s is the sideslip angle from the model, β_m is the measured sideslip angle, ω_s is the yaw rate from the model, ω_m is the measured yaw rate, and λ is a weighting factor.

Step 3: Validate the model with the identified parameters on another experimental test. If the performance is not good, carry out the test in **Step 1** again.

The experimental tests of the EGV with the navigation system are carried out at Transportation Research Center (TRC) in East Liberty, Ohio, USA. In the first step, we drive the EGV with different safe longitudinal velocities and steering angles at TRC shown in Fig. 3.1. With the collected data and measured parameters in Table 3.1, the optimization work is done in the second step and the necessary parameters are obtained. Figure 3.4 shows the sideslip angle comparison between the measured signal and the predictive value. The measured value of sideslip angle varies frequently within the range [−0.1, 0.12]rad, which indicates that the steering wheel is turned left and right several times. We can see that, after the optimization, the model-predicted sideslip angle can capture the measured signal well. Figure 3.5 is the yaw rate comparison for the identification data, in which the value is within the range of [−0.4, 0.6]rad/s. The model-predicted yaw rate with the identified parameters is

Table 3.1 Measured parameters for the EGV

Parameter	Value
m	880 kg
l_s	1.372 m
l_f	1.082 m
l_r	0.808 m

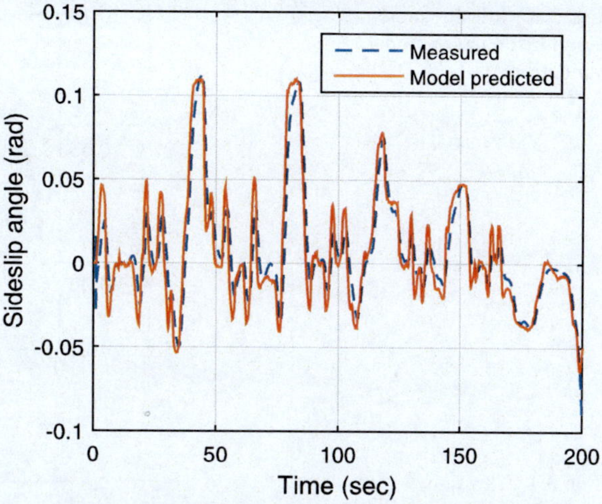

Fig. 3.4 Sideslip angle comparison in the identification

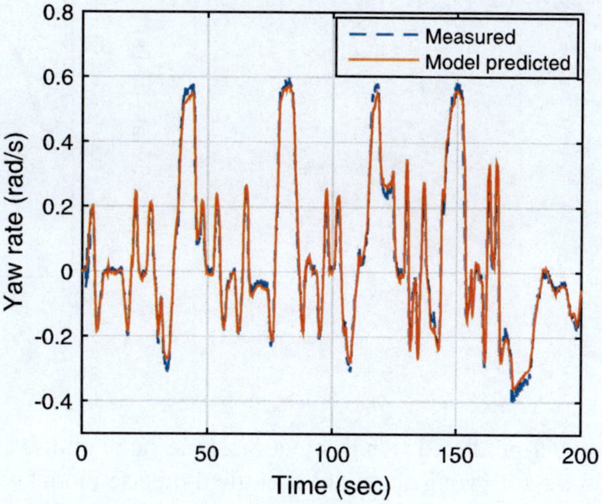

Fig. 3.5 Yaw rate comparison in the identification

identical with measured signal during most of the test. In addition, the longitudinal velocity is from 0 to 10 m/s. It infers from Figs. 3.4 and 3.5 that the lateral model with the identified parameters matches the measured signals for identification well. In order to show the lateral model with identified parameters is effective for other tests beside the identification data, another experimental test is done. The validation test lasts for 90 s and the steering wheel is operated slowly during the first 30 s. From 30 s to 80 s, the steering wheel is switched from the left to the right and from the right to the left for more than 10 times. Figures 3.6 and 3.7 depict the corresponding sideslip angle comparison and the yaw rate comparison in the model validation, respectively. We can see that the identified model is also effective for the validation test since the

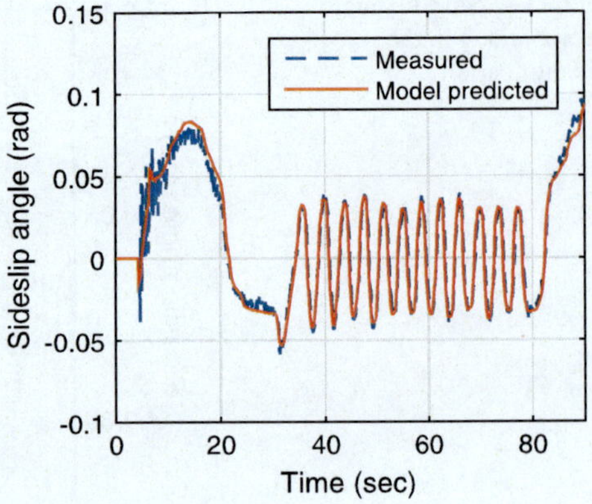

Fig. 3.6 Sideslip angle comparison in the model validation

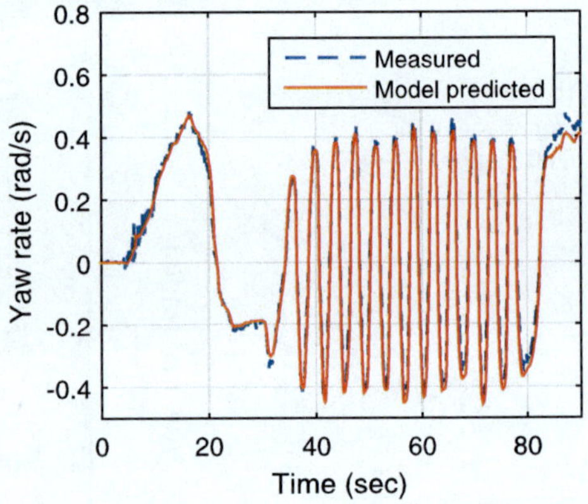

Fig. 3.7 Yaw rate comparison in the model validation

model-predicted signals are almost the same with the measured signals. Therefore, in the following study, the identified precise model will be used in the analysis and synthesis. Since the experimental tests are carried out on the dry concrete road, the identified cornering stiffness can be taken maximal one. The nominal stiffness and the minimal stiffness can be estimated based on the maximal value.

3.2.3 Model Transformation and Problem Formulation

Note that the longitudinal velocity v_x is nonconstant. Therefore, the system matrix in the lateral model is varying. In order to formulate the lateral model into the linear-parameter-varying (LPV) form, we assume that the longitudinal velocity v_x is bounded as:

$$v_x \in [\underline{v}_x,\ \bar{v}_x] \tag{3.4}$$

with the bounds $\underline{v}_x$ and $\bar{v}_x$ are known. Inspired by the work in [35], the matrix pair $(A,\ B)$ can be also written in a summation form as:

$$(A,\ B) = \sum_{l=1}^{3} \alpha_l (A_l,\ B_l), \tag{3.5}$$

where

$$(A_1,\ B_1) = \left(\begin{bmatrix} \frac{-\bar{c}_f - \bar{c}_r}{m\bar{v}_x} & \frac{l_r\bar{c}_r - l_f\bar{c}_f}{m\bar{v}_x^2} - 1 \\ \frac{l_r\bar{c}_r - l_f\bar{c}_f}{I_z} & \frac{-l_f^2\bar{c}_f - l_r^2\bar{c}_r}{I_z\bar{v}_x} \end{bmatrix},\ \begin{bmatrix} \frac{\bar{c}_f}{m\underline{\bar{v}}_x} \\ \frac{l_f\bar{c}_f}{I_z} \end{bmatrix} \right),$$

$$(A_2,\ B_2) = \left(\begin{bmatrix} \frac{-\bar{c}_f - \bar{c}_r}{m} \times \frac{\bar{v}_x + \underline{v}_x}{2\bar{v}_x\underline{v}_x} & \frac{l_r\bar{c}_r - l_f\bar{c}_f}{m\bar{v}\underline{v}_x} - 1 \\ \frac{l_r\bar{c}_r - l_f\bar{c}_f}{I_z} & \frac{-l_f^2\bar{c}_f - l_r^2\bar{c}_r}{I_z} \times \frac{\bar{v} + \underline{v}_x}{2\bar{v}\underline{v}_x} \end{bmatrix},\ \begin{bmatrix} \frac{\bar{c}_f}{m} \times \frac{\bar{v}_x + \underline{v}_x}{2\bar{v}_x\underline{v}_x} \\ \frac{l_f\bar{c}_f}{I_z} \end{bmatrix} \right),$$

$$(A_3,\ B_3) = \left(\begin{bmatrix} \frac{-\bar{c}_f - \bar{c}_r}{m\underline{v}_x} & \frac{l_r\bar{c}_r - l_f\bar{c}_f}{m\underline{v}_x^2} - 1 \\ \frac{l_r\bar{c}_r - l_f\bar{c}_f}{I_z} & \frac{-l_f^2\bar{c}_f - l_r^2\bar{c}_r}{I_z\underline{v}_x} \end{bmatrix},\ \begin{bmatrix} \frac{\bar{c}_f}{m\underline{v}_x} \\ \frac{l_f\bar{c}_f}{I_z} \end{bmatrix} \right),$$

$\sum_{l=1}^{3} \alpha_l = 1,\ \ 0 \le \alpha_l \le 1.$

With the matrix pair transformation, the lateral dynamics is rewritten as:

$$\dot{x}(t) = \sum_{l=1}^{3} \alpha_l \left(A_l x(t) + B_l \delta \right). \tag{3.6}$$

To decouple the nominal and uncertain terms, the lateral dynamics in (3.6) is further rewritten in the following form:

$$\begin{aligned} \dot{x}(t) &= \sum_{l=1}^{3} \alpha_l ((\bar{A}_l + \Delta A_l)x(t) + (\bar{B}_l + \Delta B_l)\delta) \\ &= (\bar{A}(\alpha) + \Delta A(\alpha))x(t) + (\bar{B}(\alpha) + \Delta B(\alpha))\delta, \end{aligned} \tag{3.7}$$

where

$$
\begin{aligned}
\bar{A}_1 &= \begin{bmatrix} \frac{-c_f-c_r}{m\bar{v}_x} & \frac{l_rc_r-l_fc_f}{m\bar{v}_x^2}-1 \\ \frac{l_rc_r-l_fc_f}{I_z} & \frac{-l_f^2c_f-l_r^2c_r}{I_z\bar{v}_x} \end{bmatrix}, \\
\bar{A}_2 &= \begin{bmatrix} \frac{-c_f-c_r}{m}\times\frac{\bar{v}+\underline{v}_x}{2\bar{v}\underline{v}_x} & \frac{l_rc_r-l_fc_f}{m\bar{v}\underline{v}_x}-1 \\ \frac{l_rc_r-l_fc_f}{I_z} & \frac{-l_f^2c_f-l_r^2c_r}{I_z}\times\frac{\bar{v}_x+\underline{v}_x}{2\bar{v}_x\underline{v}_x} \end{bmatrix}, \\
\bar{A}_3 &= \begin{bmatrix} \frac{-c_f-c_r}{m\underline{v}_x} & \frac{l_rc_r-l_fc_f}{m\underline{v}_x^2}-1 \\ \frac{l_rc_r-l_fc_f}{I_z} & \frac{-l_f^2c_f-l_r^2c_r}{I_z\underline{v}_x} \end{bmatrix}, \\
\Delta A_1 &= E_{1,1}\bar{M}(t)F_{1,1},\ \Delta A_2 = E_{1,2}\bar{M}(t)F_{1,2}, \\
\Delta A_3 &= E_{1,3}\bar{M}(t)F_{1,3}, \\
E_{1,1} &= \begin{bmatrix} \frac{-\Delta c_f-\Delta c_r}{m\bar{v}_x} & \frac{l_r\Delta c_r-l_f\Delta c_f}{m\bar{v}_x^2} \\ \frac{l_r\Delta c_r-l_f\Delta c_f}{I_z} & \frac{-l_f^2\Delta c_f-l_r^2\Delta c_r}{I_z\bar{v}_x} \end{bmatrix}, \\
E_{1,2} &= \begin{bmatrix} \frac{-\Delta c_f-\Delta c_r}{m}\times\frac{\bar{v}_x+\underline{v}_x}{2\bar{v}_x\underline{v}_x} & \frac{l_r\Delta c_r-l_f\Delta c_f}{m\bar{v}_x\underline{v}_x} \\ \frac{l_r\Delta c_r-l_f\Delta c_f}{I_z} & \frac{-l_f^2\Delta c_f-l_r^2\Delta c_r}{I_z}\times\frac{\bar{v}_x+\underline{v}_x}{2\bar{v}_x\underline{v}_x} \end{bmatrix}, \\
E_{1,3} &= \begin{bmatrix} \frac{-\Delta c_f-\Delta c_r}{m\underline{v}_x} & \frac{l_r\Delta c_r-l_f\Delta c_f}{m\underline{v}_x^2} \\ \frac{l_r\Delta c_r-l_f\Delta c_f}{I_z} & \frac{-l_f^2\Delta c_f-l_r^2\Delta c_r}{I_z\underline{v}_x} \end{bmatrix}, \\
\bar{M}(t) &= \begin{bmatrix} N(t) & 0 \\ 0 & N(t) \end{bmatrix},\ F_{1,1}=I,\ F_{1,2}=I,\ F_{1,3}=I, \\
B_{1,1} &= \begin{bmatrix} \frac{c_f}{m\bar{v}_x} \\ \frac{l_fc_f}{I_z} \end{bmatrix},\ B_{1,2} = \begin{bmatrix} \frac{c_f}{m}\times\frac{\bar{v}_x+\underline{v}_x}{2\bar{v}\underline{v}_x} \\ \frac{l_fc_f}{I_z} \end{bmatrix}, \\
\Delta B_{1,1} &= E_{2,1}\bar{M}(t)F_{2,1},\ \Delta B_{1,2} = E_{2,2}\bar{M}(t)F_{2,2}, \\
\Delta B_{1,3} &= E_{2,3}\bar{M}(t)F_{2,3},\ E_{2,1} = \begin{bmatrix} \frac{\Delta c_f}{m\bar{v}_x} & 0 \\ 0 & \frac{l_f\Delta c_f}{I_z} \end{bmatrix}, \\
E_{2,2} &= \begin{bmatrix} \frac{\Delta c_f}{m}\times\frac{\bar{v}_x+\underline{v}_x}{2\bar{v}_x\underline{v}_x} & 0 \\ 0 & \frac{l_f\Delta c_f}{I_z} \end{bmatrix}, \\
E_{2,3} &= \begin{bmatrix} \frac{\Delta c_f}{m\underline{v}_x} & 0 \\ 0 & \frac{l_f\Delta c_f}{I_z} \end{bmatrix},\ B_{1,3} = \begin{bmatrix} \frac{c_f}{m\underline{v}_x} \\ \frac{l_fc_f}{I_z} \end{bmatrix}, \\
F_{2,1} &= \begin{bmatrix} 1 \\ 1 \end{bmatrix},\ F_{2,2} = \begin{bmatrix} 1 \\ 1 \end{bmatrix},\ F_{2,3} = \begin{bmatrix} 1 \\ 1 \end{bmatrix}, |N(t)| \le 1.
\end{aligned}
$$

For the observer design, we assume that the yaw rate is available since the yaw rate sensor is relatively cheap. Then, the output of the lateral dynamic system is

$$y = Cx(t), \tag{3.8}$$

with $C = \begin{bmatrix} 0 & 1 \end{bmatrix}$. In order to estimate the sideslip angle with the available yaw rate measurement, the observer has the following form:

$$\dot{\hat{x}}(t) = \bar{A}(\alpha)\hat{x}(t) + \bar{B}(\alpha)\delta + L(y - C\hat{x}(t)), \tag{3.9}$$

with L as the observer gain to be determined. By defining the estimation error as:

$$e(t) = x(t) - \hat{x}(t), \tag{3.10}$$

the estimation error system has the following expression:

$$\dot{e} = (\bar{A}(\alpha) - LC)e + \Delta A(\alpha)x(t) + \Delta B(\alpha)\delta. \tag{3.11}$$

Note that not only the estimation error but also the system state $x(t)$ are involved in the estimation error system. To incorporate the system state $x(t)$, augment the estimation error system with the vehicle lateral dynamics and the following compact form is obtained:

$$\begin{aligned}\begin{bmatrix} \dot{e}(t) \\ \dot{x}(t) \end{bmatrix} = & \begin{bmatrix} \bar{A}(\alpha) - LC & \Delta A(\alpha) \\ 0 & \bar{A}(\alpha) + \Delta A(\alpha) \end{bmatrix} \begin{bmatrix} e(t) \\ x(t) \end{bmatrix} \\ & + \begin{bmatrix} \Delta B(\alpha) \\ B(\alpha) + \Delta B(\alpha) \end{bmatrix} \delta. \end{aligned} \tag{3.12}$$

In order to evaluate the uncertainties in the compact form, we have the following system as

$$\dot{\xi} = (\tilde{A}(\alpha) + \Delta\tilde{A}(\alpha))\xi + (\tilde{B}(\alpha) + \Delta\tilde{B}(\alpha))\delta, \tag{3.13}$$

where

$$\begin{aligned} \xi = & \begin{bmatrix} e(t) \\ x(t) \end{bmatrix}, \ \tilde{A}(\alpha) = \begin{bmatrix} \bar{A}(\alpha) - LC & 0 \\ 0 & \bar{A}(\alpha) \end{bmatrix}, \\ \Delta\tilde{A}(\alpha) = & \ \tilde{E}_1(\alpha)\bar{M}(t)\tilde{F}_1, \ \tilde{E}_1(\alpha) = \begin{bmatrix} E_1^{\mathrm{T}}(\alpha) \ E_1^{\mathrm{T}}(\alpha) \end{bmatrix}^{\mathrm{T}}, \\ \tilde{F}_1 = & \begin{bmatrix} 0 \ F_1 \end{bmatrix}, \ \Delta\tilde{B}(\alpha) = \tilde{E}_2(\alpha)\bar{M}(t)F_2, \\ \tilde{B}(\alpha) = & \begin{bmatrix} 0 \\ B(\alpha) \end{bmatrix}, \ \tilde{E}_2(\alpha) = \begin{bmatrix} E_2(\alpha) \\ E_2(\alpha) \end{bmatrix}, \ \tilde{F}_2 = F_2. \end{aligned}$$

3.2.4 Design Objectives

The primary objective of observer design is to ensure that the estimation error would converge to zero. Therefore, the estimation error system in (3.13) should be stable. However, beside the stability, the wheel steering angle δ would directly affect the estimation error. In order to attenuate the effect of the wheel steering angle to the estimation error, there are several strategies such as the $\mathcal{H}_\infty$ filtering and the energy-to-peak filtering. In most of the existing work on the $\mathcal{H}_\infty$ filtering, the optimization

Fig. 3.8 Wheel steering angle and the amplitude spectrum

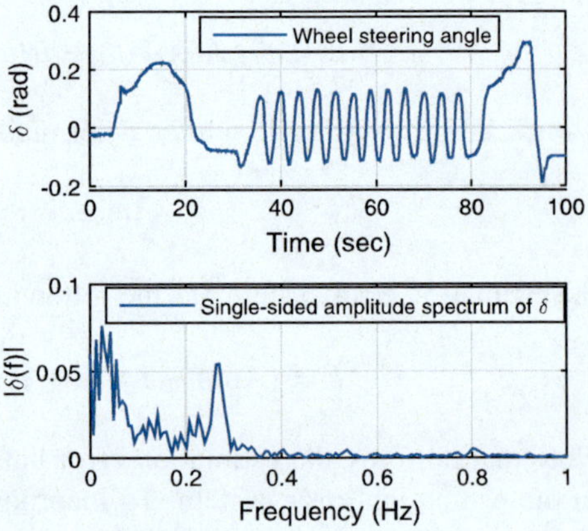

was done for the entire frequency. However, in this application, the wheel steering angle δ only works in the low frequency. An $\mathcal{H}_\infty$ observer optimized for the entire frequency may not be the optimal one for low frequency. Therefore, it is meaningful to design the $\mathcal{H}_\infty$ estimator with the optimization for the finite frequency. Figure 3.8 describes the wheel steering angle δ of an experimental test and the corresponding amplitude spectrum. We can see that the wheel steering angle δ works in the low frequency. We can get the maximal amplitude and the amplitude of frequency smaller than 0.4 Hz is smaller than 2.7% of the maximal amplitude. Therefore, it is assumed that the typical frequency of wheel steering angle δ is smaller than 0.4 Hz and the optimization should be done for the frequency range [0 0.4 Hz]. As the sideslip angle is the signal to be estimated, the signal z in the $\mathcal{H}_\infty$ filtering scheme is chosen as:

$$z = G\xi, \tag{3.14}$$

with $G = [1\ 0\ 0\ 0]$.

On the other hand side, the estimated sideslip angle may be used for the feedback control or fault diagnosis. The eigenvalue locations of $A - LC$ should be on the left-hand side of eigenvalue locations of A such that the response of the estimator is faster than the original system. Figure 3.9 illustrates the eigenvalue locations of matrix A with different longitudinal velocities v_x. When increasing the longitudinal velocity, the eigenvalue locations of matrix A become closer to the imaginary axis.

In summary, the design objectives of this work are to design the observer in (3.9) such that

O1 The compact system in (3.13) is robustly and asymptotically stable.

O2 For the finite-frequency range [0 0.4 Hz], the $\mathcal{H}_\infty$ performance in the following form is achieved:

$$\|z\|_2 < \gamma \|\delta\|_2, \tag{3.15}$$

where $\|z\|_2$ denotes the 2-norm of the signal z, $\|\delta\|_2$ is the 2 norm of the signal δ, and γ is $\mathcal{H}_\infty$ performance index.

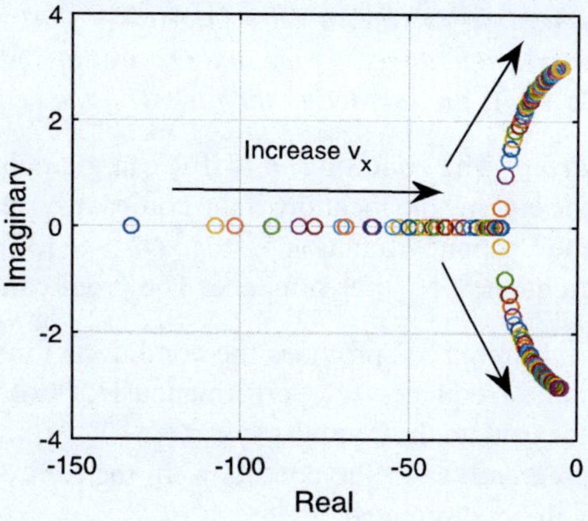

Fig. 3.9 Eigenvalue locations of matrix A with different longitudinal velocities v_x

3.3 Observer Design

In this section, we first assume that the observer gain is given and analyze the compact system dynamics. Based on the analysis results, the observer design method will be developed. Before proceeding further, we introduce the following lemmas.

Lemma 3.1 [36] *If there exist real matrices $\Omega = \Omega^{\mathrm{T}}$, $\hat{G}$ and $\hat{H}$ with compatible dimensions, and $\hat{M}(t)$ satisfying $|\hat{M}(t)| \le I$, then, the following condition*

$$\Omega + \hat{G}\hat{M}(t)\hat{H} + \hat{H}^{\mathrm{T}}\hat{M}(t)\hat{G}^{\mathrm{T}} < 0, \tag{3.16}$$

holds if and only if there exists a positive scalar $\varphi > 0$ such that the following condition

$$\begin{bmatrix} \Omega & \hat{G} & \varphi\hat{H}^{\mathrm{T}} \\ * & -\varphi I & 0 \\ * & * & -\varphi I \end{bmatrix} < 0 \tag{3.17}$$

is satisfied.

Lemma 3.2 *For a prescribed real number $\gamma > 0$, suppose that there exist a symmetric matrix P, a positive-definite matrix $Q = Q^{\mathrm{T}}$, matrices D and H such that*

$$\begin{bmatrix} -Q & P - D & 0 & 0 \\ * & \Psi_1 & G^{\mathrm{T}}H^{\mathrm{T}} & D(\tilde{B}(\alpha) + \Delta\tilde{B}(\alpha)) \\ * & * & I - H - H^{\mathrm{T}} & 0 \\ * & * & * & -\gamma^2 I \end{bmatrix} < 0, \tag{3.18}$$

$$\begin{bmatrix} -rQ & Q(\tilde{A}(\alpha) + \Delta\tilde{A}(\alpha)) + kQ \\ * & -rQ \end{bmatrix} < 0, \tag{3.19}$$

where $\Psi_1 = (0.8\pi)^2 Q + D(\tilde{A}(\alpha) + \Delta\tilde{A}(\alpha)) + (\tilde{A}(\alpha) + \Delta\tilde{A}(\alpha))^{\mathrm{T}} D^{\mathrm{T}}$ *hold. Then, the design objectives O1 and O2 are satisfied and the eigenvalues of matrix* $(\tilde{A}(\alpha) + \Delta\tilde{A}(\alpha))$ *are located in the disk with the center of* $(-k, 0)$ *and the radius of* r.

Proof The condition in (3.19) can guarantee the robust asymptotical stability and the eigenvalue location of the compact system in (3.13). It can be proved by selecting the Lyapunov function $V = \xi^{\mathrm{T}} Q \xi$; see [37]. The condition in (3.18) is for the finite-frequency $\mathcal{H}_\infty$ performance. The proof can be referred to the work [38, 39]. □

Lemma 3.2 provides the conditions for the robust asymptotical stability and the finite-frequency $\mathcal{H}_\infty$ performance. However, it is impossible to employ the conditions to evaluate the compact system in (3.13) since the varying terms $\Delta\tilde{A}(\alpha)$ and $\Delta\tilde{B}(\alpha)$ are involved in the condition. In the following theorem, the varying variable $\bar{M}(t)$ will be eliminated.

Theorem 3.1 *For prescribed real numbers* $\gamma > 0$, $k > 0$, *and* $r > 0$, *suppose that there exist a symmetric matrix* P, *a positive-definite matrix* $Q = Q^{\mathrm{T}}$, *matrices* D *and* H, *positive scalars* φ_1, φ_2, *and* φ_3 *such that*

$$\Theta = \begin{bmatrix} -Q & P-D & 0 & 0 & 0 & 0 & 0 & 0 \\ * & \Psi_2 & G^{\mathrm{T}}H^{\mathrm{T}} & D\tilde{B}(\alpha) & D\tilde{E}_1(\alpha) & \varphi_1\tilde{F}_1^{\mathrm{T}} & D\tilde{E}_2(\alpha) & 0 \\ * & * & I-H-H^{\mathrm{T}} & 0 & 0 & 0 & 0 & 0 \\ * & * & * & -\gamma^2 I & 0 & 0 & 0 & \varphi_2\tilde{F}_2^{\mathrm{T}} \\ * & * & * & * & -\varphi_1 I & 0 & 0 & 0 \\ * & * & * & * & * & -\varphi_1 I & 0 & 0 \\ * & * & * & * & * & * & -\varphi_2 I & 0 \\ * & * & * & * & * & * & * & -\varphi_2 I \end{bmatrix} < 0, \tag{3.20}$$

$$\Gamma = \begin{bmatrix} -rQ & Q\tilde{A}(\alpha)+kQ & Q\tilde{E}_1(\alpha) & 0 \\ * & -rQ & 0 & \varphi_3\tilde{F}_1^{\mathrm{T}} \\ * & * & -\varphi_3 I & 0 \\ * & * & * & -\varphi_3 I \end{bmatrix} < 0, \tag{3.21}$$

where $\Psi_2 = (0.8\pi)^2 Q + D\tilde{A}(\alpha) + \tilde{A}^{\mathrm{T}}(\alpha) D^{\mathrm{T}}$ *hold. Then, the design objectives O1 and O2 are satisfied and the eigenvalues of matrix* $(\tilde{A}(\alpha) + \Delta\tilde{A}(\alpha))$ *are located in the disk with the center of* $(-k, 0)$ *and the radius of* r.

Proof The condition in (3.19) can be rewritten as

$$\begin{aligned} &\begin{bmatrix} -rQ & Q\tilde{A}(\alpha)+kQ \\ * & -rQ \end{bmatrix} + \begin{bmatrix} Q\tilde{E}_1(\alpha) \\ 0 \end{bmatrix} \bar{M}(t) \begin{bmatrix} 0 & \tilde{F}_1 \end{bmatrix} \\ &+ \begin{bmatrix} 0 & \tilde{F}_1 \end{bmatrix}^{\mathrm{T}} \bar{M}(t) \begin{bmatrix} Q\tilde{E}_1(\alpha) \\ 0 \end{bmatrix}^{\mathrm{T}} < 0. \end{aligned} \tag{3.22}$$

In terms of Lemma 3.1, the condition (3.19) holds if and only if (3.21) is satisfied. Similarly, the equivalence between (3.20) and (3.18) can be proved. □

Since the longitudinal velocity v_x is variable, the weighting factors $\alpha_i, i = 1, 2, 3$ are varying. The dimension of conditions in Theorem 3.1 is infinite. In the following theorem, the conditions in Theorem 3.1 will be projected to the polytope vertices. Then, the conditions with finite dimensions can be used for the observer design. Different from Lemma 3.2 and Theorem 3.1 in which the observer gain is assumed to be known, the following theorem is going to provide the observer gain design method.

Theorem 3.2 *For prescribed real numbers $\gamma > 0$, $k > 0$, and $r > 0$, suppose that there exist a symmetric matrix P, a symmetric positive-definite matrix $Q = \begin{bmatrix} Q_1 & 0 \\ 0 & Q_2 \end{bmatrix}$, matrices $D_i = \begin{bmatrix} Q_1 & D_{1,i} \\ Q_1 & D_{2,i} \end{bmatrix}$, H_i and $\hat{L}_i$, positive scalars $\varphi_{1,i}$, $\varphi_{2,i}$ and $\varphi_{3,i}$ such that the following conditions are satisfied*

$$\Theta_{ij} + \Theta_{ji} < 0, \tag{3.23}$$

$$\Gamma_i = \begin{bmatrix} -rQ & \Xi_{1,i} + kQ & \Xi_{2,i} & 0 \\ * & -rQ & 0 & \varphi_{3,i} F_1^{\mathrm{T}} \\ * & * & -\varphi_{3,i} I & 0 \\ * & * & * & -\varphi_{3,i} I \end{bmatrix} < 0, \tag{3.24}$$

where

$$\Theta_{ij} = \begin{bmatrix} -Q & P - D_i & 0 & 0 & 0 & 0 & 0 & 0 \\ * & \Psi_{2,ij} & G^{\mathrm{T}} H_i^{\mathrm{T}} & \Psi_{3,ij} & \Psi_{4,ij} & \varphi_{1,i} \tilde{F}_1^{\mathrm{T}} & \Psi_{5,ij} & 0 \\ * & * & I - H_i - H_i^{\mathrm{T}} & 0 & 0 & 0 & 0 & 0 \\ * & * & * & -\gamma^2 I & 0 & 0 & 0 & \varphi_{1,i} \tilde{F}_2^{\mathrm{T}} \\ * & * & * & * & -\varphi_{1,i} I & 0 & 0 & 0 \\ * & * & * & * & * & -\varphi_{1,i} I & 0 & 0 \\ * & * & * & * & * & * & -\varphi_{2,i} I & 0 \\ * & * & * & * & * & * & * & -\varphi_{2,i} I \end{bmatrix},$$

$$\begin{aligned} \Psi_{2,ij} &= (0.8\pi)^2 Q + \begin{bmatrix} Q_1 A_i - \hat{L}_i C & D_{1,i} A_j \\ Q_1 A_i - \hat{L}_i C & D_{2,i} A_j \end{bmatrix} \\ &\quad + \begin{bmatrix} Q_1 A_i - \hat{L}_i C & D_{1,i} A_j \\ Q_1 A_i - \hat{L}_i C & D_{2,i} A_j \end{bmatrix}^{\mathrm{T}}, \\ \Psi_{3,ij} &= \begin{bmatrix} D_{1,i} B_j \\ D_{2,i} B_j \end{bmatrix}, \Psi_{4,ij} = \begin{bmatrix} Q_1 E_{1,i} + D_{1,i} E_{1,j} \\ Q_1 E_{1,i} + D_{2,i} E_{1,j} \end{bmatrix}, \end{aligned}$$

$$\Psi_{5,ij} = \begin{bmatrix} Q_1 E_{2,i} + D_{1,i} E_{2,j} \\ Q_1 E_{2,i} + D_{2,i} E_{2,j} \end{bmatrix},$$

$$\begin{aligned} \Xi_{1,i} &= \begin{bmatrix} Q_1 A_i - \hat{L}_i C & 0 \\ 0 & Q_2 A_i \end{bmatrix}, \\ \Xi_{2,i} &= \begin{bmatrix} Q_1 E_{1,i} \\ Q_2 E_{1,i} \end{bmatrix}, 1 \le i \le j \le 3. \end{aligned}$$

Then, the design objectives O1 and O2 are satisfied and the eigenvalues of matrix $(\tilde{A}(\alpha) + \Delta\tilde{A}(\alpha))$ *are located in the disk with the center of* $(-k, 0)$ *and the radius of* r. *Moreover, the observer gain can be calculated by* $L = \sum_{i=1}^{3}(\alpha_i L_i) = \sum_{i=1}^{3}(\alpha_i Q_1^{-1}\hat{L}_i)$.

Proof Define a new variable as $\hat{L} = Q_1 L$. Suppose that the observer gain L, the random variable H, and the positive scalars φ_1, φ_2, and φ_3 are linearly dependent on the weighting factors α_i, $i = 1, 2, 3$, that is,

$$
\begin{aligned}
L &= \sum_{i=1}^{3} \alpha_i L_i,\ H = \sum_{i=1}^{3} \alpha_i H_i,\ \varphi_1 = \sum_{i=1}^{3} \alpha_i \varphi_{1,i}, \\
\varphi_2 &= \sum_{i=1}^{3} \alpha_i \varphi_{2,i},\ \varphi_3 = \sum_{i=1}^{3} \alpha_i \varphi_{3,i}.
\end{aligned}
\tag{3.25}
$$

Due to the summation property, we can see that

$$
\begin{aligned}
\Theta &= \sum_{i=1}^{3} \alpha_i^2 \Theta_{ii} + \sum_{i=1}^{2} \sum_{j=i+1}^{3} \alpha_i \alpha_j (\Theta_{ij} + \Theta_{ji}), \\
\Gamma &= \sum_{i=1}^{3} \alpha_i \Gamma_i.
\end{aligned}
\tag{3.26}
$$

If the conditions in (3.23) and (3.24) are satisfied, Θ and Γ are negative definite. □

Since the performance index γ indicates the effect of the wheel steering angle to the estimated sideslip angle, the optimal observer is the one with the gain which is obtained by minimizing the value for γ. The minimization can be done in terms of the following corollary.

Corollary 3.1 *The minimum* $\mathcal{H}_\infty$ *performance index* γ^* *in Theorem 3.2 can be found by solving the following convex optimization problem:*

$$
\begin{gathered}
\gamma^* = \min \gamma. \\
s.\ t.\ (3.23), (3.24)
\end{gathered}
$$

3.4 Experimental Results

In order to show the advantage of the finite-frequency approach, we calculate the minimal γ^* for different frequency ranges. In the optimization, the 0.8π in (3.23) is replaced by different upper bounds. Table 3.2 shows the obtained minimal γ^* values for different frequency ranges. The entire frequency means that the upper bound is selected as an infinite value. The traditional $\mathcal{H}_\infty$ observer design is based on the optimization for the entire frequency. It infers from Table 3.2 that the finite-frequency approach can get smaller minimal γ^*.

Table 3.2 Minimal γ^* for different frequency ranges

Range	[0 0.1]	[0 1]	[0 0.8π]	[0 10]	Entire
γ^*	1.6056	1.6974	1.7261	2.1532	2.9719

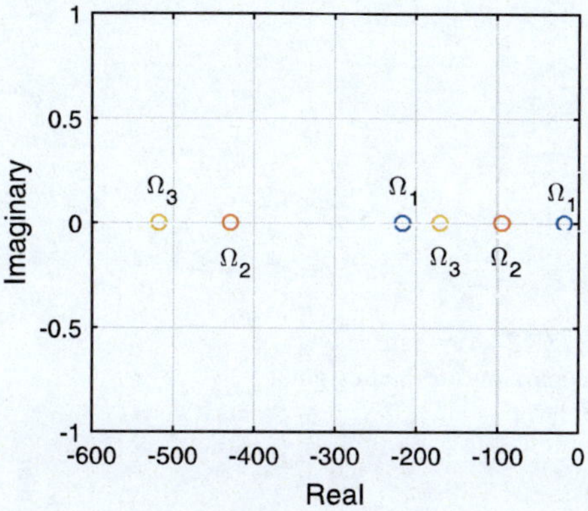

Fig. 3.10 Eigenvalue locations of the three vertices

For the finite-frequency range of [0 0.8π], the obtained gains are

$$L_1 = \begin{bmatrix} 4.76 \\ 202.91 \end{bmatrix}, L_2 = \begin{bmatrix} 8.34 \\ 356.09 \end{bmatrix}, L_3 = \begin{bmatrix} 17.10 \\ 383.57 \end{bmatrix}.$$

To analyze the eigenvalue locations of the observer, we illustrate the eigenvalue locations of matrix $(A(\alpha) - LC)$ at three vertices, as shown in Fig. 3.10. Ω_1 is the vertex with the maximal velocity. Ω_3 is the vertex with the minimal velocity. Compared to the original system, the eigenvalue locations of the observer are much farer to the imaginary axis than the corresponding eigenvalue locations of the original system. Moreover, when the longitudinal velocity is large, the eigenvalues of the observer do not have any imaginary parts, that is, the time-domain response of the observer would not have strong oscillation.

Finally, the designed observer is validated via the experimental data. Figure 3.11 shows the sideslip angle comparison between the measured signal and the estimated value. Figure 3.12 depicts the yaw rate comparison between the measured signal and the estimated value. Since the measured yaw rate is used to correct the estimation error and eigenvalues of the observer are replaced, the observed yaw rate is almost the same with the measured one (a low-pass filter is added for the measured yaw rate). For the sideslip angle comparison, the observed value can follow the shape of the measured value quite well. Small variations are caused by the modeling error. In order to future compare the results, the estimation errors are described in Figs. 3.13 and 3.14. We can see that the errors are constrained to a low level. Considering the comparisons together, the performance of the designed observer is acceptable.

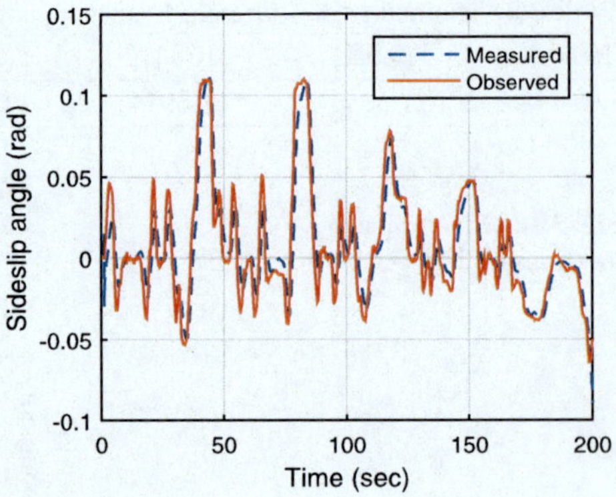

Fig. 3.11 Sideslip angle comparison for the designed observer

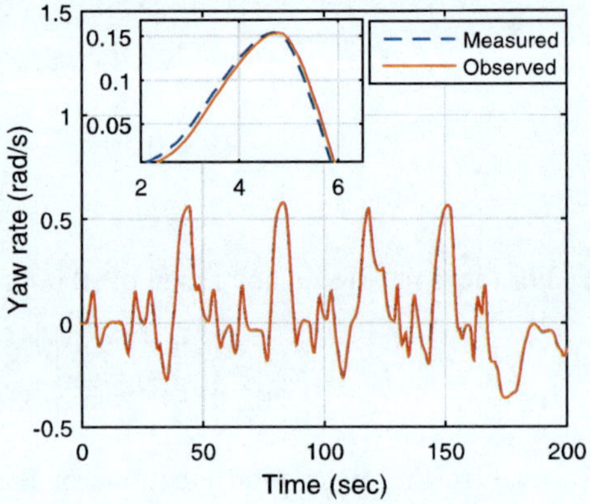

Fig. 3.12 Yaw rate comparison for the designed observer

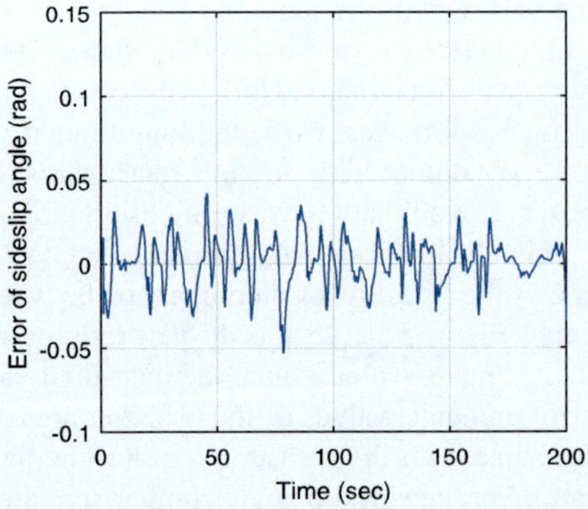

Fig. 3.13 Estimation error of sideslip angle (measured value minus the estimation value)

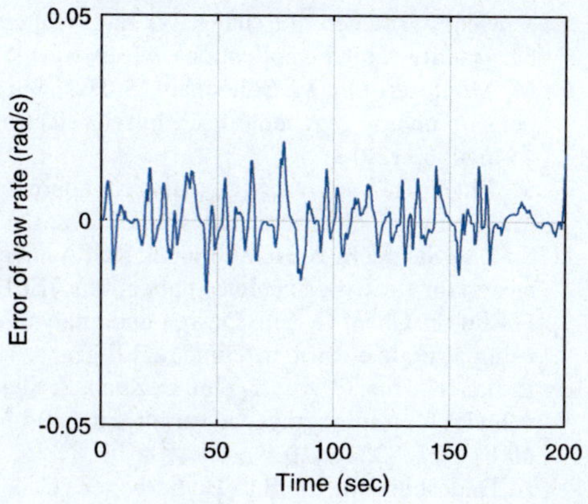

Fig. 3.14 Estimation error of yaw rate (measured value minus the estimation value)

3.5 Conclusion

In this work, we have investigated the observer design for an EGV. The work was started from the system modeling and parameter identification. The tire cornering stiffness and the inertia moment were identified with experimental data and non-linear optimization algorithm. The identified model was then validated via another experimental test. Based on the identified model, the sideslip angle observer was proposed. Considering the frequency range of the steering angle, the finite-frequency $\mathcal{H}_\infty$ observer was designed. The advantage of the finite-frequency approach was shown via a series of comparisons. At the end, experimental results were used to show the performance of the designed observer. In the future research, we would focus on the observer-based vehicle stability control.

Acknowledgements This chapter is from the previous work in [40], and some typos are corrected here.

References

1. X. Hu, R. Xiong, B. Egardt, Model-based dynamic power assessment of Lithium-Ion batteries considering different operating conditions. IEEE Trans. Ind. Inform. **10**(3), 1948–1959 (2014)
2. A. Greco, D. Cao, X. Jiang, H. Yang, A theoretical and computational study of lithium-ion battery thermal management for electric vehicles using heat pipes. J. Power Sources **257**(7), 344–355 (2014)
3. X. Hu, N. Murgovski, L.J. Mardh, B. Egardt, Comparison of three electrochemical energy buffers applied to a hybrid bus powertrain with simultaneous optimal sizing and energy management. IEEE Trans. Intell. Transp. **15**(3), 1193–1205 (2014)
4. R. Wang, C. Hu, Z. Wang, F. Yan, N. Chen, Integrated optimal dynamics control of 4WD4WS electric ground vehicle with tire-road frictional coefficient estimation. Mech. Syst. Signal Process. **60–61**, 727–741 (2015)

5. F. Meng, H. Chen, T. Zhang, X. Zhu, Clutch fill control of an automatic transmission for heavy-duty vehicle applications. Mech. Syst. Signal Process. **64–65**, 16–28 (2015)
6. M. Montazeri-Gh, M. Soleymani, S. Hashemi, Impact of traffic conditions on the active suspension energy regeneration in hybrid electric vehicles. IEEE Trans. Ind. Electron. **60**(10), 4546–4553 (2013)
7. M. Zhang, Y. Yang, C. Mi, Analytical approach for the power management of blended-mode plug-in hybrid electric vehicles. IEEE Trans. Veh. Technol. **61**(4), 1554–1566 (2012)
8. S.M.M. Sangdehi, S. Hamidifar, N. Kar, A novel bidirectional DC/AC stacked matrix converter design for electrified vehicle applications. IEEE Trans. Veh. Technol. **63**(7), 3038–3050 (2014)
9. F. Zhu, L. Chen, C. Yin, Design and analysis of a novel multimode transmission for a HEV using a single electric machine. IEEE Trans. Veh. Technol. **62**(3), 1097–1110 (2013)
10. L. Jian, H. Xue, G. Xu, X. Zhu, D. Zhao, Z. Shao, Regulated charging of plug-in hybrid electric vehicles for minimizing load variance in household smart microgrid. IEEE Trans. Ind. Electron. **60**(8), 3218–3226 (2013)
11. A. Dadashnialehi, A. Bab-Hadiashar, Z. Cao, A. Kapoor, Intelligent sensorless ABS for in-wheel electric vehicles. IEEE Trans. Ind. Electron. **61**(4), 1957–1969 (2014)
12. Y. Wang, B.M. Nguyen, H. Fujimoto, Y. Hori, Multirate estimation and control of body slip angle for electric vehicles based on onboard vision system. IEEE Trans. Ind. Electron. **61**(2), 1133–1143 (2014)
13. B. Li, H. Du, W. Li, Y. Zhang, Side-slip angle estimation based lateral dynamics control for omni-directional vehicles with optimal steering angle and traction/brake torque distribution. Mechatronics **30**, 348–362 (2015)
14. X. Jin, G. Yin, N. Chen, Gain-scheduled robust control for lateral stability of four-wheel-independent-drive electric vehicles via linear parameter-varying technique. Mechatronics **30**, 286–296 (2015)
15. Y. Sun, L. Li, B. Yan, C. Yang, G. Tang, A hybrid algorithm combining EKF and RLS in synchronous estimation of road grade and vehicle mass for a hybrid electric bus. Mech. Syst. Signal Process. **68–69**, 416–430 (2016)
16. R. Wang, H. Zhang, J. Wang, Linear parameter-varying based fault-tolerant controller design for a class of over-actuated nonlinear systems with applications to electric vehicles. IET Control Theory Appl. **8**(9), 705–717 (2014)
17. R. Wang, H. Zhang, J. Wang, Linear parameter-varying controller design for four wheel independently-actuated electric ground vehicles with active steering systems. IEEE Trans. Control Syst. Technol. 22(4), 1281–1296 (2014)
18. D.M. Bevly, J.C. Gerdes, C. Wilson, The use of GPS based velocity measurements for measurement of sideslip and wheel slip. Veh. Syst. Dyn. **38**(2), 127–147 (2002)
19. D. Piyabongkarn, R. Rajamani, J.A. Grogg, J.Y. Lew, Development and experimental evaluation of a slip angle estimator for vehicle stability control. IEEE Trans. Control Syst. Technol. **17**(1), 78–88 (2009)
20. Y.H. Hsu, S.M. Laws, J.C. Gerdes, Estimation of tire slip angle and friction limits using steering torque. IEEE Trans. Control Syst. Technol. **18**(4), 896–907 (2010)
21. K. Nam, S. Oh, H. Fujimoto, Y. Hori, Estimation of sideslip and roll angles of electric vehicles using lateral tire force sensors through RLS and Kalman filter approaches. IEEE Trans. Ind. Electron. **60**(3), 988–1000 (2013)
22. R.E. Kalman, A new approach to linear filtering and prediction problems. Trans. ASME, J Basic Eng. **82**(Series D), 35–45 (1960)
23. H. Gao, T. Chen, $\mathcal{H}_\infty$ estimation for uncertain systems with limited communication capacity. IEEE Trans. Autom. Contr. **52**(11), 2070–2084 (2007)
24. H. Gao, X. Meng, T. Chen, A parameter-dependent approach to robust $\mathcal{H}_\infty$ filtering for time-delay systems. IEEE Trans. Autom. Contr. **53**(10), 2420–2425 (2008)
25. H. Zhang, A. Saadat Mehr, Y. Shi, Improved robust energy-to-peak filtering for uncertain linear systems. Signal Process. **90**(9), 2667–2675 (2010)
26. D.-W. Ding, G. hong Yang, Fuzzy filter design for nonlinear systems in finite-frequency domain. IEEE Trans. Fuzzy Syst. **18**(5), 935–945 (2010)

27. X. Li, H. Gao, Robust finite frequency filtering for uncertain 2-D roesser systems. Automatica **48**(6), 1163–1170 (2012)
28. H. Wang, G. hong Yang, A finite frequency approach to filter design for uncertain discrete-time systems. Int. J. Adapt. Control Signal Process. 22(6), 533–550 (2008)
29. K. Nam, H. Fujimoto, Y. Hori, Lateral stability control of in-wheel-motor-driven electric vehicles based on sideslip angle estimation using lateral tire force sensors. IEEE Trans. Veh. Technol. **61**(5), 1972–1985 (2012)
30. K. Nam, S. Oh, H. Fujimoto, Y. Hori, Estimation of sideslip and roll angles of electric vehicles using lateral tire force sensors through RLS and Kalman filter approaches. IEEE Trans. Ind. Electron. **60**(3), 988–1000 (2013)
31. J.-H. Yoon, H. Peng, Robust vehicle sideslip angle estimation through a disturbance rejection filter that integrates a magnetometer with GPS. IEEE Trans. Intell. Transp. **15**(1), 191–204 (2014)
32. S. Melzi, E. Sabbioni, On the vehicle sideslip angle estimation through neural networks: Numerical and experimental results. Mech. Syst. Signal Process. **25**(6), 2005–2019 (2011)
33. L. Li, G. Jia, X. Ran, J. Song, K. Wu, A variable structure extended kalman filter for vehicle sideslip angle estimation on a low friction road. Veh. Syst. Dyn. **52**(2), 280–308 (2014)
34. R. Wang, Y. Chen, D. Feng, X. Huang, J. Wang, Development and performance characterization of an electric ground vehicle with independently actuated in-wheel motors. J. Power Sources **196**(8), 3962–3971 (2011)
35. H. Zhang, X. Huang, J. Wang, H.R. Karimi, Robust energy-to-peak sideslip angle estimation with applications to ground vehicles. Mechatronics **30**, 338–347 (2015)
36. L. Xie, Y.C. Soh, Robust control of linear systems with generalized positve real uncertainty. Automatica **33**(5), 963–967 (1997)
37. W.M. Haddad, D.S. Bernstein, Controller design with regional pole constraints. IEEE Trans. Autom. Contr. **37**(1), 54–69 (1992)
38. D.H. Lee, An improved finite frequency approach to robust $\mathcal{H}_\infty$ filter design for LTI systems with polytopic uncertainties. Int. J. Adapt. Control Signal Process. **27**(11), 944–956 (2013)
39. H. Zhang, R. Wang, J. Wang, Y. Shi, Robust finite frequency static-output-feedback control with application to vibration active control of structural systems. Mechatronics **24**(4), 354–366 (2014)
40. H. Zhang, G. Zhang, J. Wang, Sideslip angle estimation of an electric ground vehicle via finite-frequency $\mathcal{H}_\infty$ approach. IEEE Trans. Transp. Electrification **2**(2), 200–209 (2016)

Chapter 4
Active Steering Actuator Fault Detection for an Automatically Steered Electric Ground Vehicle

Abstract In this work, we investigate the actuator fault detector design problem for an electric ground vehicle (EGV) which is equipped with an active front-wheel steering system. Since the EGV can be steered by a motor automatically, it is desired to design a fault detector for the steering actuator for the safety concern. A two-degree-of-freedom (2-DOF) lateral nonlinear vehicle model is established. The nonlinear vehicle model is converted to a linear-parameter-varying (LPV) form, and the scheduling vector is related to the vehicle longitudinal velocity. Since it is not easy to measure the longitudinal velocity precisely, the uncertain measurement on the longitudinal velocity is considered and the weighting factors of LPV sub-models are subject to uncertainties. Based on the uncertain LPV model, a gain-scheduled fault detector is proposed and an augmented system is obtained. The desired steering angle and the faulty steering angle are both involved in the augmented system. As the steering angle generally has a low-frequency working range, the steering angle amplitude spectrums of three different maneuvers are studied, and the frequency working range is determined. The stability, the $\mathcal{H}_-$ performance, and the $\mathcal{H}_\infty$ performance of the augmented system are all exploited. Based on the analysis results, the mixed $\mathcal{H}_-/\mathcal{H}_\infty$ fault detector design method is developed.

4.1 Introduction

Electric vehicles (EVs) and hybrid electric vehicles (HEVs) have received much attention in the last decades due to the increased demands of fuel economy and environment protection [1–6]. Compared to the traditional engine-powered vehicles, the EVs and HEVs have new features and challenges no matter from the torque/energy management [7–11] but also from the vehicle dynamics and control [3]. At the side of EV's dynamics and control, we can see the work of EV's anti-lock braking system (ABS) in [12], the body slip angle estimation in [13], and the electric differential design in [14]. When an EV is equipped with an active-steering system, both the driving and steering motions can be controlled by electric motors. The motor actuation has various advantages such as the fast response and the high energy efficiency.

H. Zhang et al., *Robust Gain-Scheduled Estimation and Control of Electrified Vehicles via LPV Technique*, Key Technologies on New Energy Vehicles,
https://doi.org/10.1007/978-981-19-8509-6_4

However, the electric connection is not as reliable as the mechanical connection. The fault detection and fault diagnosis are two critical topics for electrified vehicles.

The primary principle of model-based fault detection is to evaluate a residual signal which is the difference between the measured signal and the observer output [15–17]. Since the detection is implemented in a software form and there is no extra hardware required, the model-based fault detection has attracted a lot of attentions. The authors in [18] studied the fault detection problem for discrete-time fuzzy system in finite-frequency domain. When the fuzzy system has unknown membership functions, the fault detection was discussed in [19]. The fault detection problem for discrete-time fuzzy system with intermittent measurements can be seen in [20]. In the fault detector design, a widely used performance is the mixed $\mathcal{H}_-/\mathcal{H}_\infty$; see [15, 17, 21, 22] and the references therein. The $\mathcal{H}_-$ index is used to denote the sensitivity of the designed detector to the fault, and the $\mathcal{H}_\infty$ performance is employed to constrain the effect of the disturbance to the designed detector. Since both $\mathcal{H}_-$ and $\mathcal{H}_\infty$ performances are related to the input frequency and the input only has finite frequency in practical applications, the fault detection with the finite-frequency technique would achieve better results than the full-frequency method. The finite-frequency technique has shown the advantages in robust control, estimation, and fault detection [23–27].

In this work, we aim to design a fault detector for an active-steering actuator of an EGV developed in The Ohio State University. The input amplitude spectrum with respect to the frequency is studied, and the finite-frequency mixed $\mathcal{H}_-/\mathcal{H}_\infty$ technique is used for the gain-scheduled observer design. To guarantee the transient response of the detector, the pole-placement skill is utilized. The original contribution of this research can be summarized as: (1) We consider the uncertainty of scheduling parameters, which makes the model more practical; (2) in the $\mathcal{D}$-stability analysis, we establish the condition with different radii and circle centers; (3) we develop new criteria for the mixed $\mathcal{H}_-/\mathcal{H}_\infty$ performance; (4) we conduct an experimental test to validate the designed detector.

Notation: The notations used in this chapter are standard. Superscript 'T' and '-1' denote matrix transposition and matrix inverse; in symmetric block matrices or long matrix expressions, $*$ is used as an ellipsis for the terms that are introduced by symmetry; and $\|X\|$ stands for the Euclidean norm for vectors and the spectral norm for matrices.

4.2 System Introduction and Problem Formulation

4.2.1 Acquisition System and Steering Actuator of EGV

The EGV studied in this work is shown in Fig. 4.1, and the experimental tests are conducted at Transportation Research Center (TRC) in East Liberty, Ohio, USA. The EGV is driven by four permanent-magnet brushless direct-current electric motors. The motors are placed in the four wheels, and each motor can be controlled separately. Therefore, the EGV is called as a four-wheel-independent-driven electric vehicle.

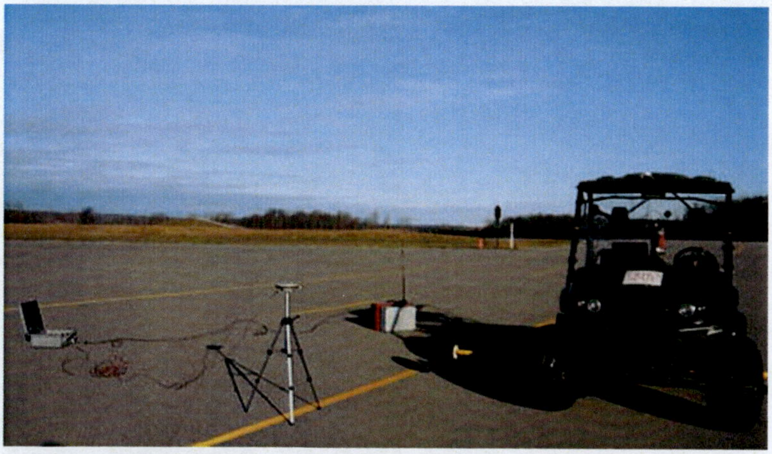

Fig. 4.1 Picture view of the EGV and the navigation system in the test field

Fig. 4.2 Active-front-wheel-steering (AFS) system for the EGV

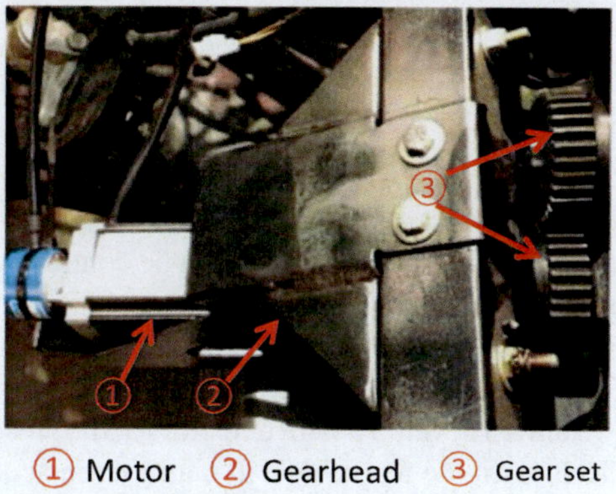

Moreover, each motor has the peak power of 7.5 KW, and the motor is powered by a 72V lithium-ion power battery pack which consists of 22 battery bricks. More details on the development of the EGV can be seen in [28]. It is necessary to mention that the EGV is equipped with an active-front-wheel-steering (AFS) system which is illustrated in Fig. 4.2. The development of the AFS system is based on the original steering system. There are three main components: a steering motor, a motor reducer, and a gear set. The type of steering motor is NEMA 23 servo motor with the controller of SV7-S-AF from Applied Motion Products Inc. The motor can provide a peak torque of 1.38 Nm, and the maximal speed is 3350 rpm. To amplify the torque and reduce the speed, a Carson 23EP055 Eliminator Planetary gearhead withe the reducing ratio of 55:1 is connected with the motor. The gearhead is from Carson Manufacturing Inc. In addition, there is one gear set with the ratio of 1:1. One gear is mounted on the steering column, and the other one is installed on the gearhead output shaft. A housing is designed to support the motor-gearhead system.

Fig. 4.3 Measuring system for the hand-wheel steering angle

The steering motor controller is connected with a dSPACE MicroBox such that the steering motion of EGV can be controlled via the dSPACE MicroBox. The actual hand-wheel steering angle can be measured by the system shown in Fig. 4.3. The angle measuring system contains a Model 260 angle encoder and a pulley system. The encoder is connected to the steering column via the pulley system. In this work, we aim to develop a fault detector for the steering actuator based on the vehicle model. To investigate the vehicle dynamics, the acquisition system for the EGV is shown in Figs. 4.4 and 4.5. The EGV is equipped with a high-end RT3003 navigation system which is from Oxford Technical Solutions. Since the navigation system includes a dual-antenna differential global positioning system (GPS) and an integrated inertial measurement system, both the vehicle lateral and longitudinal motions can be measured with a high accuracy. We can see from Fig. 4.4 that the angle encoder provides the hand-wheel steering angle δ_{HW} and the GPS system offers the longitudinal velocity (v_x), the lateral velocity (v_y), and the yaw rate (ω). The signals are transmitted to the dSPACE MicroBox and stored in the host PC. The dSPACE MicroBox sends out the motor steering angle δ_{mo} to the steering motor.

4.2.2 EGV System Modeling

In order to simplify the modeling analysis, we employ the 2-degree-of-freedom (2-DOF) bicycle model to study the vehicle lateral dynamics. The schematic diagram of

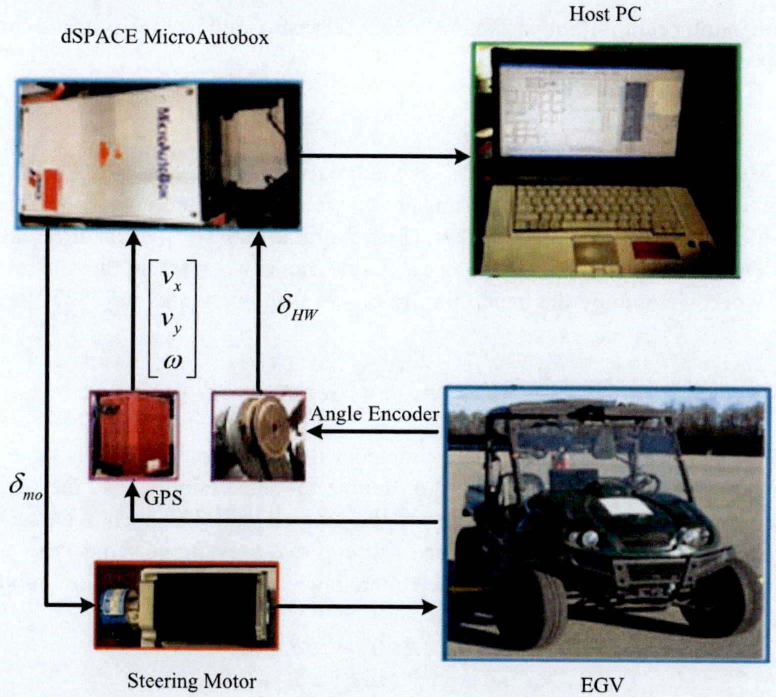

Fig. 4.4 Acquisition system and steering actuator

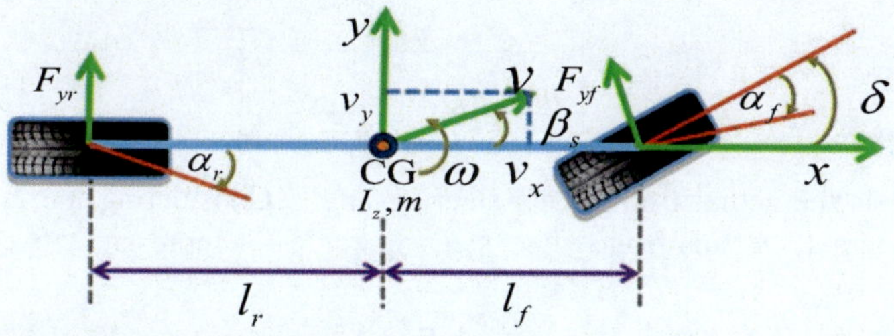

Fig. 4.5 Schematic diagram of a 2-DOF bicycle model

the 2-DOF bicycle model and the main notations are illustrated in Fig. 2.2. The center of gravity of the EGV is represented by CG. The EGV is with the total mass of m and the inertial moment about the yaw axis through the CG of I_z. l_f and l_r represent the distance of the front axis to the CG and the distance of the rear axis to the CG. σ is the actual front-wheel steering angle. α_f and α_r are the front wheel slip angle and the rear wheel steering angle, respectively. The corresponding lateral tire forces are denoted by F_{yf} and F_{yr}, respectively. Since we use the bicycle model to study the four-wheel EGV, the lateral forces are the lumped lateral forces on each axis. In addition, another important variable is vehicle sideslip angle which is approximated and represented by:

$$\beta_s = \frac{v_y}{v_x}, \tag{4.1}$$

where v_y is the lateral velocity and v_x is the longitudinal velocity.

Due to the facts that (1) the relationship between the tire lateral force and the tire slip angle is complicated and nonlinear, (2) the vehicle may be driven at different road condition, and it is quite difficult to use a simple model to represent the relationship. In this work, we employ the uncertain tire model from the works in [29, 30] as:

$$\begin{aligned} F_{\mathrm{yf}} &= \bar{c}_f \alpha_f = (c_f + \Delta c_f N)\alpha_f, \\ F_{\mathrm{yr}} &= \bar{c}_r \alpha_r = (c_r + \Delta c_r N)\alpha_r. \end{aligned} \tag{4.2}$$

where $\bar{c}_f$ and $\bar{c}_r$ represent the approximated values of cornering stiffness, c_f and c_r are the nominal values, Δc_f and Δc_r denote the uncertainties, and the scalar N satisfies $|N| \leq 1$. As the lateral forces are the lumped values on each axis, the value of cornering stiffness is also the lumped value of two tires on the same axis.

With the above formulas and Newton's law, the two-state system model is expressed as

$$\dot{x}(t) = Ax(t) + B\delta, \tag{4.3}$$

where

$$x(t) = \begin{bmatrix} \beta_s \\ \omega \end{bmatrix}, \quad B = \begin{bmatrix} \frac{\bar{c}_f}{m v_x} \\ \frac{l_f \bar{c}_f}{I_z} \end{bmatrix},$$

$$A = \begin{bmatrix} \frac{-\bar{c}_f - \bar{c}_r}{m v_x} & \frac{l_r \bar{c}_r - l_f \bar{c}_f}{m v_x^2} - 1 \\ \frac{l_r \bar{c}_r - l_f \bar{c}_f}{I_z} & \frac{-l_f^2 \bar{c}_f - l_r^2 \bar{c}_r}{I_z v_x} \end{bmatrix}.$$

Note that δ is the actual front-wheel steering angle. Considering the fault of the steering system, the actual front-wheel steering angle δ is represented by:

$$\delta = \delta_d + \delta_f, \tag{4.4}$$

where δ_d is the desired front-wheel steering angle and δ_f is the steering angle induced by the faults. In this application, the desired front-wheel steering angle is from the dSPACE MicroBox and it is available. With the fault model, the EGV state-space model is rewritten as:

$$\dot{x}(t) = Ax(t) + B\delta_d + B\delta_f. \tag{4.5}$$

The challenges for analyzing the the state-space model in (4.5) are threefold: (1) The longitudinal velocity v_x is varying such that the system in (4.5) nonlinear; (2) the measurement on the longitudinal velocity v_x by an affordable physical sensor is inaccurate; (3) the tire uncertainties are involved in the model.

To deal with the challenges, first, we convert the nonlinear model into an LPV form. Suppose that the longitudinal velocity v_x is within the range:

$$v_x \in \left[\underline{v}_x \ \bar{v}_x \right], \tag{4.6}$$

where $\underline{v}_x$ and $\bar{v}_x$ are the lower and upper bounds, respectively. Then the matrix $(A,\ B)$ can be rewritten as:

$$(A,\ B) = \sum_{i=1}^{4} \alpha_i (A_i,\ B_i), \tag{4.7}$$

where

$$(A_1,\ B_1) = \left(\begin{bmatrix} \frac{-\bar{c}_f - \bar{c}_r}{m\bar{v}_x} & \frac{l_r\bar{c}_r - l_f\bar{c}_f}{m\bar{v}_x^2} - 1 \\ \frac{l_r\bar{c}_r - l_f\bar{c}_f}{I_z} & \frac{-l_f^2\bar{c}_f - l_r^2\bar{c}_r}{I_z\bar{v}_x} \end{bmatrix},\ \begin{bmatrix} \frac{\bar{c}_f}{m\bar{v}_x} \\ \frac{l_f\bar{c}_f}{I_z} \end{bmatrix} \right),$$

$$(A_2,\ B_2) = \left(\begin{bmatrix} \frac{-\bar{c}_f - \bar{c}_r}{m\underline{v}_x} & \frac{l_r\bar{c}_r - l_f\bar{c}_f}{m\bar{v}_x^2} - 1 \\ \frac{l_r\bar{c}_r - l_f\bar{c}_f}{I_z} & \frac{-l_f^2\bar{c}_f - l_r^2\bar{c}_r}{I_z\underline{v}_x} \end{bmatrix},\ \begin{bmatrix} \frac{\bar{c}_f}{m\underline{v}_x} \\ \frac{l_f\bar{c}_f}{I_z} \end{bmatrix} \right),$$

$$(A_3,\ B_3) = \left(\begin{bmatrix} \frac{-\bar{c}_f - \bar{c}_r}{m\bar{v}_x} & \frac{l_r\bar{c}_r - l_f\bar{c}_f}{m\underline{v}_x^2} - 1 \\ \frac{l_r\bar{c}_r - l_f\bar{c}_f}{I_z} & \frac{-l_f^2\bar{c}_f - l_r^2\bar{c}_r}{I_z\bar{v}_x} \end{bmatrix},\ \begin{bmatrix} \frac{\bar{c}_f}{m\bar{v}_x} \\ \frac{l_f\bar{c}_f}{I_z} \end{bmatrix} \right),$$

$$(A_4,\ B_4) = \left(\begin{bmatrix} \frac{-\bar{c}_f - \bar{c}_r}{m\underline{v}_x} & \frac{l_r\bar{c}_r - l_f\bar{c}_f}{m\underline{v}_x^2} - 1 \\ \frac{l_r\bar{c}_r - l_f\bar{c}_f}{I_z} & \frac{-l_f^2\bar{c}_f - l_r^2\bar{c}_r}{I_z\underline{v}_x} \end{bmatrix},\ \begin{bmatrix} \frac{\bar{c}_f}{m\underline{v}_x} \\ \frac{l_f\bar{c}_f}{I_z} \end{bmatrix} \right),$$

$$\alpha_1 = \frac{\left|\frac{1}{v_x} - \frac{1}{\underline{v}_x}\right|\left|\frac{1}{v_x^2} - \frac{1}{\underline{v}_x^2}\right|}{\left|\frac{1}{\underline{v}_x} - \frac{1}{\bar{v}_x}\right|\left|\frac{1}{\underline{v}_x^2} - \frac{1}{\bar{v}_x^2}\right|},\ \alpha_2 = \frac{\left|\frac{1}{v_x} - \frac{1}{\bar{v}_x}\right|\left|\frac{1}{v_x^2} - \frac{1}{\underline{v}_x^2}\right|}{\left|\frac{1}{\underline{v}_x} - \frac{1}{\bar{v}_x}\right|\left|\frac{1}{\underline{v}_x^2} - \frac{1}{\bar{v}_x^2}\right|},$$

$$\alpha_3 = \frac{\left|\frac{1}{v_x} - \frac{1}{\underline{v}_x}\right|\left|\frac{1}{v_x^2} - \frac{1}{\bar{v}_x^2}\right|}{\left|\frac{1}{\underline{v}_x} - \frac{1}{\bar{v}_x}\right|\left|\frac{1}{\underline{v}_x^2} - \frac{1}{\bar{v}_x^2}\right|},\ \alpha_4 = \frac{\left|\frac{1}{v_x} - \frac{1}{\bar{v}_x}\right|\left|\frac{1}{v_x^2} - \frac{1}{\bar{v}_x^2}\right|}{\left|\frac{1}{\underline{v}_x} - \frac{1}{\bar{v}_x}\right|\left|\frac{1}{\underline{v}_x^2} - \frac{1}{\bar{v}_x^2}\right|}.$$

$\alpha_i,\ \forall i = 1, 2, 3, 4$ is called the weighting factor for the ith sub-model. In the formation of (4.7), $\frac{1}{v_x}$ and $\frac{1}{v_x^2}$ are selected as the scheduling parameters, that is, the longitudinal velocity v_x should be measured online precisely. However, due to the varying tire slip, the longitudinal velocity v_x cannot be measured precisely with an affordable measuring method. Let us use $\hat{v}_x$ to denote the measured longitudinal velocity. It is reasonable to assume that the measured longitudinal velocity is also within the range as:

$$\hat{v}_x \in \left[\underline{v}_x \ \bar{v}_x \right]. \tag{4.8}$$

The weighting factor for each sub-model with the measured longitudinal velocity is calculated as:

$$\hat{\alpha}_1 = \frac{\left|\frac{1}{\hat{v}_x} - \frac{1}{\underline{v}_x}\right|\left|\frac{1}{\hat{v}_x^2} - \frac{1}{\underline{v}_x^2}\right|}{\left|\frac{1}{\underline{v}_x} - \frac{1}{\bar{v}_x}\right|\left|\frac{1}{\underline{v}_x^2} - \frac{1}{\bar{v}_x^2}\right|},\ \hat{\alpha}_2 = \frac{\left|\frac{1}{\hat{v}_x} - \frac{1}{\bar{v}_x}\right|\left|\frac{1}{\hat{v}_x^2} - \frac{1}{\underline{v}_x^2}\right|}{\left|\frac{1}{\underline{v}_x} - \frac{1}{\bar{v}_x}\right|\left|\frac{1}{\underline{v}_x^2} - \frac{1}{\bar{v}_x^2}\right|},$$

$$\hat{\alpha}_3 = \frac{\left|\frac{1}{\hat{v}_x} - \frac{1}{\underline{v}_x}\right|\left|\frac{1}{\hat{v}_x^2} - \frac{1}{\bar{v}_x^2}\right|}{\left|\frac{1}{\underline{v}_x} - \frac{1}{\bar{v}_x}\right|\left|\frac{1}{\underline{v}_x^2} - \frac{1}{\bar{v}_x^2}\right|},\ \hat{\alpha}_4 = \frac{\left|\frac{1}{\hat{v}_x} - \frac{1}{\bar{v}_x}\right|\left|\frac{1}{\hat{v}_x^2} - \frac{1}{\bar{v}_x^2}\right|}{\left|\frac{1}{\underline{v}_x} - \frac{1}{\bar{v}_x}\right|\left|\frac{1}{\underline{v}_x^2} - \frac{1}{\bar{v}_x^2}\right|}.$$

It is further assumed that the

$$\alpha_i = k_i \hat{\alpha}_i, \tag{4.9}$$

and k_i is bounded as:

$$k_i \in [\underline{k}_i,\ \bar{k}_i]. \tag{4.10}$$

Then the matrix set can be reformulated as:

$$\begin{aligned} A &= \sum_{i=1}^{4} \alpha_i A_i = \sum_{i=1}^{4} k_i \hat{\alpha}_i A_i = \sum_{i=1}^{4} \hat{\alpha}_i \left(A_i + (k_i - 1)A_i\right) \\ &= \sum_{i=1}^{4} \hat{\alpha}_i \left(A_i + \Delta K_i \bar{M}_1 A_i\right), \\ B &= \sum_{i=1}^{4} \alpha_i B_i = \sum_{i=1}^{4} k_i \hat{\alpha}_i B_i = \sum_{i=1}^{4} \hat{\alpha}_i \left(B_i + (k_i - 1)B_i\right) \\ &= \sum_{i=1}^{4} \hat{\alpha}_i \left(B_i + \Delta K_i \bar{M}_1 B_i\right), \end{aligned} \tag{4.11}$$

where the diagonal matrix $\|\bar{M}_1\| \leq 1$, $\Delta K_i = \Delta k_i I$, and $\Delta k_i = \max \left\{|1 - \underline{k}_i|, |1 - \bar{k}_i|\right\}$. The uncertainty induced by the nonlinear tire involved in the matrix set. Decoupling the uncertainty, the matrix set is rewritten as:

$$\begin{aligned} A &= \bar{A}(\alpha) + \Delta\bar{A}(\alpha) = \sum_{i=1}^{4} \alpha_i (\bar{A}_i + \Delta\bar{A}_i), \\ B &= \bar{B}(\alpha) + \Delta\bar{B}(\alpha) = \sum_{i=1}^{4} \alpha_i (\bar{B}_i + \Delta\bar{B}_i), \end{aligned} \tag{4.12}$$

where

$$\bar{A}_1 = \begin{bmatrix} \frac{-c_f - c_r}{m\bar{v}_x} & \frac{l_r c_r - l_f c_f}{m\bar{v}_x^2} - 1 \\ \frac{l_r c_r - l_f c_f}{I_z} & \frac{-l_f^2 c_f - l_r^2 c_r}{I_z \bar{v}_x} \end{bmatrix},\ \bar{A}_2 = \begin{bmatrix} \frac{-c_f - c_r}{m\underline{v}_x} & \frac{l_r c_r - l_f c_f}{m\bar{v}_x^2} - 1 \\ \frac{l_r c_r - l_f c_f}{I_z} & \frac{-l_f^2 c_f - l_r^2 c_r}{I_z \underline{v}_x} \end{bmatrix},$$

$$\bar{A}_3 = \begin{bmatrix} \frac{-c_f - c_r}{m\bar{v}_x} & \frac{l_r c_r - l_f c_f}{m\underline{v}_x^2} - 1 \\ \frac{l_r c_r - l_f c_f}{I_z} & \frac{-l_f^2 c_f - l_r^2 c_r}{I_z \bar{v}_x} \end{bmatrix},\ \bar{A}_4 = \begin{bmatrix} \frac{-c_f - c_r}{m\underline{v}_x} & \frac{l_r c_r - l_f c_f}{m\underline{v}_x^2} - 1 \\ \frac{l_r c_r - l_f c_f}{I_z} & \frac{-l_f^2 c_f - l_r^2 c_r}{I_z \underline{v}_x} \end{bmatrix},$$

$$\Delta\bar{A}_1 = E_{1,1}\bar{M}F_{1,1},\ \Delta\bar{A}_2 = E_{1,2}\bar{M}F_{1,2},\ \Delta\bar{A}_3 = E_{1,3}\bar{M}F_{1,3},\ \Delta\bar{A}_4 = E_{1,4}\bar{M}F_{1,4},$$

$$E_{1,1} = \begin{bmatrix} \frac{-\Delta c_f - \Delta c_r}{m\bar{v}_x} & \frac{l_r \Delta c_r - l_f \Delta c_f}{m\bar{v}_x^2} \\ \frac{l_r \Delta c_r - l_f \Delta c_f}{I_z} & \frac{-l_f^2 \Delta c_f - l_r^2 \Delta c_r}{I_z \bar{v}_x} \end{bmatrix},\ E_{1,2} = \begin{bmatrix} \frac{-\Delta c_f - \Delta c_r}{m\underline{v}_x} & \frac{l_r \Delta c_r - l_f \Delta c_f}{m\bar{v}_x^2} \\ \frac{l_r \Delta c_r - l_f \Delta c_f}{I_z} & \frac{-l_f^2 \Delta c_f - l_r^2 \Delta c_r}{I_z \underline{v}_x} \end{bmatrix},$$

$$E_{1,3} = \begin{bmatrix} \frac{-\Delta c_f - \Delta c_r}{m\bar{v}_x} & \frac{l_r \Delta c_r - l_f \Delta c_f}{m\underline{v}_x^2} \\ \frac{l_r \Delta c_r - l_f \Delta c_f}{I_z} & \frac{-l_f^2 \Delta c_f - l_r^2 \Delta c_r}{I_z \bar{v}_x} \end{bmatrix},\ E_{1,4} = \begin{bmatrix} \frac{-\Delta c_f - \Delta c_r}{m\underline{v}_x} & \frac{l_r \Delta c_r - l_f \Delta c_f}{m\underline{v}_x^2} \\ \frac{l_r \Delta c_r - l_f \Delta c_f}{I_z} & \frac{-l_f^2 \Delta c_f - l_r^2 \Delta c_r}{I_z \underline{v}_x} \end{bmatrix},$$

$$\bar{M} = \begin{bmatrix} N_1 & 0 \\ 0 & N_1 \end{bmatrix},\ |N_1| \leq 1.\ F_{1,1} = I,\ F_{1,2} = I,\ F_{1,3} = I,\ F_{1,4} = I,$$

$$\bar{B}_1 = \begin{bmatrix} \frac{c_f}{m\bar{v}_x} \\ \frac{l_f c_f}{I_z} \end{bmatrix}, \ \bar{B}_2 = \begin{bmatrix} \frac{c_f}{m\underline{v}_x} \\ \frac{l_f c_f}{I_z} \end{bmatrix}, \ \bar{B}_3 = \begin{bmatrix} \frac{c_f}{m\bar{v}_x} \\ \frac{l_f c_f}{I_z} \end{bmatrix}, \ \bar{B}_4 = \begin{bmatrix} \frac{c_f}{m\underline{v}_x} \\ \frac{l_f c_f}{I_z} \end{bmatrix},$$

$$\Delta\bar{B}_1 = E_{2,1}\bar{M}F_{2,1}, \ \Delta\bar{B}_2 = E_{2,2}\bar{M}F_{2,2}, \ \Delta\bar{B}_3 = E_{2,3}\bar{M}F_{2,3}, \ \Delta\bar{B}_4 = E_{2,4}\bar{M}F_{2,4},$$

$$E_{2,1} = \begin{bmatrix} \frac{\Delta c_f}{m\bar{v}_x} & 0 \\ 0 & \frac{l_f \Delta c_f}{I_z} \end{bmatrix}, \ E_{2,2} = \begin{bmatrix} \frac{\Delta c_f}{m\underline{v}_x} & 0 \\ 0 & \frac{l_f \Delta c_f}{I_z} \end{bmatrix},$$

$$E_{2,3} = \begin{bmatrix} \frac{\Delta c_f}{m\bar{v}_x} & 0 \\ 0 & \frac{l_f \Delta c_f}{I_z} \end{bmatrix}, \ E_{2,4} = \begin{bmatrix} \frac{\Delta c_f}{m\underline{v}_x} & 0 \\ 0 & \frac{l_f \Delta c_f}{I_z} \end{bmatrix},$$

$$F_{2,1} = \begin{bmatrix} 1 \\ 1 \end{bmatrix}, \ F_{2,2} = \begin{bmatrix} 1 \\ 1 \end{bmatrix}, \ F_{2,3} = \begin{bmatrix} 1 \\ 1 \end{bmatrix}, \ F_{2,4} = \begin{bmatrix} 1 \\ 1 \end{bmatrix}.$$

Considering the uncertainties and the nonprecise measurements simultaneously, the matrices have the following forms:

$$\begin{aligned} A &= \sum_{i=1}^{4} \hat{\alpha}_i (\bar{A}_i + \Delta\bar{A}_i + \Delta K_i \bar{M}_1 \bar{A}_i + \Delta K_i \bar{M}_1 \Delta\bar{A}_i) \\ &= \sum_{i=1}^{4} \hat{\alpha}_i (\bar{A}_i + \Delta\bar{A}_i + \Delta K_i \bar{M}_1 \bar{A}_i + \Delta\hat{A}_i), \\ B &= \sum_{i=1}^{4} \hat{\alpha}_i (\bar{B}_i + \Delta\bar{B}_i + \Delta K_i \bar{M}_1 \bar{B}_i + \Delta K_i \bar{M}_1 \Delta\bar{B}_i) \\ &= \sum_{i=1}^{4} \hat{\alpha}_i (\bar{B}_i + \Delta\bar{B}_i + \Delta K_i \bar{M}_1 \bar{B}_i + \Delta\hat{B}_i), \end{aligned} \tag{4.13}$$

where

$$\begin{aligned} \Delta\hat{A}_i &= \Delta K_i \bar{M}_1 \Delta A_i = \Delta K_i \bar{M}_1 E_{1,i} \bar{M} F_{1,i} = \Delta K_i E_{1,i} \bar{M}_2 F_{1,i}, \\ \Delta\hat{B}_i &= \Delta K_i \bar{M}_1 \Delta B_i = \Delta K_i \bar{M}_1 E_{2,i} \bar{M} F_{2,i} = \Delta K_i E_{2,i} \bar{M}_2 F_{2,i}, \\ \|\bar{M}_2\| &\leq 1. \end{aligned}$$

The EGV state-space model becomes

$$\begin{aligned} \dot{x}(t) = (\bar{A}(\hat{\alpha}) + \Delta\tilde{A}(\hat{\alpha}))x(t) + (\bar{B}(\hat{\alpha}) + \Delta\tilde{B}(\hat{\alpha}))\delta_d \\ +(\bar{B}(\hat{\alpha}) + \Delta\tilde{B}(\hat{\alpha}))\delta_f. \end{aligned} \tag{4.14}$$

where

$$\begin{aligned} \bar{A}(\hat{\alpha}) &= \sum_{i=1}^{4} \hat{\alpha}_i \bar{A}_i, \ \bar{B}(\hat{\alpha}) = \sum_{i=1}^{4} \hat{\alpha}_i \bar{B}_i, \\ \Delta\tilde{A}(\hat{\alpha}) &= \sum_{i=1}^{4} \hat{\alpha}_i (\Delta\bar{A}_i + \Delta K_i \bar{M}_1 \bar{A}_i + \Delta\hat{A}_i), \\ \Delta\tilde{B}(\hat{\alpha}) &= \sum_{i=1}^{4} \hat{\alpha}_i (\Delta\bar{B}_i + \Delta K_i \bar{M}_1 \bar{B}_i + \Delta\hat{B}_i). \end{aligned}$$

In this work, the primary objective is to design a fault detector for the active steering actuator such that the alarm will be turned on when the fault is detected.

The fault detector is based on the model in (4.14) and the available measurements. Compared with the sideslip angle, it is much easier to obtain the vehicle yaw rate. Thus, the output is selected as the yaw rate as:

$$y = Cx, \tag{4.15}$$

where $C = \begin{bmatrix} 0 \; 1 \end{bmatrix}$. We propose the following fault detector:

$$\begin{aligned} \dot{\hat{x}}(t) &= \bar{A}(\hat{\alpha})\hat{x}(t) + \bar{B}(\hat{\alpha})\delta_d + L(\hat{\alpha})(y - \hat{y}), \\ \hat{y} &= \; C\hat{x}(t), \\ \mathrm{Re} &= \; y - \hat{y}, \end{aligned} \tag{4.16}$$

where $\hat{x}$ is the state vector of the detector, $\hat{y}$ is the estimated output, and Re is the residual signal which indicates the fault information such as the occurrence time and the location.

Defining the state estimation error as $e(t) = x(t) - \hat{x}(t)$, we have the following augmented system:

$$\begin{aligned} \dot{\xi}(t) &= (\underline{A}(\hat{\alpha}) + \Delta\underline{A}(\hat{\alpha}))\xi(t) + (\underline{B}(\hat{\alpha}) + \Delta\underline{B}(\hat{\alpha}))\delta_d \\ &\quad +(\underline{B}_1(\hat{\alpha}) + \Delta\underline{B}(\hat{\alpha}))\delta_f, \\ \mathrm{Re} &= \; \underline{C}\xi(t), \end{aligned} \tag{4.17}$$

where

$$\begin{aligned} \underline{A}(\hat{\alpha}) &= \begin{bmatrix} \bar{A}(\hat{\alpha}) & 0 \\ 0 & \bar{A}(\hat{\alpha}) - L(\hat{\alpha})C \end{bmatrix}, \xi(t) = \begin{bmatrix} x(t) \\ e(t) \end{bmatrix}, \\ \Delta\underline{A}(\hat{\alpha}) &= \begin{bmatrix} I \\ I \end{bmatrix} \Delta\tilde{A}(\hat{\alpha}) \begin{bmatrix} I \; 0 \end{bmatrix}, \\ \underline{B}(\hat{\alpha}) &= \begin{bmatrix} \bar{B}(\hat{\alpha}) \\ 0 \end{bmatrix}, \; \underline{B}_1(\hat{\alpha}) = \begin{bmatrix} \bar{B}(\hat{\alpha}) \\ \bar{B}(\hat{\alpha}) \end{bmatrix}, \\ \Delta\underline{B}(\hat{\alpha}) &= \begin{bmatrix} I \\ I \end{bmatrix} \Delta\tilde{B}(\hat{\alpha}), \\ \underline{C} &= \quad \begin{bmatrix} 0 \; 0 \; 0 \; 1 \end{bmatrix}. \end{aligned}$$

The schematic diagram of the detector principle is illustrated in Fig. 4.6. There are two inputs δ_d and δ_f for the EGV but only δ_d is available for the observer. The output of the observer is the estimated yaw rate, and the residual signal is sent to an evaluate function $J(t)$ which is defined as:

$$J(t) = \sqrt{\frac{1}{t}\int_0^t (\mathrm{Re}^{\mathrm{T}}\mathrm{Re})d\tau}. \tag{4.18}$$

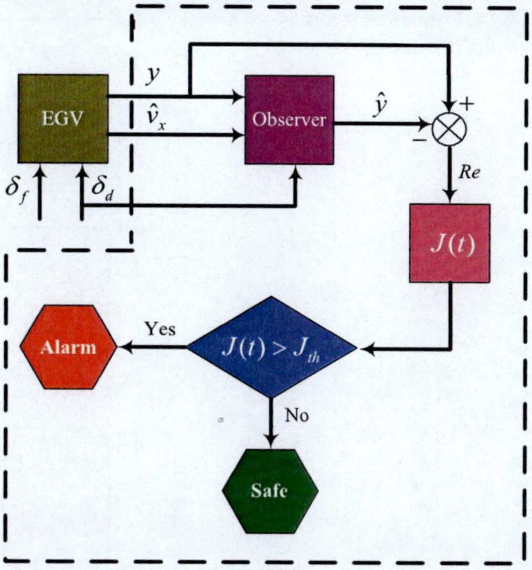

Fig. 4.6 Schematic diagram of the detector principle

The threshold Jth is defined as:

$$J_{\text{th}} = \sup_{\delta_f=0,\delta_d} J(t). \tag{4.19}$$

We can see that the threshold is the supremum of the residual evaluate function $J(t)$ when there is no fault at all. Therefore, the occurrence of fault can be detected by the following logic rule:

$$\begin{aligned} J(t) &> J_{\text{th}}, \text{ alarm} \\ J(t) &\leq J_{\text{th}}, \text{ safe and no alarm.} \end{aligned} \tag{4.20}$$

If the augmented system in (4.17) is stable, the estimation error would converge to zero. However, there are two external inputs in the augmented system (4.17) and the external inputs would affect the estimation error directly. It is desired that the effect of the desired input δ_d to the residual value is small and the effect of the fault input δ_f to the residual is large such that the designed detector is robust to the disturbance but sensitive to the faults. As the input matrices for δ_d and δ_f are different in (4.17), it is possible to achieve the two targets simultaneously. It is necessary to mention that the steering angles are operated at low frequencies. Figure 4.7 depicts the desired steering angles in three different maneuvers. The steering angles are quite different, which cover different types of operation conditions. The corresponding amplitude spectrums of three different maneuvers are illustrated in Fig. 4.8. The frequency analysis is carried out with MATLAB function *fft*. We can see that, though the typical maneuvers are different, the steering angles have quite small amplitudes when the frequency is larger than 0.5 Hz. Therefore, in the optimal detector design, we select the frequency $[-0.5\ 0.5]$ Hz=$[-\pi\ \pi]$rad for the optimization.

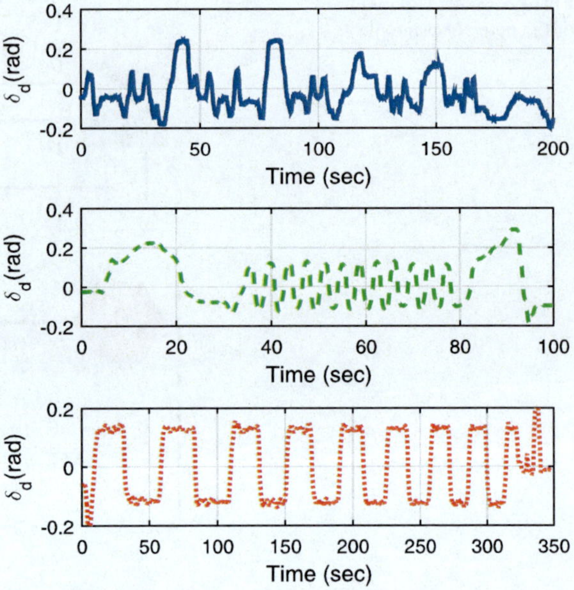

Fig. 4.7 Desired steering angles in three different maneuvers

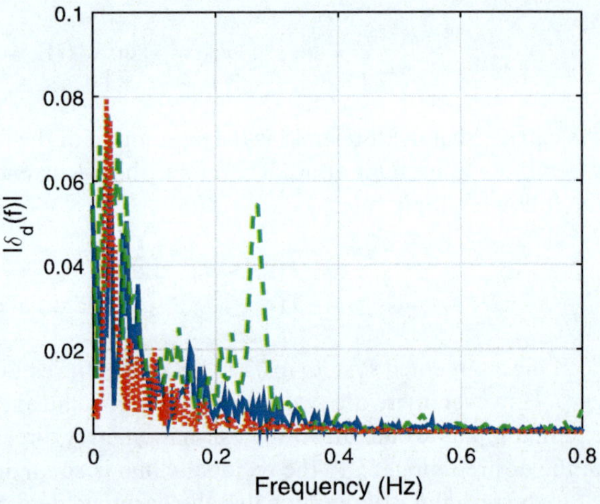

Fig. 4.8 Corresponding single-side amplitude spectrums of three different maneuvers

In order to evaluate the effects to the residual value, we introduce the following definitions.

Definition 4.1 The augmented system in (4.17) has a low-frequency ($[\,-\pi\ \ \pi\,]$rad) $\mathcal{H}_\infty$ index bound γ, if under zero initial condition, the following condition

$$\int_0^\infty \mathrm{Re}^{\mathrm{T}}\mathrm{Re}\,\mathrm{d}t < \gamma^2 \int_0^\infty \delta_d^{\mathrm{T}}\delta_d \mathrm{d}t \tag{4.21}$$

is satisfied for all solutions of (4.17) with L_2-bounded δ_d such that

$$\int_0^\infty (\pi\xi + j\dot{\xi})(-\pi\xi - j\dot{\xi})^{\mathrm{T}} < 0. \tag{4.22}$$

Here, j is the complex operator, the positive scalar γ is the $\mathcal{H}_\infty$ index bound. If the value of γ is smaller, the designed detector is more robust.

Definition 4.2 The augmented system in (4.17) has a low-frequency ($\begin{bmatrix} -\pi & \pi \end{bmatrix}$rad) $\mathcal{H}\_$ index bound β, if under zero initial condition, the following condition

$$\int_0^\infty \mathrm{Re}^{\mathrm{T}}\mathrm{Re}\,\mathrm{d}t > \beta^2 \int_0^\infty \delta_f^{\mathrm{T}}\delta_f \mathrm{d}t \tag{4.23}$$

is satisfied for all solutions of (4.17) with L_2-bounded δ_f such that the condition (4.22) holds. Here, the positive scalar β is the $\mathcal{H}\_$ index bound. If the value of γ is larger, the designed detector is more sensitive to the faults.

Then, the challenges and objectives are to design the observer gain $L(\hat{\alpha})$ such that

O1 The augmented system in (4.17) is robustly and asymptotically stable.
O2 Under zero initial condition, the $\mathcal{H}_\infty$ performance in (4.21) is achieved.
O3 Under zero initial condition, the $\mathcal{H}\_$ performance in (4.23) is achieved.

A detector in (4.16) with the gain satisfying these three objectives is called robust mixed $\mathcal{H}\_/\mathcal{H}_\infty$ detector.

4.3 Main Results

In this section, by assuming the observer gain is given, we analyze the stability and mixed $\mathcal{H}\_/\mathcal{H}_\infty$ for the augmented system dynamics. Based on the obtained results, the observer-gain design method will be developed. Before proceeding further, we introduce the following lemma to deal with uncertainties in the augmented system.

Lemma 4.1 *[31] If there exist real matrices $\Phi = \Phi^{\mathrm{T}}$, Υ, and Ω with compatible dimensions, and $\hat{M}$ satisfying $\|\hat{M}\| \le 1$, then, the following condition*

$$\Phi + \Upsilon\hat{M}\Omega + (\Upsilon\hat{M}\Omega)^{\mathrm{T}} < 0, \tag{4.24}$$

holds if and only if there exists a positive scalar $\varphi > 0$ such that the following condition

$$\begin{bmatrix} \Phi & \Upsilon & \varphi\Omega^{\mathrm{T}} \\ * & -\varphi I & 0 \\ * & * & -\varphi I \end{bmatrix} < 0 \tag{4.25}$$

is satisfied.

4.3.1 Stability Analysis and Observer Design

Besides the stability, it is desired that the observer has a fast convergence rate. In this subsystem, both the stability and the pole locations are studied.

Theorem 4.1 *Given positive scalars r_1, r_2, ς_1, and ς_2, when $r_1 < \varsigma_1$ and $r_2 < \varsigma_2$, the augmented system in (4.17) is robustly and asymptotically stable if there exist positive-definite matrices $P_1 = P_1^{\mathrm{T}} \in \mathcal{R}^{2\times 2}$ and $P_2 = P_2^{\mathrm{T}} \in \mathcal{R}^{2\times 2}$, and random matrices $G_1(\hat{\alpha}) \in \mathcal{R}^{2\times 2}$ and $G_2(\hat{\alpha}) \in \mathcal{R}^{2\times 2}$ such that the following condition is satisfied:*

$$\begin{bmatrix} \Gamma_1 & \Gamma_2 \\ * & -\Gamma_3 \end{bmatrix} < 0, \tag{4.26}$$

where

$$\Gamma_1 = \begin{bmatrix} r_1 P_1 & 0 \\ 0 & r_2 P_2 \end{bmatrix} - \begin{bmatrix} r_1 G_1(\hat{\alpha}) & 0 \\ 0 & r_2 G_2(\hat{\alpha}) \end{bmatrix} - \begin{bmatrix} r_1 G_1^{\mathrm{T}}(\hat{\alpha}) & 0 \\ 0 & r_2 G_2^{\mathrm{T}}(\hat{\alpha}) \end{bmatrix},$$

$$\Gamma_2 = \begin{bmatrix} G_1(\hat{\alpha}) & 0 \\ 0 & G_2(\hat{\alpha}) \end{bmatrix} (\underline{A}(\hat{\alpha}) + \Delta \underline{A}(\hat{\alpha})) + \begin{bmatrix} \varsigma_1 G_1(\hat{\alpha}) & 0 \\ 0 & \varsigma_2 G_2(\hat{\alpha}) \end{bmatrix},$$

$$\Gamma_3 = \begin{bmatrix} r_1 P_1 & 0 \\ 0 & r_2 P_2 \end{bmatrix}.$$

Proof Since P_1 and P_2 are positive-definite, the inverse of matrix $\begin{bmatrix} P_1 & 0 \\ 0 & P_2 \end{bmatrix}$ exists. With the fact that

$$\left(\begin{bmatrix} P_1 & 0 \\ 0 & P_2 \end{bmatrix} - \begin{bmatrix} G_1(\hat{\alpha}) & 0 \\ 0 & G_2(\hat{\alpha}) \end{bmatrix} \right)^{\mathrm{T}} \begin{bmatrix} P_1 & 0 \\ 0 & P_2 \end{bmatrix}^{-1} \left(\begin{bmatrix} P_1 & 0 \\ 0 & P_2 \end{bmatrix} - \begin{bmatrix} G_1(\hat{\alpha}) & 0 \\ 0 & G_2(\hat{\alpha}) \end{bmatrix} \right) \geq 0,$$

we have

$$\begin{aligned} & - \begin{bmatrix} G_1(\hat{\alpha}) & 0 \\ 0 & G_2(\hat{\alpha}) \end{bmatrix}^{\mathrm{T}} \begin{bmatrix} P_1 & 0 \\ 0 & P_2 \end{bmatrix}^{-1} \begin{bmatrix} G_1(\hat{\alpha}) & 0 \\ 0 & G_2(\hat{\alpha}) \end{bmatrix} \\ & \leq \begin{bmatrix} P_1 & 0 \\ 0 & P_2 \end{bmatrix} - \begin{bmatrix} G_1(\hat{\alpha}) & 0 \\ 0 & G_2(\hat{\alpha}) \end{bmatrix}^{\mathrm{T}} - \begin{bmatrix} G_1(\hat{\alpha}) & 0 \\ 0 & G_2(\hat{\alpha}) \end{bmatrix}. \end{aligned}$$

Then, the condition in (4.26) is satisfied if the following one is satisfied:

$$\begin{bmatrix} \Gamma_4 & \Gamma_2 \\ * & -\Gamma_3 \end{bmatrix} < 0, \tag{4.27}$$

where

$$\Gamma_4 = -\begin{bmatrix} r_1 I & 0 \\ 0 & r_2 I \end{bmatrix}\begin{bmatrix} G_1(\hat{\alpha}) & 0 \\ 0 & G_2(\hat{\alpha}) \end{bmatrix}^{\mathrm{T}}\begin{bmatrix} P_1 & 0 \\ 0 & P_2 \end{bmatrix}^{-1}\begin{bmatrix} G_1(\hat{\alpha}) & 0 \\ 0 & G_2(\hat{\alpha}) \end{bmatrix}. \tag{4.28}$$

It infers from (4.26) that the matrix $\begin{bmatrix} G_1(\hat{\alpha}) & 0 \\ 0 & G_2(\hat{\alpha}) \end{bmatrix}$ is nonsingular and the inverse exists. Performing a congruence transformation to (4.27) with the diagonal matrix $\begin{bmatrix} P_1 & 0 \\ 0 & P_2 \end{bmatrix}\begin{bmatrix} G_1(\hat{\alpha}) & 0 \\ 0 & G_2(\hat{\alpha}) \end{bmatrix}^{-\mathrm{T}}$, one gets

$$\begin{bmatrix} -\Gamma_3 & \Gamma_5 \\ * & -\Gamma_3 \end{bmatrix} < 0, \tag{4.29}$$

where

$$\Gamma_5 = \begin{bmatrix} P_1 & 0 \\ 0 & P_2 \end{bmatrix}(\underline{A}(\hat{\alpha}) + \Delta\underline{A}(\hat{\alpha})) + \begin{bmatrix} \varsigma_1 I & 0 \\ 0 & \varsigma_2 I \end{bmatrix}\begin{bmatrix} P_1 & 0 \\ 0 & P_2 \end{bmatrix}. \tag{4.30}$$

Since $r_1 \leq \varsigma_1$ and $r_2 \leq \varsigma_2$, in terms of the pole placement in [32] and [33], all the eigenvalues of matrix $(\underline{A}(\hat{\alpha}) + \Delta\underline{A}(\hat{\alpha}))$ have negative real parts if the condition (4.26) holds, that is, the augmented system in (4.17) is robustly and asymptotically stable. This completes the proof. □

Based on the condition in Theorem 4.1, the robust observer gain design is proposed in the following theorem.

Theorem 4.2 *Given positive scalars r_1, r_2, ς_1, and ς_2, when $r_1 < \varsigma_1$ and $r_2 < \varsigma_2$, the augmented system in (4.17) is robustly and asymptotically stable if there exist positive-definite matrices $P_1 = P_1^{\mathrm{T}} \in \mathcal{R}^{2\times 2}$ and $P_2 = P_2^{\mathrm{T}} \in \mathcal{R}^{2\times 2}$, random matrices $G_{1,i} \in \mathcal{R}^{2\times 2}$, $G_{2,i} \in \mathcal{R}^{2\times 2}$, and $\tilde{L}_i \in \mathcal{R}^{2\times 1}$, and scalars $\psi_{1,i}$, $\psi_{2,i}$, and $\psi_{3,i}$, such that the following condition is satisfied:*

$$\Theta_{1,ij} + \Theta_{1,ji} < 0, \ \forall\, 1 \leq i \leq j \leq 4, \tag{4.31}$$

where

$$\Theta_{1,ij} = \begin{bmatrix} \bar{\Gamma}_{1,i} & \bar{\Gamma}_{2,ij} & \bar{\Gamma}_{3,ij} & 0 & \bar{\Gamma}_{4,ij} & 0 & \bar{\Gamma}_{10,ij} & 0 \\ * & -\bar{\Gamma}_{5,i} & 0 & \bar{\Gamma}_{6,ij} & 0 & \bar{\Gamma}_{7,ij} & 0 & \bar{\Gamma}_{11,ij} \\ * & * & -\bar{\Gamma}_{8,i} & 0 & 0 & 0 & 0 & 0 \\ * & * & * & -\bar{\Gamma}_{8,i} & 0 & 0 & 0 & 0 \\ * & * & * & * & -\bar{\Gamma}_{9,i} & 0 & 0 & 0 \\ * & * & * & * & * & -\bar{\Gamma}_{9,i} & 0 & 0 \\ * & * & * & * & * & * & -\bar{\Gamma}_{12,i} & 0 \\ * & * & * & * & * & * & * & -\bar{\Gamma}_{12,i} \end{bmatrix},$$

$$\bar{\Gamma}_{1,i} = \begin{bmatrix} r_1 P_1 & 0 \\ 0 & r_2 P_2 \end{bmatrix} - \begin{bmatrix} r_1 G_{1,i} & 0 \\ 0 & r_2 G_{2,i} \end{bmatrix} - \begin{bmatrix} r_1 G_{1,i}^{\mathrm{T}} & 0 \\ 0 & r_2 G_{2,i}^{\mathrm{T}} \end{bmatrix},$$

$$\begin{aligned}
\bar{\Gamma}_{2,ij} &= \begin{bmatrix} G_{1,i}\bar{A}_j & 0 \\ 0 & G_{2,i}\bar{A}_j - \tilde{L}_i C \end{bmatrix} + \begin{bmatrix} \varsigma_1 G_{1,i} & 0 \\ 0 & \varsigma_2 G_{2,i} \end{bmatrix}, \\
\bar{\Gamma}_{3,ij} &= \begin{bmatrix} G_{1,i} \\ G_{2,i} \end{bmatrix} E_{1,j}, \bar{\Gamma}_{4,ij} = \begin{bmatrix} G_{1,i} \\ G_{2,i} \end{bmatrix} \Delta K_j E_{1,j}, \bar{\Gamma}_{5,i} = \begin{bmatrix} r_1 P_1 & 0 \\ 0 & r_2 P_2 \end{bmatrix}, \\
\bar{\Gamma}_{6,ij} &= \psi_{1,i} \begin{bmatrix} F_{1,j}^{\mathrm{T}} \\ 0 \end{bmatrix}, \bar{\Gamma}_{7,ij} = \psi_{2,i} \begin{bmatrix} F_{1,j}^{\mathrm{T}} \\ 0 \end{bmatrix}, \bar{\Gamma}_{10,ij} = \begin{bmatrix} G_{1,i} \\ G_{2,i} \end{bmatrix} \Delta K_j, \\
\bar{\Gamma}_{11,ij} &= \psi_{3,i} \begin{bmatrix} \bar{A}_j^{\mathrm{T}} \\ 0 \end{bmatrix}, \bar{\Gamma}_{8,i} = \psi_{1,i} I, \; \bar{\Gamma}_{9,i} = \psi_{2,i} I, \; \bar{\Gamma}_{12,i} = \psi_{3,i} I.
\end{aligned}$$

In addition, the observer gain is computed as:

$$L(\hat{\alpha}) = G_2^{-1}(\hat{\alpha})\tilde{L}(\hat{\alpha}). \tag{4.32}$$

Proof The condition in (4.26) is rewritten as:

$$\begin{aligned}
&\begin{bmatrix} \Gamma_1 & \bar{\Gamma}_2 \\ * & \Gamma_3 \end{bmatrix} + \begin{bmatrix} \begin{bmatrix} G_1(\hat{\alpha})E_1(\hat{\alpha}) \\ G_2(\hat{\alpha})E_1(\hat{\alpha}) \end{bmatrix} \\ 0 \end{bmatrix} \bar{M} \left[0 \left[F_1(\hat{\alpha}) \; 0 \right] \right] \\
&+ \begin{bmatrix} \begin{bmatrix} G_1(\hat{\alpha})\Delta K(\hat{\alpha}) \\ G_2(\hat{\alpha})\Delta K(\hat{\alpha}) \end{bmatrix} \\ 0 \end{bmatrix} \bar{M}_1 \left[0 \left[\bar{A}(\hat{\alpha}) \; 0 \right] \right] \\
&+ \begin{bmatrix} \begin{bmatrix} G_1(\hat{\alpha})\Delta \bar{K}(\hat{\alpha}) \\ G_2(\hat{\alpha})\Delta \bar{K}(\hat{\alpha}) \end{bmatrix} \\ 0 \end{bmatrix} \bar{M}_2 \left[0 \left[F_1(\hat{\alpha}) \; 0 \right] \right] \\
&+ \left(\begin{bmatrix} \begin{bmatrix} G_1(\hat{\alpha})E_1(\hat{\alpha}) \\ G_2(\hat{\alpha})E_1(\hat{\alpha}) \end{bmatrix} \\ 0 \end{bmatrix} \bar{M} \left[0 \left[F_1(\hat{\alpha}) \; 0 \right] \right] \right)^{\mathrm{T}} \\
&+ \left(\begin{bmatrix} \begin{bmatrix} G_1(\hat{\alpha})\Delta K(\hat{\alpha}) \\ G_2(\hat{\alpha})\Delta K(\hat{\alpha}) \end{bmatrix} \\ 0 \end{bmatrix} \bar{M}_1 \left[0 \left[\bar{A}(\hat{\alpha}) \; 0 \right] \right] \right)^{\mathrm{T}} \\
&+ \left(\begin{bmatrix} \begin{bmatrix} G_1(\hat{\alpha})\Delta \bar{K}(\hat{\alpha}) \\ G_2(\hat{\alpha})\Delta \bar{K}(\hat{\alpha}) \end{bmatrix} \\ 0 \end{bmatrix} \bar{M}_2 \left[0 \left[F_1(\hat{\alpha}) \; 0 \right] \right] \right)^{\mathrm{T}} < 0,
\end{aligned} \tag{4.33}$$

where

$$
\begin{aligned}
\bar{\Gamma}_2 = & \begin{bmatrix} G_1(\hat{\alpha}) & 0 \\ 0 & G_2(\hat{\alpha}) \end{bmatrix} \underline{A}(\hat{\alpha}) + \begin{bmatrix} \varsigma_1 I & 0 \\ 0 & \varsigma_2 I \end{bmatrix} \begin{bmatrix} G_1(\hat{\alpha}) & 0 \\ 0 & G_2(\hat{\alpha}) \end{bmatrix}, \\
E_1(\hat{\alpha}) = & \sum_{i=1}^{4} \hat{\alpha}_i E_{1,i}, \; F_1(\hat{\alpha}) = \sum_{i=1}^{4} \hat{\alpha}_i F_{1,i}, \\
\Delta K(\hat{\alpha}) = & \sum_{i=1}^{4} \hat{\alpha}_i \Delta K_i, \; \Delta \bar{K}(\hat{\alpha}) = \sum_{i=1}^{4} \hat{\alpha}_i \Delta K_i E_{1,i}.
\end{aligned}
$$

In terms of Lemma 4.1, the condition in (4.33) holds if and only if the following condition is satisfied:

$$
\Theta_1 = \begin{bmatrix}
\Gamma_1 & \bar{\Gamma}_2 & \begin{bmatrix} G_1(\hat{\alpha})E_1(\hat{\alpha}) \\ G_2(\hat{\alpha})E_1(\hat{\alpha}) \end{bmatrix} & 0 & \begin{bmatrix} G_1(\hat{\alpha})\Delta K(\hat{\alpha}) \\ G_2(\hat{\alpha})\Delta K(\hat{\alpha}) \end{bmatrix} & 0 & \begin{bmatrix} G_1(\hat{\alpha})\Delta \bar{K}(\hat{\alpha}) \\ G_2(\hat{\alpha})\Delta \bar{K}(\hat{\alpha}) \end{bmatrix} & 0 \\
* & \Gamma_3 & 0 & \psi_1(\hat{\alpha}) \begin{bmatrix} F_1^{\mathrm{T}}(\hat{\alpha}) \\ 0 \end{bmatrix} & 0 & \psi_3(\hat{\alpha}) \begin{bmatrix} \bar{A}^{\mathrm{T}}(\hat{\alpha}) \\ 0 \end{bmatrix} & 0 & \psi_2(\hat{\alpha}) \begin{bmatrix} F_1^{\mathrm{T}}(\hat{\alpha}) \\ 0 \end{bmatrix} \\
* & * & -\psi_1(\hat{\alpha})I & 0 & 0 & 0 & 0 & 0 \\
* & * & * & -\psi_1(\hat{\alpha})I & 0 & 0 & 0 & 0 \\
* & * & * & * & -\psi_3(\hat{\alpha})I & 0 & 0 & 0 \\
* & * & * & * & * & -\psi_3(\hat{\alpha})I & 0 & 0 \\
* & * & * & * & * & * & -\psi_2(\hat{\alpha})I & 0 \\
* & * & * & * & * & * & * & -\psi_2(\hat{\alpha})I
\end{bmatrix}.
$$

It is assumed that $\tilde{L}(\hat{\alpha}) = G_2(\hat{\alpha})L(\hat{\alpha})$ and the following matrices are linearly dependent on the weighting factor vector $\hat{\alpha}$ as:

$$
\begin{aligned}
& G_1(\hat{\alpha}) = \sum_{i=1}^{4} \hat{\alpha}_i G_{1,i}, \; G_2(\hat{\alpha}) = \sum_{i=1}^{4} \hat{\alpha}_i G_{2,i}, \; \tilde{L}(\hat{\alpha}) = \sum_{i=1}^{4} \hat{\alpha}_i \tilde{L}_i, \\
& \psi_1(\hat{\alpha}) = \sum_{i=1}^{4} \hat{\alpha}_i \psi_{1,i}, \; \psi_2(\hat{\alpha}) = \sum_{i=1}^{4} \hat{\alpha}_i \psi_{2,i}, \; \psi_3(\hat{\alpha}) = \sum_{i=1}^{4} \hat{\alpha}_i \psi_{3,i}.
\end{aligned} \tag{4.34}
$$

Then, it is easy to find that

$$
\Theta_1 = \sum_{i=1}^{4} \hat{\alpha}_i^2 \Theta_{1,ii} + \sum_{i=1}^{3} \sum_{j=i+1}^{4} \hat{\alpha}_i \hat{\alpha}_j (\Theta_{1,ij} + \Theta_{1,ji}). \tag{4.35}
$$

If the condition in (4.31) is satisfied, in terms of the Eq. (4.35), Θ_1 is negative-definite and the augmented system is asymptotically stable. Since $G_1(\hat{\alpha})$ is nonsingular, the observer gain can be calculated according to the expression in (4.32). The proof is completed. □

4.3.2 $\mathcal{H}\_$ Performance and Observer Design

Lemma 4.2 *Given a positive scalar β, the augmented system in (4.17) has the $\mathcal{H}\_$ performance in (4.23) if there exist symmetric matrices Q_1 and Q_2 such that the following condition holds:*

$$\begin{bmatrix} \underline{A}(\hat{\alpha}) + \Delta\underline{A}(\hat{\alpha}) & \underline{B}_1(\hat{\alpha}) + \Delta\underline{B}(\hat{\alpha}) \\ I & 0 \end{bmatrix}^{\mathrm{T}} \begin{bmatrix} -Q_1 & Q_2 \\ Q_2 & \pi^2 Q_1 \end{bmatrix} \begin{bmatrix} \underline{A}(\hat{\alpha}) + \Delta\underline{A}(\hat{\alpha}) & \underline{B}_1(\hat{\alpha}) + \Delta\underline{B}(\hat{\alpha}) \\ I & 0 \end{bmatrix} + \begin{bmatrix} \underline{C} & 0 \\ 0 & I \end{bmatrix}^{\mathrm{T}} \begin{bmatrix} -I & 0 \\ 0 & \beta^2 I \end{bmatrix} \begin{bmatrix} \underline{C} & 0 \\ 0 & I \end{bmatrix} < 0. \tag{4.36}$$

Before proceeding, the following Finsler's lemma, which is important for the subsequent discussion, is first introduced.

Lemma 4.3 ***(Finsler's lemma** [34, 35]): Given a vector $\bar{\xi} \in \mathbb{R}^n$, a symmetric matrix $\Theta = \Theta^{\mathrm{T}} \in \mathbb{R}^{n\times n}$, and a matrix $\mathcal{B} \in \mathbb{R}^{m\times n}$, if the rank$(\mathcal{B}) < n$, the following two conditions are equivalent:*

(i) $\xi^{\mathrm{T}}\Theta\xi < 0, \forall\, \mathcal{B}\bar{\xi} = 0, \bar{\xi} \neq 0.$
(ii) $\exists \mathcal{X} \in \mathbb{R}^{n\times m}$ *such that* $\Theta + \mathcal{X}\mathcal{B} + \mathcal{B}^{\mathrm{T}}\mathcal{X}^{\mathrm{T}} < 0.$

Theorem 4.3 *Given a positive scalar β, the augmented system in (4.17) has the $\mathcal{H}\_$ performance in (4.23) if there exist symmetric matrices Q_1 and Q_2, and random matrices $S_1(\hat{\alpha})$, $S_2(\hat{\alpha})$, $H_1(\hat{\alpha})$, $H_2(\hat{\alpha})$, $H_3(\hat{\alpha})$, and $H_4(\hat{\alpha})$ such that the following condition holds:*

$$\begin{bmatrix} -Q_1 & \Omega_1 & -H_1(\hat{\alpha}) & -S_2^{\mathrm{T}}(\hat{\alpha}) \\ * & \Omega_2 & -H_2(\hat{\alpha}) + (H_3(\hat{\alpha})\underline{C})^{\mathrm{T}} & \Omega_3 \\ * & * & -I - H_3(\hat{\alpha}) - H_3^{\mathrm{T}}(\hat{\alpha}) & -H_4^{\mathrm{T}}(\hat{\alpha}) \\ * & * & * & \Omega_4 \end{bmatrix} < 0, \tag{4.37}$$

where

$$\begin{aligned} \Omega_1 &= Q_2 + H_1(\hat{\alpha})\underline{C} - S_1^{\mathrm{T}}(\hat{\alpha}), \\ \Omega_2 &= \pi^2 Q_1 + S_1(\hat{\alpha})(\underline{A}(\hat{\alpha}) + \Delta\underline{A}(\hat{\alpha})) \\ &\quad + (S_1(\hat{\alpha})(\underline{A}(\hat{\alpha}) + \Delta\underline{A}(\hat{\alpha})))^{\mathrm{T}} + H_2(\hat{\alpha})\underline{C} + (H_2(\hat{\alpha})\underline{C})^{\mathrm{T}}, \\ \Omega_3 &= S_1(\hat{\alpha})(\underline{B}_1(\hat{\alpha}) + \Delta\underline{B}(\hat{\alpha})) + (S_2(\hat{\alpha})(\underline{A}(\hat{\alpha}) + \Delta\underline{A}(\hat{\alpha})))^{\mathrm{T}} + (H_4(\hat{\alpha})\underline{C})^{\mathrm{T}}, \\ \Omega_4 &= \beta^2 I + S_2(\hat{\alpha})(\underline{B}_1(\hat{\alpha}) + \Delta\underline{B}(\hat{\alpha})) + (S_2(\hat{\alpha})(\underline{B}_1(\hat{\alpha}) + \Delta\underline{B}(\hat{\alpha})))^{\mathrm{T}}. \end{aligned}$$

Proof The condition in (4.36) can be rewritten as:

$$\begin{bmatrix} \underline{A}(\hat{\alpha})+\Delta\underline{A}(\hat{\alpha}) & \underline{B}_1(\hat{\alpha})+\Delta\underline{B}(\hat{\alpha}) \\ I & 0 \\ \underline{C} & 0 \\ 0 & I \end{bmatrix}^{\mathrm{T}} \times \begin{bmatrix} -Q_1 & Q_2 & 0 & 0 \\ * & \pi^2 Q_1 & 0 & 0 \\ * & * & -I & 0 \\ * & * & * & \beta^2 I \end{bmatrix} \times \begin{bmatrix} \underline{A}(\hat{\alpha})+\Delta\underline{A}(\hat{\alpha}) & \underline{B}_1(\hat{\alpha})+\Delta\underline{B}(\hat{\alpha}) \\ I & 0 \\ \underline{C} & 0 \\ 0 & I \end{bmatrix} < 0. \tag{4.38}$$

It is noted that

$$\begin{bmatrix} -I & \underline{A}(\hat{\alpha})+\Delta\underline{A}(\hat{\alpha}) & 0 & \underline{B}_1(\hat{\alpha})+\Delta\underline{B}(\hat{\alpha}) \\ 0 & \underline{C} & -I & 0 \end{bmatrix} \begin{bmatrix} \underline{A}(\hat{\alpha})+\Delta\underline{A}(\hat{\alpha}) & \underline{B}_1(\hat{\alpha})+\Delta\underline{B}(\hat{\alpha}) \\ I & 0 \\ \underline{C} & 0 \\ 0 & I \end{bmatrix} = 0. \tag{4.39}$$

In terms of Finsler's lemma, the condition in (4.38) holds if the following one is satisfied:

$$\begin{bmatrix} -Q_1 & Q_2 & 0 & 0 \\ * & \pi^2 Q_1 & 0 & 0 \\ * & * & -I & 0 \\ * & * & 0 & \beta^2 I \end{bmatrix} + \begin{bmatrix} 0 & H_1(\hat{\alpha}) \\ S_1(\hat{\alpha}) & H_2(\hat{\alpha}) \\ 0 & H_3(\hat{\alpha}) \\ S_2(\hat{\alpha}) & H_4(\hat{\alpha}) \end{bmatrix} \begin{bmatrix} -I & \underline{A}(\hat{\alpha})+\Delta\underline{A}(\hat{\alpha}) & 0 & \underline{B}_1(\hat{\alpha})+\Delta\underline{B}(\hat{\alpha}) \\ 0 & \underline{C} & -I & 0 \end{bmatrix} + \left(\begin{bmatrix} 0 & H_1(\hat{\alpha}) \\ S_1(\hat{\alpha}) & H_2(\hat{\alpha}) \\ 0 & H_3(\hat{\alpha}) \\ S_2(\hat{\alpha}) & H_4(\hat{\alpha}) \end{bmatrix} \begin{bmatrix} -I & \underline{A}(\hat{\alpha})+\Delta\underline{A}(\hat{\alpha}) & 0 & \underline{B}_1(\hat{\alpha})+\Delta\underline{B}(\hat{\alpha}) \\ 0 & \underline{C} & -I & 0 \end{bmatrix} \right)^{\mathrm{T}}. \tag{4.40}$$

Expanding the condition in (4.40), one gets (4.37). Therefore, the $\mathcal{H}_{-}$ performance in (4.23) is satisfied. The proof is completed. □

The following theorem would project the condition in Theorem 4.3 to the polytope vertices. Before offering the theorem, we introduce a new matrix $H_s = \begin{bmatrix} 1 & 1 \end{bmatrix}$.

Theorem 4.4 *Given a positive scalar β and scalars λ_1, λ_2, and λ_3, the augmented system in (4.17) has the $\mathcal{H}_{-}$ performance in (4.23) if there exist symmetric matrices Q_1 and Q_2, and random matrices $S_{1,i} = \begin{bmatrix} S_{11,i} & \lambda_1 G_{2,i} \\ S_{12,i} & \lambda_2 G_{2,i} \end{bmatrix}$, $S_{2,i} = \begin{bmatrix} S_{21,i} & \lambda_3 H_s G_{2,i} \end{bmatrix}$, $\tilde{L}_i$, $\psi_{4,i}$, $\psi_{5,i}$, $\psi_{6,i}$, $\psi_{7,i}$, $\psi_{8,i}$, $\psi_{9,i}$, $H_{1,i}$, $H_{2,i}$, $H_{3,i}$, and $H_{4,i}$ such that the following condition holds:*

$$\Theta_{2,ij} + \Theta_{2,ji} < 0,\ \forall\, 1 \leq i \leq j \leq 4, \tag{4.41}$$

where

$$\Theta_{2,ij} = \left[\begin{array}{cccccccccccccccc}
-Q_1 & \Omega_{1,i} & -H_{1,i} & -S_{2,i}^{\mathrm{T}} & 0 & 0 & 0 & 0 & 0 & 0 & 0 & 0 & 0 & 0 & 0 & 0 \\
* & \Omega_{2,ij} & -H_{2,i} + (H_{3,i}\underline{C})^{\mathrm{T}} & \Omega_{3,ij} & \Omega_{5,ij} & \Omega_{8,ij} & \Omega_{9,ij} & \Omega_{12,ij} & \Omega_{13,ij} & \Omega_{16,ij} & \Omega_{17,ij} & 0 & \Omega_{20,ij} & 0 & \Omega_{13,ij} & 0 \\
* & * & -I - H_{3,i} - H_{3,i}^{\mathrm{T}} & -H_{4,i}^{\mathrm{T}} & 0 & 0 & 0 & 0 & 0 & 0 & 0 & 0 & 0 & 0 & 0 & 0 \\
* & * & * & \Omega_{4,ij} & \Omega_{6,ij} & 0 & \Omega_{10,ij} & 0 & \Omega_{14,ij} & 0 & \Omega_{18,ij} & \Omega_{19,ij} & \Omega_{21,ij} & \Omega_{22,ij} & \Omega_{14,ij} & \Omega_{23,ij} \\
* & * & * & * & -\Omega_{7,i} & 0 & 0 & 0 & 0 & 0 & 0 & 0 & 0 & 0 & 0 & 0 \\
* & * & * & * & * & -\Omega_{7,i} & 0 & 0 & 0 & 0 & 0 & 0 & 0 & 0 & 0 & 0 \\
* & * & * & * & * & * & -\Omega_{11,i} & 0 & 0 & 0 & 0 & 0 & 0 & 0 & 0 & 0 \\
* & * & * & * & * & * & * & -\Omega_{11,i} & 0 & 0 & 0 & 0 & 0 & 0 & 0 & 0 \\
* & * & * & * & * & * & * & * & -\Omega_{15,i} & 0 & 0 & 0 & 0 & 0 & 0 & 0 \\
* & * & * & * & * & * & * & * & * & -\Omega_{15,i} & 0 & 0 & 0 & 0 & 0 & 0 \\
* & * & * & * & * & * & * & * & * & * & -\Omega_{24,i} & 0 & 0 & 0 & 0 & 0 \\
* & * & * & * & * & * & * & * & * & * & * & -\Omega_{24,i} & 0 & 0 & 0 & 0 \\
* & * & * & * & * & * & * & * & * & * & * & * & -\Omega_{25,i} & 0 & 0 & 0 \\
* & * & * & * & * & * & * & * & * & * & * & * & * & -\Omega_{25,i} & 0 & 0 \\
* & * & * & * & * & * & * & * & * & * & * & * & * & * & -\Omega_{26,i} & 0 \\
* & * & * & * & * & * & * & * & * & * & * & * & * & * & * & -\Omega_{26,i}
\end{array}\right],$$

$$\Omega_{1,i} = Q_2 + H_{1,i}\underline{C} - \begin{bmatrix} S_{11,i} & \lambda_1 G_{2,i} \\ S_{12,i} & \lambda_2 G_{2,i} \end{bmatrix},$$

$$\Omega_{2,ij} = \pi^2 Q_1 + \begin{bmatrix} S_{11,i}\bar{A}_j & \lambda_1 G_{2,i}\bar{A}_j - \lambda_1 \tilde{L}_i C \\ S_{12,i}\bar{A}_j & \lambda_2 G_{2,i}\bar{A}_j - \lambda_2 \tilde{L}_i C \end{bmatrix},$$

$$+ \begin{bmatrix} S_{11,i}\bar{A}_j & \lambda_1 G_{2,i}\bar{A}_j - \lambda_1 \tilde{L}_i C \\ S_{12,i}\bar{A}_j & \lambda_2 G_{2,i}\bar{A}_j - \lambda_2 \tilde{L}_i C \end{bmatrix}^{\mathrm{T}} + H_{2,i}\underline{C} + (H_{2,i}\underline{C})^{\mathrm{T}},$$

$$\Omega_{3,ij} = \begin{bmatrix} S_{11,i}\bar{B}_j + \lambda_1 G_{2,i}\bar{B}_j \\ S_{12,i}\bar{B}_j + \lambda_2 G_{2,i}\bar{B}_j \end{bmatrix}$$

$$+ \begin{bmatrix} S_{21,i}\bar{A}_j & \lambda_3 H_s G_{2,i}\bar{A}_j - \lambda_3 H_s \tilde{L}_i C \end{bmatrix}^{\mathrm{T}} + (H_{4,i}\underline{C})^{\mathrm{T}},$$

$$\Omega_{4,ij} = \beta^2 I + \left[S_{21,i}\bar{B}_j + \lambda_3 H_s G_{2,i}\bar{B}_j \right] + \left[S_{21,i}\bar{B}_j + \lambda_3 H_s G_{2,i}\bar{B}_j \right]^{\mathrm{T}},$$

$$\Omega_{5,ij} = \begin{bmatrix} S_{11,i}E_{1,j} + \lambda_1 G_{2,i}E_{1,j} \\ S_{12,i}E_{1,j} + \lambda_2 G_{2,i}E_{1,j} \end{bmatrix},$$

$$\Omega_{6,ij} = \left[S_{21,i}E_{1,j} + \lambda_3 H_s G_{2,i}E_{1,j} \right], \Omega_{7,i} = \psi_{4,i} I,$$

$$\Omega_{8,ij} = \psi_{4,i} \begin{bmatrix} F_{1,j}^{\mathrm{T}} \\ 0 \end{bmatrix},$$

$$\Omega_{9,ij} = \begin{bmatrix} S_{11,i}\Delta K_j E_{1,j} + \lambda_1 G_{2,i}\Delta K_j E_{1,j} \\ S_{12,i}\Delta K_j E_{1,j} + \lambda_2 G_{2,i}\Delta K_j E_{1,j} \end{bmatrix},$$

$$\Omega_{10,ij} = \left[S_{21,i}\Delta K_j E_{1,j} + \lambda_3 H_s G_{2,i}\Delta K_j E_{1,j} \right], \Omega_{11,i} = \psi_{5,i} I,$$

$$\Omega_{12,ij} = \psi_{5,i} \begin{bmatrix} F_{1,j}^{\mathrm{T}} \\ 0 \end{bmatrix},$$

$$\Omega_{13,ij} = \begin{bmatrix} S_{11,i}\Delta K_j + \lambda_1 G_{2,i}\Delta K_j \\ S_{12,i}\Delta K_j + \lambda_2 G_{2,i}\Delta K_j \end{bmatrix},$$

$$\Omega_{14,ij} = \left[S_{21,i}\Delta K_j + \lambda_3 H_s G_{2,i}\Delta K_j \right], \Omega_{16,i} = \psi_{9,i} I,$$

$$\Omega_{16,ij} = \psi_{6,i} \begin{bmatrix} \bar{A}_j^{\mathrm{T}} \\ 0 \end{bmatrix},$$

$$\Omega_{17,ij} = \begin{bmatrix} S_{11,i}E_{2,j} + \lambda_1 G_{2,i}E_{2,j} \\ S_{12,i}E_{2,j} + \lambda_2 G_{2,i}E_{2,j} \end{bmatrix},$$

$$\Omega_{18,ij} = \left[S_{21,i}E_{2,j} + \lambda_3 H_s G_{2,i}E_{2,j} \right],$$

$$\Omega_{19,ij} = \psi_{7,i} F_{2,j}^{\mathrm{T}},$$

$$\Omega_{20,ij} = \begin{bmatrix} S_{11,i}\Delta K_j E_{2,j} + \lambda_1 G_{2,i}\Delta K_j E_{2,j} \\ S_{12,i}\Delta K_j E_{2,j} + \lambda_2 G_{2,i}\Delta K_j E_{2,j} \end{bmatrix},$$

$$\Omega_{21,ij} = \left[S_{21,i}\Delta K_j E_{2,j} + \lambda_3 H_s G_{2,i}\Delta K_j E_{2,j} \right],$$

$$\Omega_{22,ij} = \psi_{8,i} F_{2,j}^{\mathrm{T}}, \ \Omega_{23,ij} = \psi_{9,i} \bar{B}_j^{\mathrm{T}},$$

$$\Omega_{24,i} = \psi_{7,i} I, \ \Omega_{25,i} = \psi_{8,i} I, \ \Omega_{26,i} = \psi_{9,i} I.$$

In addition, the observer gain is computed in terms of (4.32).

Proof By applying Lemma 4.1 three times, the condition in (4.37) holds if and only if the following one is satisfied:

$$
\Theta_2 = \left[\begin{array}{cccccccccccccccc}
-Q_1 & \Omega_1 & -H_1(\hat{\alpha}) & -S_2^{\mathrm{T}}(\hat{\alpha}) & 0 & 0 & 0 & 0 & 0 & 0 & 0 & 0 & 0 & 0 & 0 & 0 \\
* & \bar{\Omega}_2 & -H_2(\hat{\alpha}) + (H_3(\hat{\alpha})\underline{C})^{\mathrm{T}} & \bar{\Omega}_3 & \bar{\Omega}_5 & \bar{\Omega}_8 & \bar{\Omega}_9 & \bar{\Omega}_{12} & \bar{\Omega}_{13} & \bar{\Omega}_{16} & \bar{\Omega}_{17} & 0 & \bar{\Omega}_{20} & 0 & \bar{\Omega}_{13} & 0 \\
* & * & -I - H_3(\hat{\alpha}) - H_3^{\mathrm{T}}(\hat{\alpha}) & -H_4^{\mathrm{T}}(\hat{\alpha}) & 0 & 0 & 0 & 0 & 0 & 0 & 0 & 0 & 0 & 0 & 0 & 0 \\
* & * & * & \bar{\Omega}_4 & \bar{\Omega}_6 & 0 & \bar{\Omega}_{10} & 0 & \bar{\Omega}_{14} & 0 & \bar{\Omega}_{18} & \bar{\Omega}_{19} & \bar{\Omega}_{21} & \bar{\Omega}_{22} & \bar{\Omega}_{14} & \bar{\Omega}_{23} \\
* & * & * & * & -\bar{\Omega}_7 & 0 & 0 & 0 & 0 & 0 & 0 & 0 & 0 & 0 & 0 & 0 \\
* & * & * & * & * & -\bar{\Omega}_7 & 0 & 0 & 0 & 0 & 0 & 0 & 0 & 0 & 0 & 0 \\
* & * & * & * & * & * & -\bar{\Omega}_{11} & 0 & 0 & 0 & 0 & 0 & 0 & 0 & 0 & 0 \\
* & * & * & * & * & * & * & -\bar{\Omega}_{11} & 0 & 0 & 0 & 0 & 0 & 0 & 0 & 0 \\
* & * & * & * & * & * & * & * & -\bar{\Omega}_{15} & 0 & 0 & 0 & 0 & 0 & 0 & 0 \\
* & * & * & * & * & * & * & * & * & -\bar{\Omega}_{15} & 0 & 0 & 0 & 0 & 0 & 0 \\
* & * & * & * & * & * & * & * & * & * & -\bar{\Omega}_{24} & 0 & 0 & 0 & 0 & 0 \\
* & * & * & * & * & * & * & * & * & * & * & -\bar{\Omega}_{24} & 0 & 0 & 0 & 0 \\
* & * & * & * & * & * & * & * & * & * & * & * & -\bar{\Omega}_{25} & 0 & 0 & 0 \\
* & * & * & * & * & * & * & * & * & * & * & * & * & -\bar{\Omega}_{25} & 0 & 0 \\
* & * & * & * & * & * & * & * & * & * & * & * & * & * & -\bar{\Omega}_{26} & 0 \\
* & * & * & * & * & * & * & * & * & * & * & * & * & * & * & -\bar{\Omega}_{26}
\end{array}\right] < 0, \qquad (4.42)
$$

where

$$
\begin{aligned}
\bar{\Omega}_2 &= \pi^2 Q_1 + S_1(\hat{\alpha})\underline{A}(\hat{\alpha}) \\
&\quad + (S_1(\hat{\alpha})\underline{A}(\hat{\alpha}))^{\mathrm{T}} + H_2(\hat{\alpha})\underline{C} + (H_2(\hat{\alpha})\underline{C})^{\mathrm{T}}, \\
\bar{\Omega}_3 &= S_1(\hat{\alpha})\underline{B}_1(\hat{\alpha}) + (S_2(\hat{\alpha})\underline{A}(\hat{\alpha}))^{\mathrm{T}} + (H_4(\hat{\alpha})\underline{C})^{\mathrm{T}}, \\
\bar{\Omega}_4 &= \beta^2 I + S_2(\hat{\alpha})\underline{B}_1(\hat{\alpha}) + (S_2(\hat{\alpha})\underline{B}_1(\hat{\alpha}))^{\mathrm{T}},
\end{aligned}
$$

$$
\bar{\Omega}_5 = S_1(\hat{\alpha}) \begin{bmatrix} E_1(\hat{\alpha}) \\ E_1(\hat{\alpha}) \end{bmatrix}, \ \bar{\Omega}_6 = S_2(\hat{\alpha}) \begin{bmatrix} E_1(\hat{\alpha}) \\ E_1(\hat{\alpha}) \end{bmatrix},
$$

$$
\bar{\Omega}_8 = \psi_4(\hat{\alpha}) \begin{bmatrix} F_1^{\mathrm{T}}(\hat{\alpha}) \\ 0 \end{bmatrix}, \ \bar{\Omega}_9 = S_1(\hat{\alpha}) \begin{bmatrix} \Delta\bar{K}(\hat{\alpha}) \\ \Delta\bar{K}(\hat{\alpha}) \end{bmatrix},
$$

$$\bar{\Omega}_{10} = S_2(\hat{\alpha})\begin{bmatrix} \Delta\bar{K}(\hat{\alpha}) \\ \Delta\bar{K}(\hat{\alpha}) \end{bmatrix},\ \bar{\Omega}_{12} = \psi_5(\hat{\alpha})\begin{bmatrix} F_1^{\mathrm{T}}(\hat{\alpha}) \\ 0 \end{bmatrix},$$

$$\Omega_7 = \psi_4(\hat{\alpha})I,\ \Omega_{11} = \psi_5(\hat{\alpha})I, \Omega_{15}(\hat{\alpha}) = \psi_6(\hat{\alpha})I,$$

$$\bar{\Omega}_{13} = S_1(\hat{\alpha})\begin{bmatrix} \Delta K(\hat{\alpha}) \\ \Delta K(\hat{\alpha}) \end{bmatrix},\ \bar{\Omega}_{14} = S_2(\hat{\alpha})\begin{bmatrix} \Delta K(\hat{\alpha}) \\ \Delta K(\hat{\alpha}) \end{bmatrix},$$

$$\bar{\Omega}_{16} = \psi_6(\hat{\alpha})\begin{bmatrix} \bar{A}^{\mathrm{T}}(\hat{\alpha}) \\ 0 \end{bmatrix},\ \bar{\Omega}_{17} = S_1(\hat{\alpha})\begin{bmatrix} E_2(\hat{\alpha}) \\ E_2(\hat{\alpha}) \end{bmatrix},$$

$$\bar{\Omega}_{18} = S_2(\hat{\alpha})\begin{bmatrix} E_2(\hat{\alpha}) \\ E_2(\hat{\alpha}) \end{bmatrix},\ \bar{\Omega}_{19} = \psi_7(\hat{\alpha})F_2^{\mathrm{T}}(\hat{\alpha}),$$

$$\bar{\Omega}_{20} = S_1(\hat{\alpha})\begin{bmatrix} \Delta\bar{K}_2(\hat{\alpha}) \\ \Delta\bar{K}_2(\hat{\alpha}) \end{bmatrix},\ \bar{\Omega}_{21} = S_2(\hat{\alpha})\begin{bmatrix} \Delta\bar{K}_2(\hat{\alpha}) \\ \Delta\bar{K}_2(\hat{\alpha}) \end{bmatrix},$$

$$\bar{\Omega}_{22} = \psi_8(\hat{\alpha})F_2^{\mathrm{T}}(\hat{\alpha}),\ \Omega_{23} = \psi_9(\hat{\alpha})\bar{B}^{\mathrm{T}}(\hat{\alpha}),$$

$$\bar{\Omega}_{24} = \psi_7(\hat{\alpha})I, \bar{\Omega}_{25} = \psi_8(\hat{\alpha})I, \bar{\Omega}_{26} = \psi_9(\hat{\alpha})I,$$

$$\Delta\bar{K}_2(\hat{\alpha}) = \sum_{i=1}^{4}\hat{\alpha}_i\Delta K_i E_{2,i}.$$

It is assumed that

$$S_1(\hat{\alpha}) = \sum_{i=1}^{4}\hat{\alpha}_i\begin{bmatrix} S_{11,i} & \lambda_1 G_{2,i} \\ S_{12,i} & \lambda_2 G_{2,i} \end{bmatrix},\ S_2(\hat{\alpha}) = \sum_{i=1}^{4}\hat{\alpha}_i\begin{bmatrix} S_{21,i} & \lambda_3 H_s G_{2,i} \end{bmatrix},$$

$$\psi_4(\hat{\alpha}) = \sum_{i=1}^{4}\hat{\alpha}_i\psi_{4,i},\ \psi_5(\hat{\alpha}) = \sum_{i=1}^{4}\hat{\alpha}_i\psi_{5,i},\ \psi_6(\hat{\alpha}) = \sum_{i=1}^{4}\hat{\alpha}_i\psi_{6,i},$$

$$\psi_7(\hat{\alpha}) = \sum_{i=1}^{4}\hat{\alpha}_i\psi_{7,i},\ \psi_8(\hat{\alpha}) = \sum_{i=1}^{4}\hat{\alpha}_i\psi_{8,i},\ \psi_9(\hat{\alpha}) = \sum_{i=1}^{4}\hat{\alpha}_i\psi_{9,i},$$

$$\tilde{L}(\hat{\alpha}) = G_2(\hat{\alpha})L(\hat{\alpha}) = \sum_{i=1}^{4}\hat{\alpha}_i\tilde{L}_i,\ H_1(\hat{\alpha}) = \sum_{i=1}^{4}\hat{\alpha}_i H_{1,i},\ H_2(\hat{\alpha}) = \sum_{i=1}^{4}\hat{\alpha}_i H_{2,i},$$

$$H_3(\hat{\alpha}) = \sum_{i=1}^{4}\hat{\alpha}_i H_{3,i},\ H_4(\hat{\alpha}) = \sum_{i=1}^{4}\hat{\alpha}_i H_{4,i}.$$

With the matrix expressions, one gets:

$$\Theta_2 = \sum_{i=1}^{4}\hat{\alpha}_i^2\Theta_{2,ii} + \sum_{i=1}^{3}\sum_{j=i+1}^{4}\hat{\alpha}_i\hat{\alpha}_j(\Theta_{2,ij} + \Theta_{2,ji}). \tag{4.43}$$

In terms of the summation property in (4.43), if the condition in (4.41) holds, the matrix Θ_2 is negative-definite. Therefore, the $\mathcal{H}\_$ performance in (4.23) is guaranteed. This completes the proof. □

Remark 4.1 The proposed condition with pole constraints in Theorem 4.1 is inspired by the work in [32] and [33]. However, we have two circle constraints instead of one in [33], that is, some eigenvalues of the augmented system matrix are constrained into one circle and the other eigenvalues are restricted into the other circle. Since the observer gain to be designed cannot change the eigenvalue of the EGV dynamics, the multi-circle constraints are quite useful for the augmented system in (4.17). This is one contribution of the work. In addition, with the assist of Finsler's lemma, we obtain new criterion for the finite-frequency $\mathcal{H}\_$ performance in Theorem 4.3. Gain-scheduled slack matrices are introduced to reduce the conservativeness of the criterion. It is necessary to mention that there are three prescribed scalars λ_1, λ_2, and λ_3 in Theorem 4.4. In the observer design, these scalars can be optimized with the function *fminsearch* of MATLAB.

4.3.3 $\mathcal{H}_\infty$ Performance and Observer Design

Lemma 4.4 *Given a positive scalar γ, the augmented system in (4.17) has the $\mathcal{H}_\infty$ performance in (4.21) if there exist symmetric matrices Q_3 and Q_4 such that the following condition holds:*

$$\begin{bmatrix} \underline{A}(\hat{\alpha}) + \Delta\underline{A}(\hat{\alpha}) & \underline{B}(\hat{\alpha}) + \Delta\underline{B}(\hat{\alpha}) \\ I & 0 \end{bmatrix}^{\mathrm{T}} \begin{bmatrix} -Q_3 & Q_4 \\ Q_4 & \pi^2 Q_3 \end{bmatrix} \begin{bmatrix} \underline{A}(\hat{\alpha}) + \Delta\underline{A}(\hat{\alpha}) & \underline{B}(\hat{\alpha}) + \Delta\underline{B}(\hat{\alpha}) \\ I & 0 \end{bmatrix} + \begin{bmatrix} \underline{C} & 0 \\ 0 & I \end{bmatrix}^{\mathrm{T}} \begin{bmatrix} I & 0 \\ 0 & -\gamma^2 I \end{bmatrix} \begin{bmatrix} \underline{C} & 0 \\ 0 & I \end{bmatrix} < 0. \tag{4.44}$$

Theorem 4.5 *Given a positive scalar γ, the augmented system in (4.17) has the $\mathcal{H}_\infty$ performance in (4.21) if there exist symmetric matrices Q_3 and Q_4, and random matrices $S_3(\hat{\alpha})$, $S_4(\hat{\alpha})$, $H_5(\hat{\alpha})$, $H_6(\hat{\alpha})$, $H_7(\hat{\alpha})$, and $H_8(\hat{\alpha})$ such that the following condition holds:*

$$\begin{bmatrix} -Q_3 & \Lambda_1 & -H_5(\hat{\alpha}) & -S_4^{\mathrm{T}}(\hat{\alpha}) \\ * & \Lambda_2 & -H_6(\hat{\alpha}) + (H_7(\hat{\alpha})\underline{C})^{\mathrm{T}} & \Lambda_3 \\ * & * & I - H_7(\hat{\alpha}) - H_7^{\mathrm{T}}(\hat{\alpha}) & -H_8^{\mathrm{T}}(\hat{\alpha}) \\ * & * & * & \Lambda_4 \end{bmatrix} < 0, \tag{4.45}$$

where

$$
\begin{aligned}
\Lambda_1 &= Q_4 + H_5(\hat{\alpha})\underline{C} - S_3^{\mathrm{T}}(\hat{\alpha}),\\
\Lambda_2 &= \pi^2 Q_3 + S_3(\hat{\alpha})(\underline{A}(\hat{\alpha}) + \Delta\underline{A}(\hat{\alpha}))\\
&\quad +(S_3(\hat{\alpha})(\underline{A}(\hat{\alpha}) + \Delta\underline{A}(\hat{\alpha})))^{\mathrm{T}} + H_6(\hat{\alpha})\underline{C} + (H_6(\hat{\alpha})\underline{C})^{\mathrm{T}},\\
\Lambda_3 &= S_3(\hat{\alpha})(\underline{B}(\hat{\alpha}) + \Delta\underline{B}(\hat{\alpha}))\\
&\quad +(S_4(\hat{\alpha})(\underline{A}(\hat{\alpha}) + \Delta A(\hat{\alpha})))^{\mathrm{T}} + (H_8(\hat{\alpha})\underline{C})^{\mathrm{T}},\\
\Lambda_4 &= -\gamma^2 I + S_4(\hat{\alpha})(\underline{B}(\hat{\alpha}) + \Delta\underline{B}(\hat{\alpha}))\\
&\quad +(S_4(\hat{\alpha})(\underline{B}(\hat{\alpha}) + \Delta\underline{B}(\hat{\alpha})))^{\mathrm{T}}.
\end{aligned}
$$

Proof The condition in (4.44) can be rewritten as:

$$
\begin{aligned}
&\begin{bmatrix} \underline{A}(\hat{\alpha}) + \Delta\underline{A}(\hat{\alpha}) & \underline{B}(\hat{\alpha}) + \Delta\underline{B}(\hat{\alpha}) \\ I & 0 \\ \underline{C} & 0 \\ 0 & I \end{bmatrix}^{\mathrm{T}}\\
&\times \begin{bmatrix} -Q_3 & Q_4 & 0 & 0 \\ * & \pi^2 Q_3 & 0 & 0 \\ * & * & I & 0 \\ * & * & * & -\gamma^2 I \end{bmatrix}\\
&\times \begin{bmatrix} \underline{A}(\hat{\alpha}) + \Delta\underline{A}(\hat{\alpha}) & \underline{B}(\hat{\alpha}) + \Delta\underline{B}(\hat{\alpha}) \\ I & 0 \\ \underline{C} & 0 \\ 0 & I \end{bmatrix} < 0.
\end{aligned}
\tag{4.46}
$$

It is noted that

$$
\begin{aligned}
&\begin{bmatrix} -I & \underline{A}(\hat{\alpha}) + \Delta\underline{A}(\hat{\alpha}) & 0 & \underline{B}(\hat{\alpha}) + \Delta\underline{B}(\hat{\alpha}) \\ 0 & \underline{C} & -I & 0 \end{bmatrix}\\
&\times \begin{bmatrix} \underline{A}(\hat{\alpha}) + \Delta\underline{A}(\hat{\alpha}) & \underline{B}(\hat{\alpha}) + \Delta\underline{B}(\hat{\alpha}) \\ I & 0 \\ \underline{C} & 0 \\ 0 & I \end{bmatrix} = 0.
\end{aligned}
\tag{4.47}
$$

In terms of Finsler's lemma, the condition in (4.46) holds if the following one is satisfied:

$$
\begin{aligned}
&\begin{bmatrix} -Q_3 & Q_4 & 0 & 0 \\ * & \pi^2 Q_3 & 0 & 0 \\ * & * & I & 0 \\ * & * & 0 & -\gamma^2 I \end{bmatrix} + \begin{bmatrix} 0 & H_5(\hat{\alpha}) \\ S_3(\hat{\alpha}) & H_6(\hat{\alpha}) \\ 0 & H_7(\hat{\alpha}) \\ S_4(\hat{\alpha}) & H_8(\hat{\alpha}) \end{bmatrix}\\
&\times \begin{bmatrix} -I & \underline{A}(\hat{\alpha}) + \Delta\underline{A}(\hat{\alpha}) & 0 & \underline{B}(\hat{\alpha}) + \Delta\underline{B}(\hat{\alpha}) \\ 0 & \underline{C} & -I & 0 \end{bmatrix}
\end{aligned}
$$

$$
+\left(\begin{bmatrix} 0 & H_5(\hat{\alpha}) \\ S_3(\hat{\alpha}) & H_6(\hat{\alpha}) \\ 0 & H_7(\hat{\alpha}) \\ S_4(\hat{\alpha}) & H_8(\hat{\alpha}) \end{bmatrix}\right. \tag{4.48}
$$

$$
\left.\times\begin{bmatrix} -I & \underline{A}(\hat{\alpha})+\Delta\underline{A}(\hat{\alpha}) & 0 & \underline{B}(\hat{\alpha})+\Delta\underline{B}(\hat{\alpha}) \\ 0 & \underline{C} & -I & 0 \end{bmatrix}\right)^{\mathrm{T}}.
$$

Expanding the condition in (4.48), one gets (4.45). Therefore, the $\mathcal{H}_\infty$ performance in (4.21) is satisfied. The proof is completed. □

Theorem 4.6 *Given a positive scalar γ and scalars λ_4, λ_5, and λ_6, the augmented system in (4.17) has the $\mathcal{H}_\infty$ performance in (4.21) if there exist symmetric matrices Q_3 and Q_4, and random matrices $S_{3,i} = \begin{bmatrix} S_{31,i} & \lambda_4 G_{2,i} \\ S_{32,i} & \lambda_5 G_{2,i} \end{bmatrix}$, $S_{4,i} = \begin{bmatrix} S_{41,i} & \lambda_6 H_s G_{2,i} \end{bmatrix}$, $\tilde{L}_i$, $\psi_{10,i}$, $\psi_{11,i}$, $\psi_{12,i}$, $\psi_{13,i}$, $\psi_{14,i}$, $\psi_{15,i}$, $H_{5,i}$, $H_{6,i}$, $H_{7,i}$, and $H_{8,i}$ such that the following condition holds:*

$$
\Theta_{3,\mathrm{ij}} + \Theta_{3,\mathrm{ji}} < 0, \ \forall\, 1 \le i \le j \le 4, \tag{4.49}
$$

where

$$
\Theta_{3,ij} = \left[\begin{array}{cccccccccc}
-Q_3 & \Lambda_{1,i} & -H_{5,i} & -S_{4,i}^{\mathrm{T}} & 0 & 0 & 0 & 0 & 0 & 0 \\
* & \Lambda_{2,ij} & -H_{6,i}+(H_{7,i}\underline{C})^{\mathrm{T}} & \Lambda_{3,ij} & \Lambda_{5,ij} & \Lambda_{8,ij} & \Lambda_{9,ij} & \Lambda_{12,ij} & \Lambda_{13,ij} & \Lambda_{16,ij} \\
* & * & I-H_{7,i}-H_{7,i}^{\mathrm{T}} & -H_{8,i}^{\mathrm{T}} & 0 & 0 & 0 & 0 & 0 & 0 \\
* & * & * & \Lambda_{4,ij} & \Lambda_{6,ij} & 0 & \Lambda_{10,ij} & 0 & \Lambda_{14,ij} & 0 \\
* & * & * & * & -\Lambda_{7,i} & 0 & 0 & 0 & 0 & 0 \\
* & * & * & * & * & -\Lambda_{7,i} & 0 & 0 & 0 & 0 \\
* & * & * & * & * & * & -\Lambda_{11,i} & 0 & 0 & 0 \\
* & * & * & * & * & * & * & -\Lambda_{11,i} & 0 & 0 \\
* & * & * & * & * & * & * & * & -\Lambda_{15,i} & 0 \\
* & * & * & * & * & * & * & * & * & -\Lambda_{15,i} \\
* & * & * & * & * & * & * & * & * & * \\
* & * & * & * & * & * & * & * & * & * \\
* & * & * & * & * & * & * & * & * & * \\
* & * & * & * & * & * & * & * & * & * \\
* & * & * & * & * & * & * & * & * & * \\
* & * & * & * & * & * & * & * & * & *
\end{array}\right.
$$

$$\begin{array}{cccccc} 0 & 0 & 0 & 0 & 0 & 0 \\ \Lambda_{17,ij} & 0 & \Lambda_{20,ij} & 0 & \Lambda_{13,ij} & 0 \\ 0 & 0 & 0 & 0 & 0 & 0 \\ \Lambda_{18,ij} & \Lambda_{19,ij} & \Lambda_{21,ij} & \Lambda_{22,ij} & \Lambda_{14,ij} & \Lambda_{23,ij} \\ 0 & 0 & 0 & 0 & 0 & 0 \\ 0 & 0 & 0 & 0 & 0 & 0 \\ 0 & 0 & 0 & 0 & 0 & 0 \\ 0 & 0 & 0 & 0 & 0 & 0 \\ 0 & 0 & 0 & 0 & 0 & 0 \\ 0 & 0 & 0 & 0 & 0 & 0 \\ -\Lambda_{24,i} & 0 & 0 & 0 & 0 & 0 \\ * & -\Lambda_{24,i} & 0 & 0 & 0 & 0 \\ * & * & -\Lambda_{25,i} & 0 & 0 & 0 \\ * & * & * & -\Lambda_{25,i} & 0 & 0 \\ * & * & * & * & -\Lambda_{26,i} & 0 \\ * & * & * & * & * & -\Lambda_{26,i} \end{array}\right],$$

$$\Lambda_{1,i} = Q_4 + H_{5,i}\underline{C} - \begin{bmatrix} S_{31,i} & \lambda_4 G_{2,i} \\ S_{32,i} & \lambda_5 G_{2,i} \end{bmatrix},$$

$$\Lambda_{2,ij} = \pi^2 Q_3 + \begin{bmatrix} S_{31,i}\bar{A}_j & \lambda_4 G_{2,i}\bar{A}_j - \lambda_4 \tilde{L}_i C \\ S_{32,i}\bar{A}_j & \lambda_5 G_{2,i}\bar{A}_j - \lambda_5 \tilde{L}_i C \end{bmatrix}$$

$$+ \begin{bmatrix} S_{31,i}\bar{A}_j & \lambda_4 G_{2,i}\bar{A}_j - \lambda_4 \tilde{L}_i C \\ S_{32,i}\bar{A}_j & \lambda_5 G_{2,i}\bar{A}_j - \lambda_5 \tilde{L}_i C \end{bmatrix}^{\mathrm{T}} + H_{6,i}\underline{C} + (H_{6,i}\underline{C})^{\mathrm{T}},$$

$$\Lambda_{3,ij} = \begin{bmatrix} S_{31,i}\bar{B}_j \\ S_{32,i}\bar{B}_j \end{bmatrix} + \begin{bmatrix} S_{41,i}\bar{A}_j & \lambda_6 G_{2,i}\bar{A}_j - \lambda_6 H_s \tilde{L}_i C \end{bmatrix}^{\mathrm{T}} + (H_{8,i}\underline{C})^{\mathrm{T}},$$

$$\Lambda_{4,ij} = -\gamma^2 I + S_{41,i}\bar{B}_j + (S_{41,i}\bar{B}_j)^{\mathrm{T}},\ \Lambda_{5,ij} = \begin{bmatrix} S_{31,i}E_{1,j} + \lambda_4 G_{2,i}E_{1,j} \\ S_{32,i}E_{1,j} + \lambda_5 G_{2,i}E_{1,j} \end{bmatrix},$$

$$\Lambda_{6,ij} = \begin{bmatrix} S_{41,i}E_{1,j} + \lambda_6 H_s G_{2,i}E_{1,j} \end{bmatrix},\ \Lambda_{7,i} = \psi_{10,i} I,$$

$$\Lambda_{8,ij} = \psi_{10,i}\begin{bmatrix} F_{1,j}^{\mathrm{T}} \\ 0 \end{bmatrix},\ \Lambda_{9,ij} = \begin{bmatrix} S_{31,i}\Delta K_j E_{1,j} + \lambda_4 G_{2,i}\Delta K_j E_{1,j} \\ S_{32,i}\Delta K_j E_{1,j} + \lambda_5 G_{2,i}\Delta K_j E_{1,j} \end{bmatrix},$$

$$\Lambda_{10,ij} = \begin{bmatrix} S_{41,i}\Delta K_j E_{1,j} + \lambda_6 H_s G_{2,i}\Delta K_j E_{1,j} \end{bmatrix},$$

$$\Lambda_{12,ij} = \psi_{11,i}\begin{bmatrix} F_{1,j}^{\mathrm{T}} \\ 0 \end{bmatrix},\ \Lambda_{11,i} = \psi_{11,i} I,$$

$$\Lambda_{13,ij} = \begin{bmatrix} S_{31,i}\Delta K_j + \lambda_4 G_{2,i}\Delta K_j \\ S_{32,i}\Delta K_j + \lambda_5 G_{2,i}\Delta K_j \end{bmatrix},$$

$$\Lambda_{14,ij} = \begin{bmatrix} S_{41,i}\Delta K_j + \lambda_6 H_s G_{2,i}\Delta K_j \end{bmatrix},\ \Lambda_{15,i} = \psi_{12,i} I,$$

$$\Lambda_{16,ij} = \psi_{12,i}\begin{bmatrix} \bar{A}_j^{\mathrm{T}} \\ 0 \end{bmatrix}, \Lambda_{17,ij} = \begin{bmatrix} S_{31,i}E_{2,j} + \lambda_4 G_{2,i}E_{2,j} \\ S_{32,i}E_{2,j} + \lambda_5 G_{2,i}E_{2,j} \end{bmatrix},$$

$$\Lambda_{18,ij} = \begin{bmatrix} S_{41,i}E_{2,j} + \lambda_6 H_s G_{2,i}E_{2,j} \end{bmatrix}, \Lambda_{19,ij} = \psi_{13,i}F_{2,j}^{\mathrm{T}},$$

$$\Lambda_{20,ij} = \begin{bmatrix} S_{31,i}\Delta K_j E_{2,j} + \lambda_4 G_{2,i}\Delta K_j E_{2,j} \\ S_{32,i}\Delta K_j E_{2,j} + \lambda_5 G_{2,i}\Delta K_j E_{2,j} \end{bmatrix},$$

$$\Lambda_{21,ij} = \begin{bmatrix} S_{41,i}\Delta K_j E_{2,j} + \lambda_6 G_{2,i}\Delta K_j E_{2,j} \end{bmatrix},$$

$$\Lambda_{22,ij} = \psi_{14,i}F_{2,j}^{\mathrm{T}}, \quad \Lambda_{23,ij} = \psi_{15,i}\bar{B}_{2,j}^{\mathrm{T}},$$

$$\Lambda_{24,i} = \psi_{13,i}I, \Lambda_{25,i} = \psi_{14,i}I, \Lambda_{26,i} = \psi_{15,i}I.$$

In addition, the observer gain is computed in terms of (4.32).

Proof The proof is similar to the one of Theorem 4.4. It is omitted for the simplicity. □

4.3.4 Mixed $\mathcal{H}\_/\mathcal{H}_\infty$ Observer Design

It infers from the design objectives that the augmented system should be stable and it is desired that the $\mathcal{H}\_$ performance index β should be as large as possible and the $\mathcal{H}_\infty$ performance index γ should be as small as possible. However, the largest $\mathcal{H}\_$ performance index β and the smallest $\mathcal{H}_\infty$ performance index γ cannot be satisfied simultaneously. An alternative approach is to study the optimal mixed $\mathcal{H}\_/\mathcal{H}_\infty$ performance in which one performance index is constrained and the other one is minimized. In this study, we desire the fault detector is sensitive to the actuator fault. Therefore, in the optimal mixed $\mathcal{H}\_/\mathcal{H}_\infty$ observer design, the $\mathcal{H}_\infty$ performance index γ is constrained and the $\mathcal{H}\_$ performance index β is to be maximized. For a given $\mathcal{H}_\infty$ performance index γ, the maximal $\mathcal{H}\_$ performance index β can be obtained from the following corollary.

Corollary 4.1 *For a given $\mathcal{H}_\infty$ performance index γ, the maximum $\mathcal{H}\_$ performance index β in Theorem 4.4 can be found by solving the following convex optimization problem:*

$$\max \beta^2,$$
$$\text{s.t. } (4.31)(4.41)\text{and}(4.49)$$

$\forall i = 1, \cdots, 4, \ j = i, \cdots, 4.$

Remark 4.2 Uncertainties from the system parameters or structures are typical in practical engineering [36]. In order to deal with the uncertainties in this work, Lemmas 4.2 and 4.4 are adopted from the work in [37] in which the studied setup is a Takagi–Sugeno fuzzy system. Due to the nonlinearities of tire model and the uncertainties in the scheduling vector, the fault detection work in this chapter is more

Table 4.1 Maximum $\mathcal{H}_-$ performance index β for different frequency ranges

range(Hz)	[–0.1 0.1]	[–0.5 0.5]	[–10 10]
β	0.2049	0.1993	0.0935

challenging. In addition, we can see from Fig. 4.6 that, when the fault is detected, an alarm system will be set off. If the vehicle is with a fault-tolerant controller, the vehicle is still functional. But it is better to repair the steering system to avoid further damage on the steering. The alarm is very helpful for the on-board diagnostics in this case. If the vehicle is not with a fault-tolerant controller, when the alarm is on, pull it over for the safety.

Remark 4.3 The steering angles shown in Fig. 4.7 at different maneuvers are obtained at the TRC test field when the EVG is randomly driven at relatively low speeds. The purpose of the tests is to derive Fig. 4.8 and the low frequency range. Though high-speed maneuvers are not covered for the safety concern, the steering angle is generally small and the steering wheel is not operated too fast at high-speed maneuvers. The low-frequency range $\left[-0.5\ 0.5\right]$Hz = $\left[-\pi\ \pi\right]$ rad adequately reflects the steering characteristics of typical maneuvers.

4.4 Experiment-Based Simulation Results

The proposed fault-detector design method is applied to the steering system of an EGV. For the safety consideration, the longitudinal velocity is selected in the range of $\left[1\ 10\right]$m/s. In order to derive less-conservative results, different from the paper in [38], by analyzing the pole locations of the EGV dynamics in (4.3), the circle disks are selected as $\varsigma_1 = 1000$, $r_1 = 1000$, $\varsigma_2 = 800$, and $r_2 = 800$. Meanwhile, the value for $k_i,\ i = 1, 2, 3, 4$ is assumed to be within the range of $\left[0.9\ 1.1\right]$. The $\mathcal{H}_\infty$ performance index β is constrained to be 100. To show the advantage of the finite-frequency technique, when the values for $\lambda_i,\ i = 1, \cdots, 6$ are selected as $-1, 220, -110, -1, 15, 000, -110$, the maximum $\mathcal{H}_-$ performance index β for different frequency ranges is listed in Table 4.1. We can see that the maximum $\mathcal{H}_-$ performance index β decreases when increasing the bounds of the frequency. The entire-frequency range is the case when the bounds of the frequency are infinity. Since a larger maximum $\mathcal{H}_-$ performance index indicates that the designed detector is more sensitive to the fault, the finite-frequency approach can lead to better results than the entire-frequency approach and the proposed method is better than the exiting work. In addition, when the circle disks are expanded, the maximum $\mathcal{H}_-$ performance index β in this work is much larger than the one in [38].

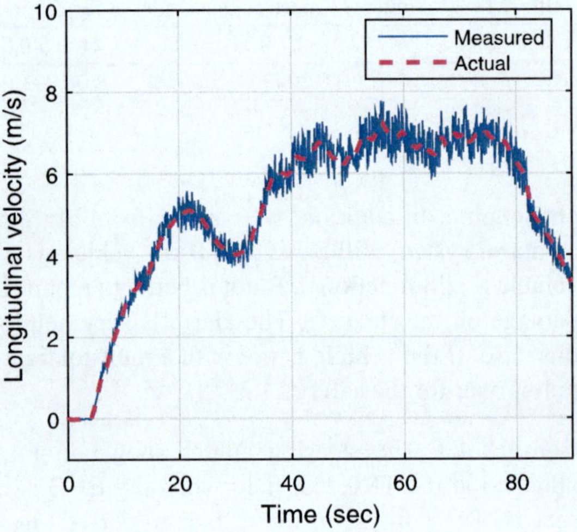

Fig. 4.9 Longitudinal velocity during the test for the EGV

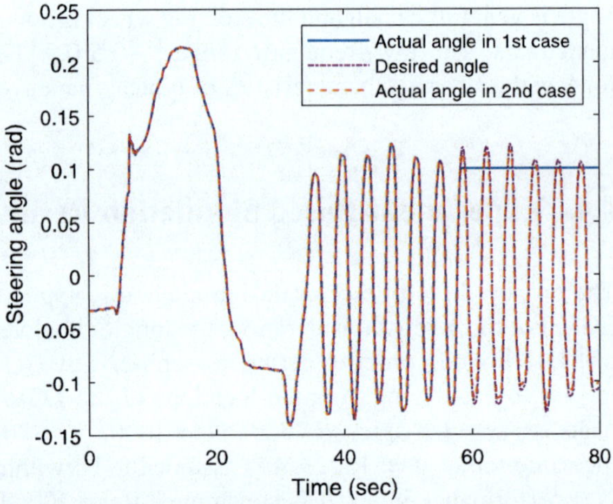

Fig. 4.10 Steering angles with two kinds of faults

The data from an experimental test is used to evaluate the designed fault detector. The test is carried out at TRC in East Liberty, Ohio, USA, with a good road condition. Figure 4.9 depicts the longitudinal velocity during the test. The dash curve is the measured velocity from the accurate GPS system. The solid curve is the longitudinal velocity used for the implementation of designed detector. It is obvious that the weighting factor vector $\hat{\alpha}$ is different from α. As shown in Fig. 4.10, we inject two kinds of faults to the desired front-wheel steering angle. In the first case, the faulty steering angle is stuck from 58.3 s, which is a significant fault. In the second case, the value of the faulty steering angle is 95% of the desired value from 58.3 s.

The detector is designed based on the range of [−0.5 0.5]HZ. The values of fault evaluate function for both cases are shown in Figs. 4.11 and 4.12. It infers from

Fig. 4.11 Value of fault evaluate function for the first case

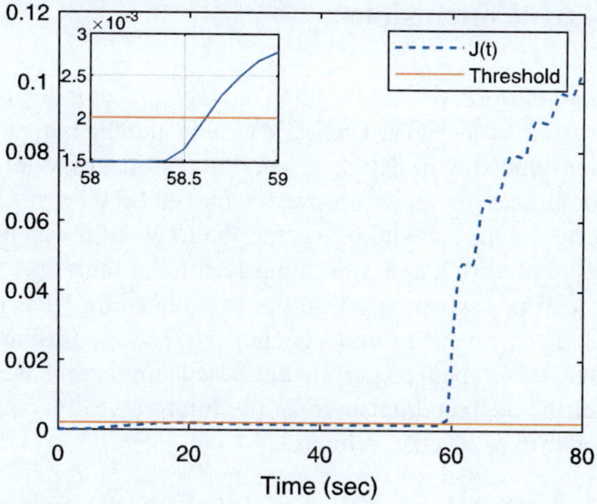

Fig. 4.12 Value of fault evaluate function for the second case

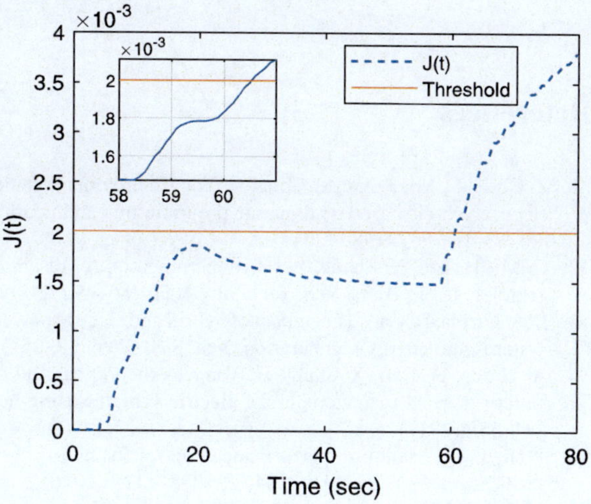

Fig. 4.11 that the designed detector can detect the fault instantaneously for the significant fault. Compared to the detector in [38], the designed detector in this work is more effective and the performance comparison is consistent with the comparison of maximum $\mathcal{H}_-$ performance index β. We can see from Fig. 4.12 that the designed detector can detect the fault at 60.3 s which is 2 s after the fault occurrence. Since the fault in the second case is not so significant, we can claim that the designed detector is sensitive to the fault.

4.5 Conclusions

In this work, we have studied the fault detector design problem for the active-steering system of an EGV. The EGV system and the corresponding acquisition system were introduced with details. Then, the nonlinear model of EGV was developed, and the nonlinear model was converted into an LPV form. Considering the online measuring error of the scheduling vector, the LPV form was transferred into an uncertain LPV system with a new weighting factor. An fault detector was proposed for the active-steering system based on the new uncertain LPV model. By studying the stability and the mixed finite-frequency $\mathcal{H}\_/\mathcal{H}_\infty$ performance, the detector design method was developed. Experimental-based simulations were used to show the performance of the designed detector. In the future research, we will focus on the fault-tolerant control of electric vehicles.

Acknowledgements This chapter is from the previous work in [38], and some typos are corrected here.

References

1. Z. Chen, C. Mi, J. Xu, X. Gong, C. You, Energy management for a power-split plug-in hybrid electric vehicle based on dynamic programming and neural networks. IEEE Trans. Veh. Technol. **63**(4), 1567–1580 (2014)
2. C. Mi, H. Lin, Y. Zhang, Iterative learning control of antilock braking of electric and hybrid vehicles. IEEE Trans. Veh. Technol. **54**(2), 486–494 (2005)
3. D.A. Crolla, D. Cao, The impact of hybrid and electric powertrains on vehicle dynamics, control systems and energy regeneration. Veh. Syst. Dyn. **50**, 95–109 (2012)
4. A. Greco, D. Cao, X. Jiang, H. Yang, A theoretical and computational study of lithium-ion battery thermal management for electric vehicles using heat pipes. J. Power Sources **257**(7), 344–355 (2014)
5. J. Ni, J. Hu, Handling performance control for hybrid 8-wheel-drive vehicle and simulation verification. Veh. Syst. Dyn. **54**(6), 1098–1119 (2016)
6. F. Meng, H. Zhang, D. Cao, H. Chen, System modeling and pressure control of a clutch actuator for heavy-duty automatic transmission systems. IEEE Trans. Veh. Technol. **65**(7), 4865–4874 (2016)
7. F. Yan, J. Wang, K. Huang, Hybrid electric vehicle model predictive control torque-split strategy incorporating engine transient characteristics. IEEE Trans. Veh. Technol. **61**(6), 2458–2467 (2012)
8. C. Yang, X. Jiao, L. Li, Y. Zhang, L. Zhang, J. Song, Robust coordinated control for hybrid electric bus with single-shaft parallel hybrid powertrain. IET Control Theory Appl. **9**(2), 270–282 (2015)
9. C. Yang, J. Song, L. Li, S. Li, D. Cao, Economical launching and accelerating control strategy for a single-shaft parallel hybrid electric bus. Mech. Syst. Signal Process. **76–77**, 649–664 (2016)
10. X. Hu, S.J. Moura, N. Murgovski, B. Egardt, D. Cao, Integrated optimization of battery sizing, charging, and power management in plug-in hybrid electric vehicles. IEEE Trans. Control Syst. Technol. **24**(3), 1036–1043 (2016)
11. X. Hu, S.E. Li, Y. Yang, Advanced machine learning approach for lithium-ion battery state estimation in electric vehicles. IEEE Trans. Transp. Electrification **2**(2), 140–149 (2016)

12. A. Dadashnialehi, A. Bab-Hadiashar, Z. Cao, A. Kapoor, Intelligent sensorless ABS for in-wheel electric vehicles. IEEE Trans. Ind. Electron. **61**(4), 1957–1969 (2014)
13. Y. Wang, B.M. Nguyen, H. Fujimoto, Y. Hori, Multirate estimation and control of body slip angle for electric vehicles based on onboard vision system. IEEE Trans. Ind. Electron. **61**(2), 1133–1143 (2014)
14. Y. Chen, J. Wang, Design and evaluation on electric differentials for overactuated electric ground vehicles with four independent in-wheel motors. IEEE Trans. Veh. Technol. **61**(4), 1534–1542 (2012)
15. M. Chadli, A. Abdo, S.X. Ding, $\mathcal{H}_-/\mathcal{H}_\infty$ fault detection filter design for discrete-time Takagi-Sugeno fuzzy system. Automatica **49**(7), 1996–2005 (2013)
16. M. Zhong, S.X. Ding, J. Lam, H. Wang, An LMI approach to design robust fault detection filter for uncertain LTI systems. Automatica **39**(3), 543–550 (2003)
17. J.L. Wang, G.H. Yang, J. Liu, An LMI approach to $\mathcal{H}_-$ index and mixed $\mathcal{H}_-/\mathcal{H}_\infty$ fault detection observer design. Automatica **43**(9), 1656–1665 (2007)
18. H. Yang, Y. Xia, B. Liu, Fault detection for T-S fuzzy discrete systems in finite-frequency domain. IEEE Trans. Syst., Man, Cybern. B., **41**,(4), 911–920 (2011)
19. X.-J. Li, G.H. Yang, Fault detection for T-S fuzzy systems with unknown membership functions. IEEE Trans. Fuzzy Syst. **22**(1), 139–152 (2014)
20. Y. Zhao, J. Lam, H. Gao, Fault detection for fuzzy systems with intermittent measurements. IEEE Trans. Fuzzy Syst. **17**(2), 398–410 (2009)
21. B. Boulkroune, S. Halabi, A. Zemouche, $\mathcal{H}_-/\mathcal{H}_\infty$ fault detection filter for a class of nonlinear descriptor systems. Int. J. Control **86**(2), 253–262 (2013)
22. I.M. Jaimoukha, Z. Li, V. Papakos, A matrix factorization solution to the $\mathcal{H}_-/\mathcal{H}_\infty$ fault detection problem. Automatica **42**(11), 1907–1912 (2006)
23. W. Sun, H. Gao, O. Kaynak, Finite frequency $\mathcal{H}_\infty$ control for vehicle active suspension systems. IEEE Trans. Control Syst. Technol. **19**(2), 416–422 (2011)
24. H. Wang, G.H. Yang, A finite frequency approach to filter design for uncertain discrete-time systems. Int. J. Adapt. Control Signal Process. **22**(6), 533–550 (2008)
25. X. Li, H. Gao, Robust finite frequency filtering for uncertain 2-D roesser systems. Automatica **48**(6), 1163–1170 (2012)
26. D.-W. Ding and G. hong Yang, "Fuzzy filter design for nonlinear systems in finite-frequency domain," IEEE Trans. Fuzzy Syst., vol. 18, no. 5, pp. 935–945, 2010
27. J. Chen, Y.-Y. Cao, A stable fault detection observer design in finite frequency domain. Int. J. Control **86**(2), 290–298 (2013)
28. R. Wang, Y. Chen, D. Feng, X. Huang, J. Wang, Development and performance characterization of an electric ground vehicle with independently actuated in-wheel motors. J. Power Sources **196**(8), 3962–3971 (2011)
29. H. Zhang, X. Huang, J. Wang, H.R. Karimi, Robust energy-to-peak sideslip angle estimation with applications to ground vehicles. Mechatronics **30**, 338–347 (2015)
30. H. Zhang, X. Zhang, J. Wang, Robust gain-scheduled energy-to-peak control of vehicle lateral dynamics stabilisation. Veh. Syst. Dyn. **52**(3), 309–340 (2014)
31. L. Xie, Y.C. Soh, Robust control of linear systems with generalized positve real uncertainty. Automatica **33**(5), 963–967 (1997)
32. W.M. Haddad, D.S. Bernstein, Controller design with regional pole constraints. IEEE Trans. Autom. Contr. **37**(1), 54–69 (1992)
33. H. Gao, X. Yang, P. Shi, Multi-objective robust $\mathcal{H}_\infty$ control of spacecraft rendezvous. IEEE Trans. Contr. Sys. Techn. **17**(4), 794–802 (2009)
34. M.C. de Oliveira, Investigating duality on stability conditions. Syst. Contr. Lett. **52**(1), 1–6 (2004)
35. J. Zhang, Y. Xia, P. Shi, Parameter-dependent robust $\mathcal{H}_\infty$ filtering for uncertain discrete-time systems. Automatica **45**(2), 560–565 (2009)
36. W. Sun, H. Gao, O. Kaynak, Vibration isolation for active suspensions with performance constraints and actuator saturation. IEEE/ASME Trans. Mechatron. **20**(2), 675–683 (2015)

37. X.-J. Li, G.H. Yang, Fault detection in finite frequency domain for Takagi-Sugeno fuzzy systems with sensor faults. IEEE Trans. Cybern. **44**(8), 1446–1458 (2014)
38. H. Zhang, J. Wang, Active steering actuator fault detection for an automatically-steered electric ground vehicle. IEEE Trans. Veh. Technol. **66**(5), 3685–3702 (2017)

Chapter 5
Robust $\mathcal{H}_\infty$ Output-Feedback Yaw Control for In-Wheel Motor-Driven Electric Vehicles with Differential Steering

Abstract This chapter investigates the yaw control issue for in-wheel-motor (IWM) electric ground vehicles (EGVs) based on the differential steering in the presence of the complete failure of the active front-wheel steering. Differential steering is an emerging steering mechanism, generated from the differential torque between the left and right wheels in IWM EGVs. In case that the regular steering system is defective, differential steering can be utilized to act as the sole steering power and thus avoid dangerous consequences for vehicles. For this purpose, a robust $\mathcal{H}_\infty$ output-feedback controller based on differential steering is designed to achieve yaw stabilization, considering that the desired steering angle is uncertain and hard to obtain. Parameter uncertainties for the cornering stiffnesses and the external disturbances are considered to make vehicle robust to different driving conditions. CarSim-Simulink joint simulation results based on a high-fidelity and full-car model verify the effectiveness of the proposed controller to guarantee the equal vehicle handling and stability.

5.1 Introduction

In recent decades, electric ground vehicles (EGVs) have showed great advantages in emissions reductions, fuel economy, and energy security and thus attracted worldwide research attentions [1]. Especially, the emerging and development of the independent electric drive for EGVs have been proved relatively effective in improving the vehicle handling and stability [2–4]. The in-wheel motors (IWMs) (or hub motors) mounted in EGVs can lead to flexibility actuation due to the electric motors' fast and precise torque responses [5]. Numerous studies have revealed that IWM EGV is a promising vehicle architecture for its capability in considerably enhancing the vehicle maneuverability, stability and safety in defective driving conditions [6].

The actuator redundancies in IWM EGVs can greatly contribute to yaw motion control owing to the direct yaw moment provided by the differential driving forces. Additionally, the generated yaw moment can produce a differential steering, which can be utilized to steer the vehicle in emergency situations. Extensive previous research works were proposed to study the vehicle motion control with the differential steering. Skid steering was discussed in [7] and applied to four-wheel drive

H. Zhang et al., *Robust Gain-Scheduled Estimation and Control of Electrified Vehicles via LPV Technique*, Key Technologies on New Energy Vehicles,
https://doi.org/10.1007/978-981-19-8509-6_5

electric vehicles, where steering was achieved by differentially varying the speeds of the lines of wheels on different sides of the vehicle. In [8], the usefulness of a brake steer system (BSS) was examined, which used differential brake forces for steering intervention. In [9], an indirect power steering measure was proposed named as differential drive torque-assisted steering, and its feasibility of assisting steering, the capability of road feel keeping, and the effect of the torque distribution control system were validated. Nozaki [10] and Wu and Yeh [11] proposed similar ideas to generate assisted steering force by differential driving of two front wheels of electric vehicle, which had electrical steering system according to the steering geometry. Nevertheless, the above literatures mainly investigated the differential drive as the assisted power source for steering. Actually, differential steering can be a sole steering power source to provide emergency steering control against active steering system failure. When the steering motor breaks down, the regular active steering system (e.g., steering-by-wire, hydraulic power steering) will be out of control, triggering dangerous consequences to vehicles [12, 13]. In this situation, the differential steering generated by the differential drive torque of the front axle through the steering system mechanism can be utilized to realize the normal steering function. This principle can be employed for automatic steering without the external steering power in faulted-steering condition, but few researchers investigated that benefit. In the latest few decades, many control strategies have been well studied and implemented in engineering field, i.e., robust $\mathcal{H}_\infty$ control [2–4], data-driven control [14–17], adaptive robust control [18], adaptive backstepping control [19], output-feedback control [20], and fault-tolerant control [21]. Moreover, the fault detection and isolation for vehicle suspension system problem was proposed in [22], where both fault diagnosis and fuzzy positivistic C-means clustering was adopted in the control schema. As the parameter uncertainties for the tire cornering stiffness and the external disturbances always exist in the real application, and the reference steering angle is hard to calculate, the control strategy of differential steering should be effective to handle the above problem. The robust $\mathcal{H}_\infty$ output-feedback control strategy has been proved effective on handling the parameter uncertainties, external disturbances, and the unmeasurable variables. Consequently, this chapter adopts the robust $\mathcal{H}_\infty$ output-feedback yaw control for IWM EGVs using the differential steering in the presence of the complete failure of the regular steering system.

What is more, the parameter uncertainties for the tire cornering stiffness and the external disturbances in differential steering system have been taken into consideration. It should be noted that when the regular active steering system completely breaks, the differential steering will be activated and generated by the differential torque. Therefore, the front steering angle is not a control input any more, but a state controlled by the longitudinal tire force difference of the front wheels. The yaw stabilization in this work is achieved by the tracking control of the yaw rate toward the desired values. Note that the desired yaw rate is calculated from the driver's steering command [23], but the desired steering angle is unknown. That can be explained that there exist uncertainties for the desired steering angle, which is related with the road adhesion, tire cornering stiffness, and vehicle states; also, the yaw moment contributes to the steering of the vehicle which further complicates the constitution of the desired steering angle. Considering the reference steering angle is hard to calculate, a dynamic output-feedback strategy is employed to handle that problem based on linear matrix inequalities (LMIs) approach. The proposed controller is supposed to maintain the equal vehicle handling and stability in case that the active steering completely fails.

As a novel mechanism, differential steering has not been deeply investigated in vehicle dynamics and control. This chapter employs differential steering as a sole steering power in the presence of the complete failure of the regular steering system, which is more innovated and very few papers have been published on the issue. The main contributions of the chapter are listed as follows: (1) The yaw control problem for IWM EGVs based on the differential steering is presented in case of the complete failure of the active front-wheel steering, and thus dangerous consequences for vehicles can be avoided; (2) parameter uncertainties for the cornering stiffness and the external disturbances are considered to make vehicle robust to different driving conditions; and (3) considering the reference steering angle is hard to calculate, a dynamic output-feedback strategy is employed to handle that problem.

5.2 System Modeling and Problem Formulation

5.2.1 *Vehicle Dynamics with Differential Steering*

The bicycle model is used for the simplification of controller design. As shown in Fig. 5.1, l_f and l_r are the distances between the center of gravity (CG) to the front and rear axles, respectively. The vehicle has the moment inertia I_z, yaw rate Ω_z, longitudinal velocity v_x, and lateral velocity v_y at the CG point. F_{yf} and F_{yf} are the lateral tire forces of the front and rear wheel, respectively. α_f and α_r are the tire slip angles of the front and rear wheel, respectively. δ is the front-wheel steering angle. The vehicle yaw motion can be modeled as [9]

$$I_z \dot{\Omega}_z = F_{yf} l_f - F_{yr} l_r + \Delta M_1, \tag{5.1}$$

where ΔM_1 is the external yaw moment generated with the longitudinal tire force difference between the two sides of the front wheels. Assuming the front-wheel steering angle is small, ΔM_1 can be written as

$$\Delta M_1 = (F_{xfr} - F_{xfl}) l_s, \tag{5.2}$$

with F_{xfj} $(j = r,\ l)$ being the longitudinal tire force of front-right and front-left wheels, respectively, and l_s being the half of wheel track. Given a small tire slip angle, the tire lateral force can be written in linear form

$$F_{yi} = C_i \alpha_i, \quad (i = f,\ r), \tag{5.3}$$

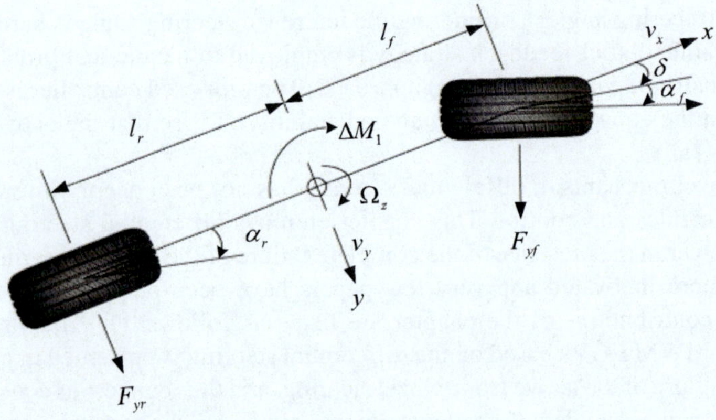

Fig. 5.1 Schematic of the bicycle model for vehicle yaw control

where C_i is the tire corning stiffness, α_i is the tire slip angle which can be expressed as

$$\alpha_f = -\left(\frac{v_y + l_f \Omega_z}{v_x} - \delta\right), \quad \alpha_r = -\frac{v_y - l_r \Omega_z}{v_x}. \tag{5.4}$$

Utilizing (5.3) and (5.4), the vehicle yaw motion (5.1) can be rewritten as

$$\dot{\Omega}_z = \frac{C_f l_f \delta}{I_z} - \frac{C_f l_f^2 + C_r l_r^2}{v_x I_z}\Omega_z + \frac{\Delta M_1}{I_z} - \frac{(C_f l_f - C_r l_r) v_y}{I_z v_x}. \tag{5.5}$$

The configuration of a steer-by-wire system of the EGV with in-wheel motors is presented in Fig. 5.2. Note that there is no mechanical connection between steering wheel and road wheel. Normally, the steering wheel provides steering torque and angle signals to ECU, which will then realize the steering function by an electric steering motor. This chapter deals with the occasion when the steering motor completely fails, i.e., the steering motor torque is zero. Once there is no torque transmitted to the Rack-and-Pin in Fig. 5.2, the conventional steering mechanism fails to function the desired steering action. At the time, the differential steering control is activated and functioned as a sole steering power source. Compared with the conventional vehicle, each wheel of the IWM EGVs can be driven independently; thus, the driven torque of the left and right wheels can be different. As shown in Fig. 5.2, if the driven torque of left wheel is bigger than that of the right ones, the steering mechanism will definitely force the two wheels to rotate around its Kingpin in certain direction, simultaneously.

Consequently, the above principle is adopted by differential steering control schema to complete the driver's steering attempt in following steps: When the conventional steering system fails down, the differential steering system is activated and an external moment ΔM_1 is generated corresponding to the δ_{sw}, to regulate the

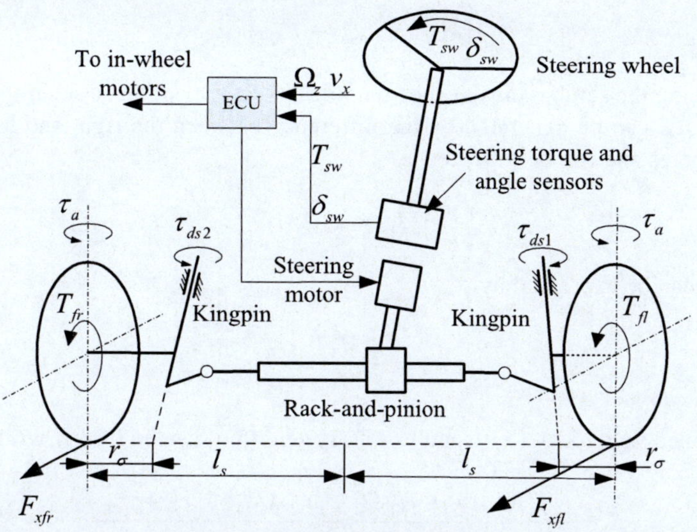

Fig. 5.2 System model with differential steering

vehicle yaw motion. As discussed above, the steering action can be proceeded with the help of the different driven torque of the right and left wheels.

The dynamic equation of the steering system can be written as

$$J_{\mathrm{eff}}\ddot{\delta} + b_{\mathrm{eff}}\dot{\delta} = \tau_a + \Delta M_1' - \tau_f, \tag{5.6}$$

where J_{eff} and b_{eff} are the effective moment of inertia and effective damping of the steering system, respectively. τ_f is the friction torque, and τ_a means the tire self-aligning moment as shown in Fig. 5.2. According to the brush model [24], when the front tire slip angle α_f is smaller than the sliding slip angle α_{sl}, the self-aligning moment can be expressed as

$$\begin{aligned} \tau_a =& \mu F_z l \theta_y \tan(\alpha_f)\{1 - 3|\theta_y \tan(\alpha_f)| \\ &+ 3[\theta_y \tan(\alpha_f)]^2 - |\theta_y \tan(\alpha_f)|^3\}, \end{aligned} \tag{5.7}$$

where $\alpha_{sl} = \arctan(1/\theta_y)$, $\theta_y = \frac{2c_p l^2}{3\mu F_z}$, μ is the tire-road adhesion coefficient, F_z is the normal force, l is the half of the tire contact length, and c_p is the stiffness coefficient of tire tread in unit length with $c_p = C_f/(2l)$. Under the small tire slip angle assumption, we have

$$\tau_a = \kappa_1 C_f \alpha_f, \tag{5.8}$$

with $\kappa_1 = l^2/3$ being a constant. $\Delta M_1'$ in equation (5.6) is the moment on steering system which comes from ΔM_1, and can be written as

$$\Delta M_1' = (F_{xfr} - F_{xfl})r_\sigma, \tag{5.9}$$

with r_σ being the scrub radius as shown in Fig. 5.2. It is clear that a steering-assisted torque $\Delta M_1{}'$ can be generated by the difference between the right and left wheel driving forces and can be rewritten as

$$\Delta M_1{}' = \kappa_2 \Delta M_1, \tag{5.10}$$

where κ_2 can be expressed by

$$\kappa_2 = \frac{\Delta M_1'}{\Delta M_1} = \frac{r_\sigma}{l_s}. \tag{5.11}$$

Combined with (5.8), (5.9), and (5.11), system (5.6) can be further written as

$$\dot{\delta} = \frac{\kappa_1 C_f \delta}{b_{\text{eff}}} - \frac{\kappa_1 C_f l_f \Omega_z}{v_x b_{\text{eff}}} + \frac{\kappa_2 \Delta M_1}{b_{\text{eff}}} - \frac{\tau_f + J_{\text{eff}}\ddot{\delta}}{b_{\text{eff}}}. \tag{5.12}$$

Define the control input as $u = \Delta M_1$, the vehicle model can be written as

$$\begin{cases} \dot{\delta} = \dfrac{\kappa_1 C_f}{b_{\text{eff}}}\delta - \dfrac{\kappa_1 C_f l_f}{v_x b_{\text{eff}}}\Omega_z + \dfrac{\kappa_2}{b_{\text{eff}}}u + d_1, \\ \dot{\Omega}_z = \dfrac{C_f l_f}{I_z}\delta - \dfrac{C_f l_f^2 + C_r l_r^2}{v_x I_z}\Omega_z + \dfrac{1}{I_z}u + d_2, \end{cases} \tag{5.13}$$

where $d_1 = \frac{1}{b_{\text{eff}}}(\kappa_1 C_f \frac{v_y}{v_x} - \tau_f - J_{\text{eff}}\ddot{\delta})$ and $d_2 = -\frac{(C_f l_f - C_r l_r)v_y}{I_z v_x}$ are the disturbance terms. Define the following state variables: $x_1 = \delta$, $x_2 = \Omega_z$, $x(t) = [x_1, x_2]^{\mathrm{T}}$, and $d(t) = [d_1, d_2]^{\mathrm{T}}$, and the state space form of the dynamic model (5.13) can be given as

$$\dot{x}(t) = Ax(t) + Bu(t) + B_1 d(t), \tag{5.14}$$

where

$$A = \begin{bmatrix} \frac{\kappa_1 C_f}{b_{\text{eff}}} & -\frac{\kappa_1 C_f l_f}{v_x b_{\text{eff}}} \\ \frac{C_f l_f}{I_z} & -\frac{C_f l_f^2 + C_r l_r^2}{v_x I_z} \end{bmatrix}, \quad B = \begin{bmatrix} \frac{\kappa_2}{b_{\text{eff}}} \\ \frac{1}{I_z} \end{bmatrix}, \quad B_1 = \begin{bmatrix} 1 & 0 \\ 0 & 1 \end{bmatrix}.$$

Remark 5.1 The vehicle lateral velocity v_y can usually be assumed small in normal driving; meanwhile, the value of v_y is hard to measure with low cost sensor, so in this work, v_y is regarded as a bounded disturbance. $\ddot{\delta}$ is the second derivative of steering angle δ, which is relatively small based on the small steering assumption, and thus can be neglected here. The friction torque of steering system τ_f, which is difficult to measure, can also be assumed small. Based on the above analysis, d_1 and d_2 can be regarded as bounded disturbances to facilitate the control law design. Robust controller will be designed to attenuate the uncertainties and disturbances.

5.2.2 Vehicle Modeling with Parameter Uncertainties

Due to the change of road conditions and the vehicle states, the tire cornering stiffness is always varying and unknown. Thus, the cornering stiffness of front and rear wheels C_i $(i = f, r)$ can be presented as follows

$$C_f = C_{f0} + \lambda_f \tilde{C}_f, \quad C_r = C_{r0} + \lambda_r \tilde{C}_r, \tag{5.15}$$

with

$$C_{f0} = \frac{C_{f\max} + C_{f\min}}{2}, \quad C_{r0} = \frac{C_{r\max} + C_{r\min}}{2},$$

where $C_{i\max}$ and $C_{i\min}$ are the maximum value and minimum value of C_i $(i = f, r)$, respectively; $\tilde{C}_f$ and $\tilde{C}_r$ can be presented by

$$\tilde{C}_f = C_{f\max} - C_{f0}, \quad \tilde{C}_r = C_{r\max} - C_{r0}, \tag{5.16}$$

λ_f and λ_r are time-varying parameters and bounded in the range $[-1, 1]$. Since $|\lambda_f| \leq 1$ and $|\lambda_r| \leq 1$, based on (5.15) the varying C_f/b_{eff} can be further written as follows:

$$\frac{C_f}{b_{\text{eff}}} = \frac{C_{f0}}{b_{\text{eff}}} + \frac{\lambda_f \tilde{C}_f}{b_{\text{eff}}} = \frac{C_{f0}}{b_{\text{eff}}} + \lambda_1 \varphi_1,$$

where $\lambda_1 = \lambda_f$ and $\varphi_1 = \tilde{C}_f/b_{\text{eff}}$. Similarly, we have $C_f/I_z = C_{f0}/I_z + \lambda_2\varphi_2$, $C_r/I_z = C_{r0}/I_z + \lambda_3\varphi_3$, where $\lambda_2 = \lambda_f$, $\lambda_3 = \lambda_r$, $\varphi_2 = \tilde{C}_f/I_z$, and $\varphi_3 = \tilde{C}_r/I_z$. So we have

$$A = A_0 + \Delta A, \tag{5.17}$$

with

$$A_0 = \begin{bmatrix} \frac{\kappa_1 C_{f0}}{b_{\text{eff}}} & -\frac{\kappa_1 C_{f0} l_f}{v_x b_{\text{eff}}} \\ \frac{C_{f0} l_f}{I_z} & -\frac{C_{f0} l_f^2 + C_{r0} l_r^2}{v_x I_z} \end{bmatrix}, \quad \Delta A = \begin{bmatrix} \kappa_1 \lambda_1 \varphi_1 & -\frac{\kappa_1 l_f}{v_x} \lambda_1 \varphi_1 \\ l_f \lambda_2 \varphi_2 & -\frac{l_f^2}{v_x} \lambda_2 \varphi_2 - \frac{l_r^2}{v_x} \lambda_3 \varphi_3 \end{bmatrix}.$$

Note that there still exist time-varying parameters $1/v_x$ in the system matrices. Since the vehicle longitudinal speed v_x is actually bounded in $[v_{x\min}, \ v_{x\max}]$, we can assume that $1/v_x$ is varying in the range of $[1/v_{x\max}, \ 1/v_{x\min}]$. Therefore, the vertices' coordinates can be written as $\tilde{\vartheta}_1 = \frac{1}{v_{x\max}}$, $\tilde{\vartheta}_2 = \frac{1}{v_{x\min}}$. The varying parameter $1/v_x$ can be represented by a summation of the vertices' coordinates as

$$\frac{1}{v_x} = \sum_{i=1}^{2} h_i(t) \tilde{\vartheta}_i, \tag{5.18}$$

with

$$h_1 = \frac{1/v_{x\,\min} - 1/v_x}{1/v_{x\,\min} - 1/v_{x\,\max}}, h_2 = \frac{1/v_x - 1/v_{x\,\max}}{1/v_{x\,\min} - 1/v_{x\,\max}}. \tag{5.19}$$

Note that based on the $h_i(t)$ definitions in (5.19), we have

$$\sum_{i=1}^{2} h_i(t) = 1, \quad h_i(t) \geq 0. \tag{5.20}$$

Based on (5.18), the vehicle model (5.13) can be rewritten in the polytopic form as

$$\begin{aligned} \dot{x}(t) &= \sum_{i=1}^{2} h_i(t)[A_i x(t) + Bu(t) + B_1 d(t)] \\ &= \sum_{i=1}^{2} h_i(t)\big[(A_{0i} + \Delta A_i)x(t) + Bu(t) + B_1 d(t)\big]. \end{aligned} \tag{5.21}$$

where the matrices A_{0i} and ΔA_i are obtained by replacing $1/v_x$ with $\tilde{\vartheta}_i$ in A_0 and ΔA, respectively, and can be written as

$$A_{0i} = \begin{bmatrix} \frac{\kappa_1 C_{f0}}{b_{\text{eff}}} & -\frac{\kappa_1 C_{f0} l_f}{b_{\text{eff}}} \tilde{\vartheta}_i \\ \frac{C_{f0} l_f}{I_z} & -\frac{C_{f0} l_f^2 + C_{r0} l_r^2}{I_z} \tilde{\vartheta}_i \end{bmatrix}, \quad \Delta A_i = \begin{bmatrix} \kappa_1 \lambda_1 \varphi_1 & -\kappa_1 l_f \lambda_1 \varphi_1 \tilde{\vartheta}_i \\ l_f \lambda_2 \varphi_2 & -l_f^2 \lambda_2 \varphi_2 \tilde{\vartheta}_i - l_r^2 \lambda_3 \varphi_3 \tilde{\vartheta}_i \end{bmatrix}.$$

Remark 5.2 The number of the uncertain parameters in matrices ΔA and ΔB should be decreased to avoid computational burden and possible design conservation. It is noted that the road conditions are usually uniform to the front and rear wheels; thus, it is reasonable to assume that the road coherent coefficients for both tires are identical, which means that $\lambda_f = \lambda_r$ can be assumed. Therefore, we have λ_1, λ_2, and λ_3 in ΔA_i are identical, and ΔA_i can be rewritten as

$$\Delta A_i = H \Lambda_1 E_{1i}, \tag{5.22}$$

where

$$H = \begin{bmatrix} 1\,0\,0 \\ 0\,1\,1 \end{bmatrix}, \quad \Lambda_1 = \text{diag}\{\lambda_1, \lambda_1, \lambda_1\},$$

$$E_{1i} = \begin{bmatrix} \kappa_1 \varphi_1 & l_f \varphi_2 & 0 \\ -\kappa_1 l_f \varphi_1 \tilde{\vartheta}_i & -l_f^2 \varphi_2 \tilde{\vartheta}_i & -l_r^2 \varphi_3 \tilde{\vartheta}_i \end{bmatrix}^{\mathrm{T}}.$$

5.2.3 Problem Statement

As mentioned earlier, we want to realize the yaw control solely through the differential steering produced by differential torque of front wheels. The yaw stabilization is achieved by the tracking control of the yaw rate, a robust dynamic output-feedback controller is presented to handle that problem. System (5.21) can be transformed into the linear form as

$$\begin{cases} \dot{x}(t) = \sum_{i=1}^{2} h_i(t)\big[(A_{0i} + \Delta A_i)x(t) \\ \qquad + Bu(t) + B_1 d(t)\big] \\ y(t) = C_1 x(t) \\ z(t) = C_2 x(t) \end{cases} \tag{5.23}$$

where $C_1 = C_2 = [0 \ \ 1]$. The control objective is to design a robust dynamic output-feedback controller to generate a control signal $u(t)$, such that the system in equation (5.23) is robustly asymptotically stable and has the following $\mathcal{H}_\infty$ disturbance attenuation performance in the presence of parameter uncertainties and external disturbances,

$$\int_0^t z^{\mathrm{T}}(t)z(t)dt < \gamma^2 \int_0^t d^{\mathrm{T}}(t)d(t)dt, \tag{5.24}$$

where γ is the prescribed attenuation level.

Remark 5.3 It should be emphasized that the front-wheel steering angle δ is a state variable rather than a control input in this work, since we investigate the failure situation of the active steering system in this study. When the steering motor breaks down, the differential steering will be activated. The yaw control in this work is realized through making the yaw rate and the steering angle track their respective values, where the reference yaw rate can be obtained based on the driver's steering command; however, the desired steering angle is uncertain and hard to obtained. That is because the desired steering angle for generating a certain yaw rate is related with the vehicle parameters and states, tire-road friction coefficient, as well as the yaw-moment control input. Thus, it is difficult to calculate the reference steering angle only by the reference yaw rate. That also explains why we use the output-feedback approach in this study.

5.3 Robust Controller Design

The dynamic output-feedback controller is designed as

$$\begin{cases} \dot{\hat{x}}(t) = \sum_{i=1}^{2}\sum_{j=1}^{2} h_i(t)h_j(t)A_{kij}\hat{x}(t) \\ \quad + \sum_{j=1}^{2} h_j(t)B_{kj}y(t) \\ u(t) = \sum_{j=1}^{2} h_j(t)C_{kj}\hat{x}(t) + D_k y(t) \end{cases}, \tag{5.25}$$

where $\hat{x}(t)$ is the states of the controller, A_{kij}, B_{kj}, C_{kj}, and D_k $(i = 1, 2;\ j = 1, 2)$ are the control matrices to be designed. Define $\bar{x} = [x \quad \hat{x}]^{\mathrm{T}}$, by combining (5.23) and (5.25), the closed-loop system is given as follows

$$\begin{cases} \dot{\bar{x}}(t) = \sum_{i=1}^{2}\sum_{j=1}^{2} h_i(t)h_j(t)\left[(\bar{A}_{ij} + \bar{H}\Lambda_1\bar{E}_{ij})\bar{x}(t) + \bar{B}d(t)\right] \\ z = \bar{C}\bar{x} \end{cases}, \tag{5.26}$$

where

$$\begin{aligned} &\bar{x} = \begin{bmatrix} x \\ \hat{x} \end{bmatrix}, \quad \bar{B} = \begin{bmatrix} B_1 \\ 0 \end{bmatrix}, \quad \bar{H} = \begin{bmatrix} H \\ 0 \end{bmatrix}, \\ &\bar{A}_{ij} = \begin{bmatrix} A_{0i} + BD_kC_1 & BC_{kj} \\ B_{kj}C_1 & A_{kij} \end{bmatrix}, \\ &\bar{E}_i = [E_{1i} \quad 0], \ \bar{C} = [C_2 \quad 0]. \end{aligned} \tag{5.27}$$

In order to deal with the parameter uncertainties and external disturbances, we introduce the following lemma.

Lemma 5.1 [23] *Let $Y = Y^{\mathrm{T}}$, Γ, and Ψ be the real matrices with proper dimensions and make $\tilde{N}$ satisfy $\Lambda^{\mathrm{T}}\Lambda < I$, then the following condition:*

$$Y + \Gamma\Lambda\Psi + \Psi^{\mathrm{T}}\Lambda^{\mathrm{T}}\Gamma^{\mathrm{T}} < 0, \tag{5.28}$$

holds if and only if there exists a positive scalar $\varepsilon > 0$ such that

$$Y + \epsilon\Gamma\Gamma^{\mathrm{T}} + \epsilon^{-1}\Psi^{\mathrm{T}}\Psi < 0. \tag{5.29}$$

By using Schur complement, the above inequality can be rewritten as

$$\begin{bmatrix} Y & \epsilon\Gamma & \Psi^{\mathrm{T}} \\ * & -\epsilon I & 0 \\ * & * & -\epsilon I \end{bmatrix} < 0, \tag{5.30}$$

where in this work $*$ denotes the symmetric elements in a symmetric matrix. Now, we are in the position to give the robust dynamic output-feedback controller. □

Theorem 5.1 *Given a positive constant γ, the closed-loop system in (5.26) is asymptotically stable with $d(t) = 0$ and satisfies the $\mathcal{H}_\infty$ performance index (5.24) for all $d(t) \in [0, \infty)$, if there exists a symmetric positive-definite matrix P and positive scalars ϵ_1, such that*

$$\bar{\Pi}_{1ii} < 0, \quad i = 1, 2 \tag{5.31a}$$

$$\bar{\Pi}_{1ij} + \bar{\Pi}_{1ji} < 0, \quad 1 \le i < j \le 2 \tag{5.31b}$$

where

$$\bar{\Pi}_{1ij} = \begin{bmatrix} \bar{A}_{ij}^{\mathrm{T}} P + P\bar{A}_{ij} & P\bar{B} & \bar{C}^{\mathrm{T}} & \epsilon_1 P\bar{H} & \bar{E}_i^{\mathrm{T}} \\ * & -I & 0 & 0 & 0 \\ * & * & -\gamma^2 I & 0 & 0 \\ * & * & * & -\epsilon_1 I & 0 \\ * & * & * & * & -\epsilon_1 I \end{bmatrix}.$$

Proof Define a Lyapunov function for the system in (5.26) as

$$V = \bar{x}^{\mathrm{T}}(t) P \bar{x}(t), \tag{5.32}$$

where P is a positive matrix. The time derivative of the above Lyapunov function is

$$\begin{aligned} \dot{V} &= \dot{\bar{x}}^{\mathrm{T}}(t) P \bar{x}(t) + \bar{x}^{\mathrm{T}}(t) P \dot{\bar{x}}(t) \\ &= \sum_{i=1}^{2}\sum_{j=1}^{2} h_i(t)h_j(t)\Big(\big((\bar{A}_{ij} + \bar{H}\Lambda_1\bar{E}_i)\bar{x}(t) + \bar{B}d(t)\big)^{\mathrm{T}} P\bar{x}(t) \\ &\quad + \bar{x}^{\mathrm{T}}(t) P\big((\bar{A}_{ij} + \bar{H}\Lambda_1\bar{E}_i)\bar{x}(t) + \bar{B}d(t)\big)\Big) \\ &= \sum_{i=1}^{2}\sum_{j=1}^{2} h_i(t)h_j(t)\Big(\bar{x}^{\mathrm{T}}T(t)\big(\bar{A}_{ij}^{T} P + P\bar{A}_{ij} \\ &\quad + \bar{H}\Lambda_1\bar{E}_i + \bar{E}_{ij}^{\mathrm{T}}\Lambda_1\bar{H}^{\mathrm{T}}\big)\bar{x}(t) + d^{\mathrm{T}}(t)\bar{B}^{\mathrm{T}} P\bar{x}(t) + \bar{x}^{\mathrm{T}}(t) P\bar{B}d(t)\Big). \end{aligned} \tag{5.33}$$

Denoting $\Theta_{ij} = \bar{A}_{ij}^{\mathrm{T}} P + P\bar{A}_{ij} + P\bar{H}\Lambda_1\bar{E}_i + \bar{E}_i^{\mathrm{T}}\Lambda_1\bar{H}^{\mathrm{T}} P$ and by adding $z^{\mathrm{T}}(t)z(t) - \gamma^2 d^{\mathrm{T}}(t)d(t)$ with $\gamma > 0$ for both sides of (5.33), we have

$$
\begin{aligned}
&\dot{V} + z^{\mathrm{T}}(t)z(t) - \gamma^2 d^{\mathrm{T}}(t)d(t) \\
=&\dot{V} + \sum_{i=1}^{3}\sum_{j=1}^{3} h_i(t)h_j(t)\Big(z^{\mathrm{T}}(t)z(t) - \gamma^2 d^{\mathrm{T}}(t)d(t)\Big) \\
=&\sum_{i=1}^{2}\sum_{j=1}^{2} h_i(t)h_j(t)\Big(\bar{x}^{\mathrm{T}}(t)(\Theta_{ij} + \bar{C}^{\mathrm{T}}\bar{C})x(t) \\
&+ d^{\mathrm{T}}(t)\bar{B}P\bar{x}(t) + \bar{x}^{\mathrm{T}}(t)P\bar{B}d(t) - \gamma^2 d^{\mathrm{T}}(t)d(t)\Big) \\
=&\sum_{i=1}^{2}\sum_{j=1}^{2} h_i(t)h_j(t)\begin{bmatrix}\bar{x}(t)\\ d(t)\end{bmatrix}^{\mathrm{T}} \Pi_{ij}\begin{bmatrix}\bar{x}(t)\\ d(t)\end{bmatrix},
\end{aligned} \tag{5.34}
$$

where

$$
\Pi_{ij} = \begin{bmatrix}\Theta_{ij} + \bar{C}^{\mathrm{T}}\bar{C} & P\bar{B} \\ \bar{B}^{\mathrm{T}}P & -\gamma^2 I\end{bmatrix}. \tag{5.35}
$$

It can be seen that if the following inequality holds,

$$
\sum_{i=1}^{2}\sum_{j=1}^{2} h_i(t)h_j(t)\Pi_{1ij} < 0, \tag{5.36}
$$

then the controlled system (5.26) is robustly asymptotically stable and the $\mathcal{H}_\infty$ performance defined in (5.24) can be satisfied. Note that (5.36) can be rewritten as

$$
\begin{aligned}
&\sum_{i=1}^{2}\sum_{j=1}^{2} h_i(t)h_j(t)\Pi_{1ij} \\
=&\sum_{i=1}^{2} h_i^2(t)\Pi_{1ii} + \sum_{i=1}^{2}\sum_{i<j}^{2} h_i(t)h_j(t)(\Pi_{1ij} + \Pi_{1ji}) \\
<&\ 0.
\end{aligned} \tag{5.37}
$$

Therefore, the controlled system is robustly stable with an $\mathcal{H}_\infty$ performance index (5.24) if,

$$
\Pi_{1ii} < 0, \quad i = 1, 2 \tag{5.38a}
$$

$$
\Pi_{1ij} + \Pi_{1ji} < 0, \quad 1 \le i < j \le 2 \tag{5.38b}
$$

Using Schur complement, $\Pi_{1ij} < 0$ is equivalent to

$$
\begin{bmatrix}\mathrm{sys}\{\bar{A}_{ij}^{\mathrm{T}}P + P\bar{H}\Lambda_1\bar{E}_i\} & P\bar{B} & \bar{C}^{\mathrm{T}} \\ * & -I & 0 \\ * & * & -\gamma^2 I\end{bmatrix} < 0, \tag{5.39}
$$

where here and everywhere in the sequel, sys{•} denotes $\bullet + \bullet^{\mathrm{T}}$. Then, the above inequality can be further written as

$$\begin{bmatrix} \mathrm{sys}\{\bar{A}_{ij}^{\mathrm{T}}P + P\bar{H}\Lambda_1\bar{E}_i\} & P\bar{B} & \bar{C}^{\mathrm{T}} \\ * & -I & 0 \\ * & * & -\gamma^2 I \end{bmatrix}$$

$$= \begin{bmatrix} \mathrm{sys}\{\bar{A}_{ij}^{\mathrm{T}}P\} & P\bar{B} & \bar{C}^{\mathrm{T}} \\ * & -I & 0 \\ * & * & -\gamma^2 I \end{bmatrix} + \mathrm{sys}\left\{ \begin{bmatrix} P\bar{H} \\ 0 \\ 0 \end{bmatrix} \Lambda_1 [\bar{E}_i \; 0 \; 0] \right\} < 0.$$

As $\Lambda_1^T\Lambda_1 < I$, it follows from Lemma 5.1 that the above condition is equivalent to $\bar{\Pi}_{1ij} < 0$ as shown in (5.31a). Similarly, we can prove that the condition (5.31b) is equivalent to (5.38b). So the inequalities (5.31a) and (5.31b) ensure that the closed-loop system (5.26) is asymptotically stable and the $\mathcal{H}_\infty$ performance (5.24) can be satisfied. This completes the proof. □

There are nonlinear terms involved in (5.31a) and (5.31b), and these nonlinear terms cannot be removed by the change of variable which is usually used in the robust $\mathcal{H}_\infty$ state-feedback controllers . Since the matrix P is nonsingular, we can partition P and its inverse as

$$P = \begin{bmatrix} R & N \\ N^{\mathrm{T}} & W \end{bmatrix}, \; P^{-1} = \begin{bmatrix} S & M \\ M^{\mathrm{T}} & V \end{bmatrix}.$$

As $PP^{-1} = I$, we have

$$P \begin{bmatrix} S & I \\ M^{\mathrm{T}} & 0 \end{bmatrix} = \begin{bmatrix} I & R \\ 0 & N^{\mathrm{T}} \end{bmatrix},$$

Without loss of generality, we can assume that both M and N are full rank matrices. Let

$$F_1 = \begin{bmatrix} S & I \\ M^{\mathrm{T}} & 0 \end{bmatrix}, \quad F_2 = \begin{bmatrix} I & R \\ 0 & N^{\mathrm{T}} \end{bmatrix}.$$

Theorem 5.2 *Given positive constant γ, the closed-loop system in (5.26) is asymptotically stable with $d(t) = 0$ and satisfies the $\mathcal{H}_\infty$ performance index (5.24) for all $d(t) \in [0, \infty)$, if there exists symmetric positive-definite matrix R and S, positive scalars ϵ_1, and general matrices $\hat{A}_{ij}$, $\hat{B}_j$, $\hat{C}_j$, $\hat{D}$ such that*

$$\bar{\Xi}_{1ii} < 0, \quad i = 1, 2 \tag{5.40a}$$

$$\bar{\Xi}_{1ij} + \bar{\Xi}_{1ji} < 0, \quad 1 \le i < j \le 2 \tag{5.40b}$$

where

$$
\begin{aligned}
\bar{\Xi}_{1ij} &= \begin{bmatrix} \bar{\Xi}^1_{1ij} & \bar{\Xi}^2_{1ij} \\ * & \bar{\Xi}^3_{1ij} \end{bmatrix}, \\
\bar{\Xi}^1_{1ij} &= \begin{bmatrix} \text{sys}\{A_{0i}S + B\hat{C}_j\} & \hat{A}^{\mathrm{T}}_{ij} + (A_{0i} + B\hat{D}C_1) \\ * & \text{sys}\{RA_{0i} + \hat{B}_j C_1\} \end{bmatrix}, \\
\bar{\Xi}^2_{1ij} &= \begin{bmatrix} B_1 & SC_2^{\mathrm{T}} & \epsilon_1 H & SE^{\mathrm{T}}_{1i} \\ RB_1 & C_2^{\mathrm{T}} & \epsilon_1 RH & E^{\mathrm{T}}_{1i} \end{bmatrix}, \\
\bar{\Xi}^3_{1ij} &= \text{diag}\left[-I \ \ -\gamma^2 I \ \ -\epsilon_1 I \ \ -\epsilon_1 I\right],
\end{aligned}
$$

with

$$
\begin{cases}
\hat{A}_{ij} = R(A_{0i} + BD_k C_1)S + NB_{kj}C_1 S \\
\qquad\quad + RBC_{kj}M^{\mathrm{T}} + NA_{kij}M^{\mathrm{T}}, \\
\hat{B}_j = RBD_k + NB_{kj}, \\
\hat{C}_j = D_k C_1 S + C_{kj}M^{\mathrm{T}}, \\
\hat{D} = D_k.
\end{cases}
$$

Proof Performing a congruence transformation with $\Gamma = \text{diag}\{F_1 \ I \ I \ I \ I\}$ to the matrix $\bar{\Xi}_{1ij}$, we have

$$
\begin{bmatrix}
F_1^{\mathrm{T}}\bar{A}^{\mathrm{T}}_{ij}PF_1 & F_1^{\mathrm{T}}P\bar{B} & F_1^{\mathrm{T}}\bar{C}^{\mathrm{T}} & \epsilon_1 F_1^{\mathrm{T}}PF_1 & F_1^{\mathrm{T}}\bar{E}_i^{\mathrm{T}} \\
* & -I & 0 & 0 & 0 \\
* & * & -\gamma^2 I & 0 & 0 \\
* & * & * & -\epsilon_1 I & 0 \\
* & * & * & * & -\epsilon_1 I
\end{bmatrix} < 0.
$$

Based on the definitions of F_1 and F_2, we have,

$$
\begin{aligned}
&PF_1 = F_2, \\
&F_1^{\mathrm{T}}P\bar{A}_{ij}F_1 = \begin{bmatrix} A_{0i}S + B\hat{C} & A_{0i} + B\hat{D}C_1 \\ \hat{A}_{ij} & RA_{0i} + \hat{B}_j C_1 \end{bmatrix}, \\
&F_1^{\mathrm{T}}P\bar{B} = \begin{bmatrix} B_1 \\ RB_1 \end{bmatrix}, \ F_1^{\mathrm{T}}\bar{C}^{\mathrm{T}} = \begin{bmatrix} SC_2^{\mathrm{T}} \\ C_2^{\mathrm{T}} \end{bmatrix}, \\
&F_1^{\mathrm{T}}P\bar{H} = \begin{bmatrix} H \\ RH \end{bmatrix}, \ F_1^{\mathrm{T}}\bar{E}_i^{\mathrm{T}} = \begin{bmatrix} SE^{\mathrm{T}}_{1i} \\ E^{\mathrm{T}}_{1i} \end{bmatrix}.
\end{aligned}
$$

Therefore, the condition (5.31a) and (5.31b) are equivalent to (5.40a) and (5.40b), respectively. This completes the proof.

In order to get A_{kij}, B_{kj}, C_{kj}, and D_k, we need to find out matrices M and N. According to the equation $PP^{-1} = I$, we can get $MN^{\mathrm{T}} = I - SR$. Making singular value decomposition for $(I - SR)$, we can get M and N, and then, the matrices A_{kij}, B_{kj}, C_{kj}, and D_k can be further obtained as

$$\begin{cases} D_k = \hat{D}, \\ C_{kj} = (\hat{C}_j - D_k C_1 S)(M^{-\mathrm{T}}), \\ B_{kj} = N^{-1}(\hat{B}_j - RBD_k), \\ A_{kij} = N^{-1}[\hat{A}_{ij} - R(A_{0i} + BD_k C_1)S]M^{-\mathrm{T}} \\ \qquad - B_{kj} C_1 S M^{-\mathrm{T}} - N^{-1} R B_{0i} C_{kj}. \end{cases} \tag{5.41}$$

The finial control law can be written as:

$$\begin{cases} A_k = \sum_{i=1}^{2}\sum_{j=1}^{2} h_i(t) h_j(t) A_{kij}, \\ B_k = \sum_{j=1}^{2} h_j(t) B_{kj}, \\ C_k = \sum_{j=1}^{2} h_j(t) C_{kj}, \\ D_k = D_k. \end{cases} \tag{5.42}$$

Obviously, the final control is dependent with the $h_1(t)$ and $h_2(t)$, which can be online computed by using v_x. Thus, it is easy to used in real application, especially in the real-time test.

5.4 Simulation Results

In this section, simulations for two driving maneuvers, J-turn and double-lane change, are conducted to validate the effectiveness of the proposed control method. The steering motor is assumed to completely fail in the simulations. The control objective is to make the yaw rate of the vehicle track its desired value with only the differential steering. The desired yaw rate can be generated from the driver's steering angle and vehicle longitudinal speed as [25]

$$r_r = \frac{v_x}{l(1 + k_{us} v_x^2)} \delta(t), \tag{5.43}$$

where k_{us} is the stability factor and $l = l_f + l_s$ is the distance between the front and rear axles. The vehicle parameters are listed in Table 5.1. The uncertainties for the tire cornering stiffness C_f and C_r are both assumed as 40% of the normal values. The simulations are based on a high-fidelity and full-vehicle model via CarSim.

Table 5.1 Vehicle parameters used in the simulation

Symbol	Definition	Value
m	Vehicle mass	1500 kg
I_z	Inertia moment of the vehicle about yaw axis	$2000\ \text{kg} \cdot \text{m}^2$
b_{eff}	Effective damping of the steering system	$100\ \text{N} \cdot \text{s}/(\text{m} \cdot \text{rad})$
l_f	Distance of CG from front axle	1.3 m
l_r	Distance of CG from rear axle	1.4 m
C_f	Cornering stiffness of front tires	80000 N/rad
C_r	Cornering stiffness of rear tires	80000 N/rad
k_{us}	Stability factor	0.001
hline $v_{x\min}$	Minimum velocity	20 m/s
$v_{x\max}$	Maximum velocity	40 m/s

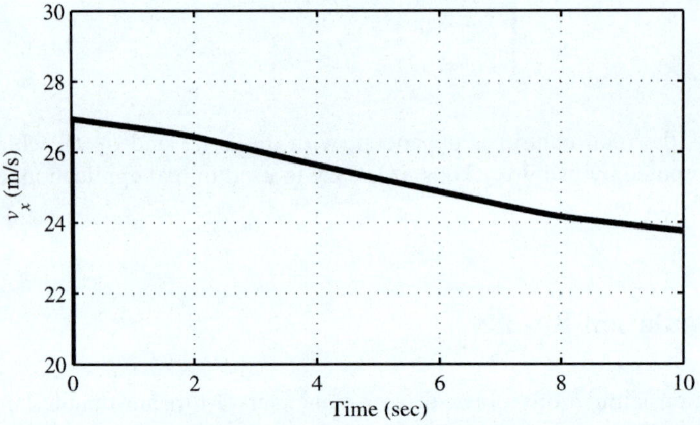

Fig. 5.3 Vehicle velocity in the J-turn simulation case

5.4.1 J-Turn Simulation

In this simulation, the vehicle is controlled to make a J-turn with the longitudinal speed shown in Fig. 5.3. The steering command from the driver is shown in Fig. 5.4. The actual steering angle of the front wheels is presented in Fig. 5.5, from which one can see that the differential steering angle is maintained in reasonable region, and has a highly similar changing trend compared with the driver's steering angle command. The fluctuation in the steering angle is caused by the unmodeled dynamics and friction in the differential steering mechanism. The result of the response for the steering angle verifies the effectiveness of the proposed controller on the steering

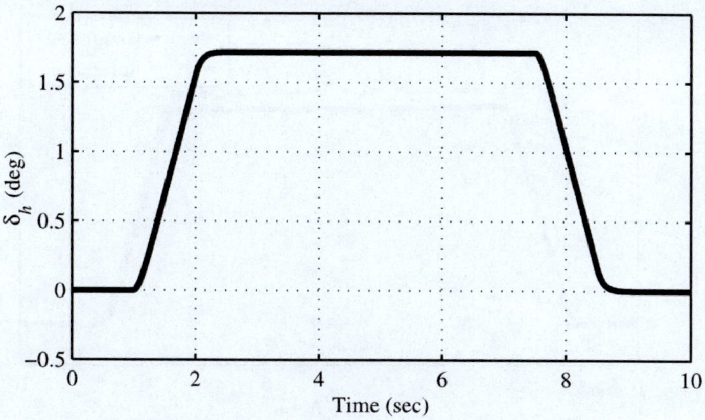

Fig. 5.4 The driver' steering command in the J-turn simulation case

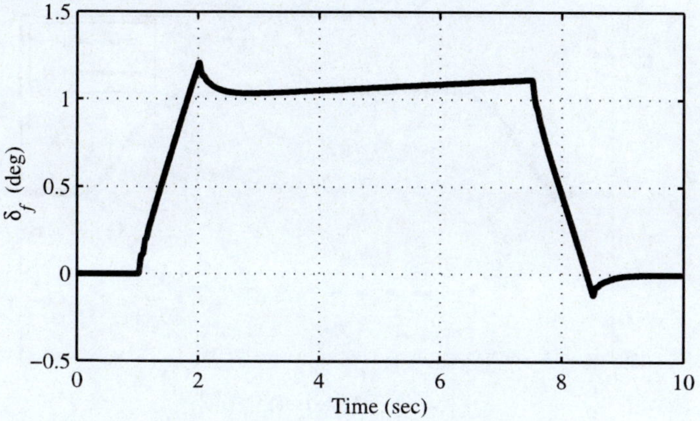

Fig. 5.5 Real steering angle of the front wheel in the J-turn simulation case

angle control and proves that the differential steering mechanism can successfully perform the steering function in the emergency situation.

The yaw rate is presented in Fig. 5.6, and we find that the tracking control of the real yaw rate toward the reference value is achieved satisfactorily, although there exists some tracking error. That is caused by the parameter uncertainties and the external disturbances in the control system. The result for the yaw stabilization indicates that with the differential steering generated by the front in-wheel motors, the vehicle can still obtain acceptable handling performance, in the presence of the steering motor's complete failure. Figure 5.7 shows the longitudinal forces of left and right front wheels. It can be observed that the two forces are equal in the magnitude but opposite in the direction and are both maintained in reasonable region. The fluctuations in the figures are caused by the uncertainties and disturbances investigated in this study.

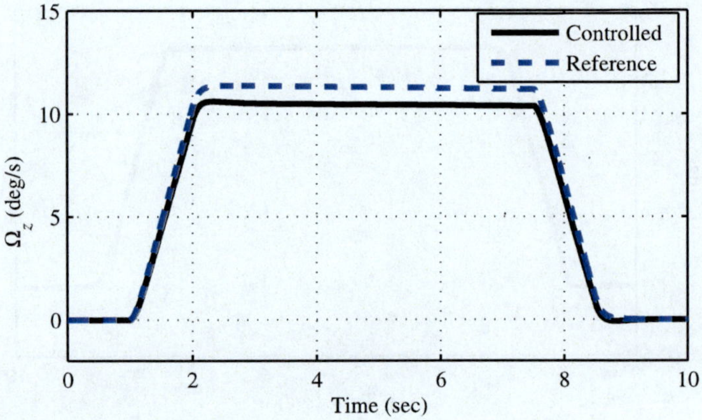

Fig. 5.6 Yaw rate in the J-turn simulation case

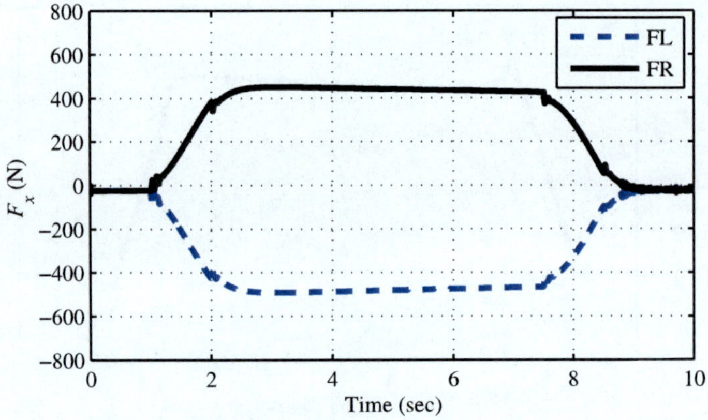

Fig. 5.7 Tire forces in the J-turn simulation case

5.4.2 *Double-Lane Change*

In this simulation, the vehicle is made to complete a double-lane change with higher speed shown in Fig. 5.8. The driver's steering angle command is presented in Fig. 5.9. In Fig. 5.10, the real steering angle for the front wheel is plotted, which is also controlled in reasonable level, and has an analogous variation trend compared to the driver's steering command. This verifies the effectiveness of the proposed controller on the steering angle control in high speed driving. The fluctuation in the steering angle is caused by the uncertainties and friction in the modeling of the differential steering mechanism.

The yaw rate control result is plotted in Fig. 5.11. We can see the real yaw rate tracks the reference value well, with small tracking error. That demonstrates the

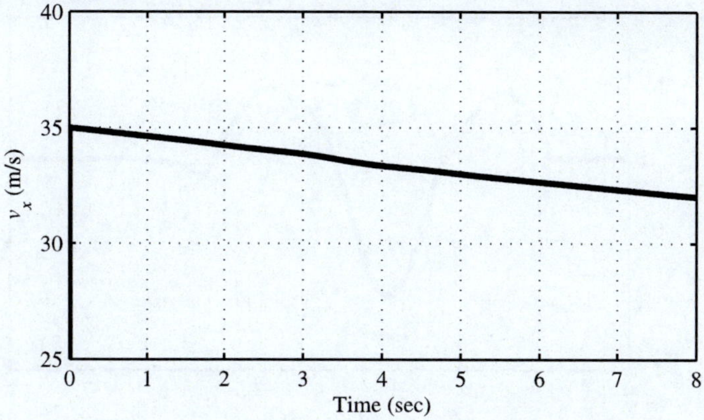

Fig. 5.8 Vehicle velocity in the double-lane change simulation case

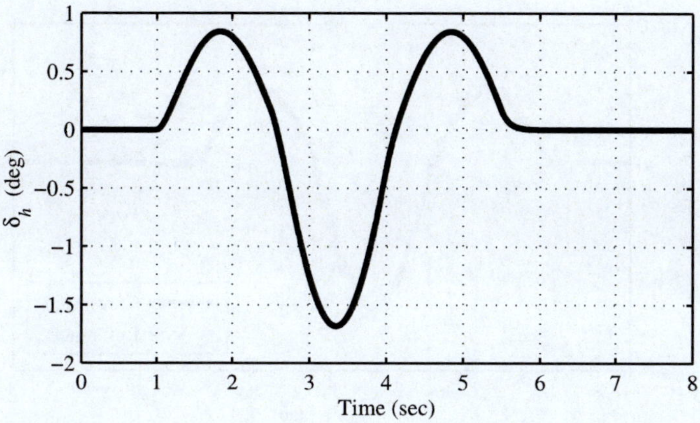

Fig. 5.9 The driver' steering command in the double-lane change simulation case

effectiveness of the differential steering mechanism in performing normal steering function and maintaining good handling performance in the emergency situation. The longitudinal forces of left and right front wheels are presented in Fig. 5.12, from which it can be found that the two forces change equally in the magnitude but opposite in direction, and are both maintained in reasonable level. It is proved that as an inherent steering actuation redundancy, the differential steering can lend a strong support for the driving safety in emergency conditions. Note that there exists tracking error in the yaw rate control result, this might be caused by the unmodeled nonlinear terms or disturbances. In the future study, differential steering control with more advanced control methods such as robust fuzzy control [26, 27] and model predictive control [28, 29] will be investigated.

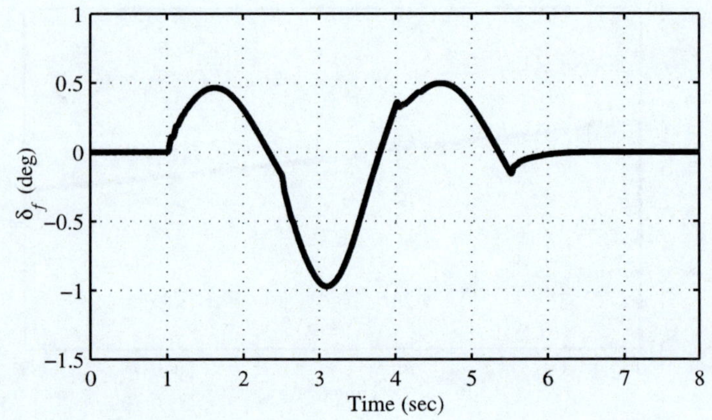

Fig. 5.10 Real steering angle of the front wheel in the double-lane change simulation case

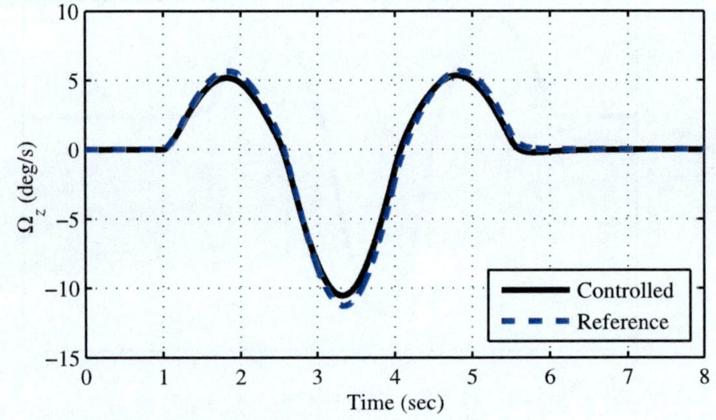

Fig. 5.11 Yaw rate in the double-lane change simulation case

5.5 Conclusion

In this study, the yaw control problem for IWM driven EGVs is investigated, with only the differential steering generated by the independently actuated front in-wheel motors. Since the steering motor is assumed in complete fault, the steering angle is not a control input; instead, it is produced by the differential torque between the left and right front wheels and thus becomes a state in the system model. Considering the desired front-wheel steering angle is hard to obtain, a robust $\mathcal{H}_\infty$ output-feedback controller is designed to ensure the equal vehicle handling based on the differential steering mechanism. Tire cornering stiffness uncertainty, vehicle longitudinal speed variation and the external disturbances are considered, making the vehicle robust to different driving conditions. Simulation results based on a high-fidelity CarSim model validate the effectiveness of the proposed controller.

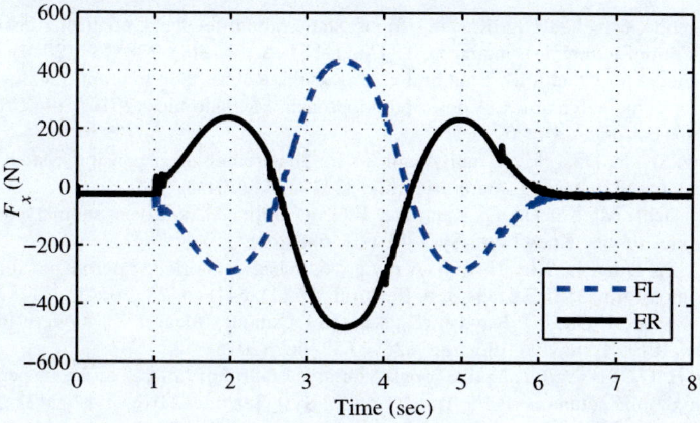

Fig. 5.12 Tire force in the double-lane change simulation case

Acknowledgements This chapter is from the previous work in [30], and some typos are corrected here.

References

1. C. Chan, A. Bouscayrol, K. Chen, Electric, hybrid, and fuel-cell vehicles: architectures and modeling. IEEE Trans. Veh. Technol. **59**(2), 589–598 (2010)
2. R. Wang, Y. Chen, D. Feng, X. Huang, J. Wang, Development and performance characterization of an electric ground vehicle with independently actuated in-wheel motors. J. Power Sources **196**(8), 3962–3971 (2011)
3. H. Zhang, J. Wang, Vehicle lateral dynamics control through AFS/DYC and robust gain-scheduling approach. IEEE Trans. Veh. Technol. (in press) (2015)
4. H. Zhang, X. Zhang, J. Wang, Robust gain-scheduling energy-to-peak control of vehicle lateral dynamics stabilisation. Vehicle Sys. Dyn. **52**(3), 309–340 (2014)
5. K. Nam, H. Fujimoto, Y. Hori, Lateral stability control of in-wheel-motor-driven electric vehicles based on sideslip angle estimation using lateral tire force sensors. IEEE Trans. Veh. Technol. **61**(5), 1972–1985 (2012)
6. D. Li, Y. Song, D. Huang, H. Chen, Model-independent adaptive fault-tolerant output tracking control of 4WS4WD road vehicles. IEEE Trans. Intell. Transp. Syst. **14**(1), 169–179 (2013)
7. S. Gao, N. Cheung, E. Cheng, L. Dong, X. Liao. Skid steering in 4-wheel-drive electric vehicle. in *Proceedings of the 7th international conference on power electronics and drive systems (PEDS)* (2007), pp. 1548–1554
8. B. Jang, Y. Yun, S. Lee, Simulation of vehicle steering control through differential braking. Int. J. Precis. Eng. Manuf. **5**(3), 26–34 (2009)
9. J. Wang, Q. Wang, L. Jin, C. Song, Independent wheel torque control of 4WD electric vehicle for differential drive assisted steering. Mechatronics **21**(1), 63–76 (2011)
10. H. Nozaki, Effect of differential steering assist on drift running performance. SAE technical paper, (2005) 2005-01-3472
11. F. Wu, T. Yeh, A control strategy for an electrical vehicle using two in-wheel motors and steering mechanism. Proc. of AVEC **8**, 796–801 (2008)

12. S. Aouaouda, M. Chadli, H. Karimi, Robust static output-feedback controller design against sensor failure for vehicle dynamics. IET Control Theory A. **8**(9), 728–737 (2014)
13. S. Aouaouda, M. Chadli, M. Boukhnifer, H. Karimi, Robust fault tolerant tracking controller design for vehicle dynamics: A descriptor approach. Mechatronics (2014). https://doi.org/10.1016/j.mechatronics.2014.09.011
14. G. Wang, S. Yin, Data-driven fault diagnosis for an automobile suspension system by using a clustering based method. J. Franklin I. **351**, 3231–3244 (2014)
15. E. Khalastchi, M. Kalech, G. Kaminka, R. Lin, Online data-driven anomaly detection in autonomous robots. Knowl. Inf. Syst. **43**, 657–688 (2015)
16. S. Yin, S.X. Ding, X. Xie, H. Luo, A review on basic data-driven approaches for industrial process monitoring. IEEE Trans. Ind. Electron. **61**(11), 6418–6428 (2014)
17. S. Yin, X. Li, H. Gao, O. Kaynak, Data-based techniques focused on modern industry: an overview. IEEE Trans. Ind. Electron. **62**(1), 657–667 (2015)
18. W. Sun, H. Gao, B. Yao, Adaptive robust vibration control of full-car active suspensions with electrohydraulic actuators. IEEE Trans. Control Syst. Technol. **21**(6), 2417–2422 (2013)
19. W. Sun, H. Gao, O. Kaynak, Adaptive backstepping control for active suspension systems with hard constraints. IEEE/ASME Trans. Mech. **18**(3), 1072–1079 (2013)
20. H. Li, X. Jing, H. Karimi, Output-feedback based $\mathcal{H}_\infty$ control for active suspension systems with control delay. IEEE Trans. Ind. Electron. **61**(1), 436–446 (2014)
21. H. Li, H. Gao, H. Liu, M. Liu, Fault-tolerant $\mathcal{H}_\infty$ control for active suspension vehicle systems with actuator faults. Proc. I. Mech. Eng. I-J. Sys. **226**, 348–363 (2012)
22. S. Yin and Z. Huang, Performance monitoring for vehicle suspension system via fuzzy positivistic C-means clustering based on accelerometer measurements. IEEE/ASME Trans. Mech., https://doi.org/10.1109/TMECH.2014.2358674
23. H. Du, N. Zhang, F. Naghdy, Velocity-dependent robust control for improving vehicle lateral dynamics. Transport Res. C-Emer. **19**(3), 454–468 (2011)
24. C. Ahn, H. Peng, H. Tseng, Robust estimation of road frictional coefficient. IEEE Trans. Control Syst. Technol. **21**(1), 1–13 (2013)
25. H. Du, N. Zhang, W. Li, Robust tracking control of vehicle lateral dynamics. Int. J. of Vehicle Design **65**(4), 314–335 (2014)
26. H. Li, C. Wu, L. Wu, H. Lam, Y. Gao, Filtering of interval type-2 fuzzy systems with intermittent measurements. IEEE Trans. Cybernetics. https://doi.org/10.1109/TCYB.2015.2413134
27. H. Li, X.Sun, L. Wu, and H. Lam, State and output feedback control of a class of fuzzy systems with mismatched membership functions, IEEE Trans. Fuzzy Systems. https://doi.org/10.1109/TFUZZ.2014.2387876
28. A. Mozaffari, M. Vajedi, N. Azad, A robust safety-oriented autonomous cruise control scheme for electric vehicles based on model predictive control and online sequential extreme learning machine with a hyper-level fault tolerance-based supervisor. Neurocomputing **151**(2), 845–856 (2015)
29. J. Shin, H. Kim, S. Park, Y. Kim, Model predictive flight control using adaptive support vector regression. Neurocomputing **73**(4–6), 1031–1037 (2010)
30. R. Wang, H. Jing, C. Hu, M. Chadli, F. Yan, Robust $\mathcal{H}_\infty$ output-feedback yaw control for in-wheel motor driven electric vehicles with differential steering. Neurocomputing **173**(3), 676–684 (2016)

Chapter 6
Robust $\mathcal{H}_\infty$ Path Following Control for Autonomous Ground Vehicles with Delay and Data Dropout

Abstract This chapter presents a robust $\mathcal{H}_\infty$ path following control strategy for autonomous ground vehicles (AGV) with delays and data dropouts. The state measurements and signal transmissions usually suffer from inevitable delays and data packet dropouts, which may degrade the control performance or even deteriorate the system stability. A robust $\mathcal{H}_\infty$ state-feedback controller is proposed to achieve the path following and vehicle lateral control simultaneously. A generalized delay representation is formulated to include the delays and data dropouts in the measurement and transmission. The uncertainties of the tire cornering stiffness and the external disturbances are also considered to enhance the robustness of the proposed controller. Two simulation cases are presented with a high-fidelity and full-car model based on the CarSim-Simulink joint platform, and the results verify the effectiveness and robustness of the proposed control approach.

6.1 Introduction

Autonomous ground vehicle (AGV) has been a rapidly developing intelligent transportation means with application from dangerous battlefields to highways. Path following is a fundamental capability for an AGV, where the destination or the desired path is predefined and the path following task needs to be completed by the control system. Generally, the control system is required to be able to make the vehicle track the desired path with zero steady-state path following errors (i.e., the lateral offset and the heading error) [1, 2], while handling the different road characteristics (e.g., varying road curvatures and terrain types).

Numerous control strategies have been presented for the traditional vehicle handling and stability control, such as learning control [3], fuzzy control [4], adaptive robust control [5], genetical algorithm-based control [6], sliding mode control (SMC) [7], phase portrait analysis [8], output-feedback control [9], model predictive control (MPC) [10], and linear matrix inequality (LMI) [11]. Nevertheless, with the arrival of the car-networking age, the emerging research topic on the vehicle motion control with delays and data dropouts in the process of the information transmission is increasingly crucial. As far as the authors have observed, there are few previ-

H. Zhang et al., *Robust Gain-Scheduled Estimation and Control of Electrified Vehicles via LPV Technique*, Key Technologies on New Energy Vehicles,
https://doi.org/10.1007/978-981-19-8509-6_6

ous literatures which investigated the delay and data dropout problem in the motion control for ground vehicles, letting alone the autonomous vehicles. For the path following control of AGVs, a critical point is to obtain the vehicle states and position information to evaluate the tracking errors which is then used to feed the control system. In last few decades, the global positioning system (GPS) has become one of the standard real-time measurement tools in the vehicle path following control. The GPS receiver is always equipped on vehicle to communicate with the GPS satellites or the GPS base stations [12]. Then, the vehicle states can be measured and sent to the controller via the in-vehicle network such as controller area network (CAN). Once the controller receives the vehicle states, the corresponding control signals are generated and sent to the actuators of the vehicle (e.g., the steering motors, or the in-wheel motors). However, there are usually some unavoidable time delays or data dropouts during the signal transmissions, which are likely to considerably deteriorate the control effects or destroy the system stability. Therefore, from a practical perspective, the path following control issue of AGVs needs to be treated as a networked control system (NCS) in the presence of the network-induced delays and data packet dropouts [13].

How to ensure the vehicle path following performance for AGVs in the presence of the delays and data packet dropouts is challenging in both the industry and academia. There are actually some well-studied theoretical achievements to deal with systems with delays and data dropouts, which can lend some useful inspirations for the path following control. In [13], a controller combined active front steering (AFS) and direct yaw-moment control (DYC) for four-wheel-independent-drive electric vehicles was presented, and the CAN network with time-varying delays was handled by Taylor series expansion. Gao et al. [14] proposed a control approach with multiple successive delay components in network-based control problem by exploiting a novel Lyapunov-Krasovskii functional and taking advantage of new techniques for time delay systems. The system with state and input delays was investigated in [15], where a delay-dependent approach was proposed to robustly $\mathcal{H}_\infty$ control the uncertain stochastic systems. In [16], the NCS with data packet dropout and transmission delays was handled via LMI approach, where the dropout and delay were lumped together by parallel distributed compensation (PDC). In [17], the authors investigated the tracking problem in NCS with delay and handled the discrete-time nonlinear systems with integral and predictive actions by $\mathcal{H}_\infty$ step tracking control schema. In [18, 19], a deterministic approach was chosen to deal with the delays and packet dropout in NCS, while in [20] the delay and data packet dropout were assumed to obey certain probability distributions. In [21], a fuzzy dynamic output-feedback control strategy was presented to regulate the nonlinear networked discrete-time system with missing measurements. Srinivasan and Ayyagari [22] handled the communication links (vehicle-to-vehicle) with random packet dropouts based on a driver assistance system for automated highway system (AHS) with graph technology. In [23], a remote wireless path tracking controller for an autonomous guided vehicle was given by using Kalman filter to estimate the optimal delay.

To handle the delay and data dropout in path following, this chapter proposes a robust $\mathcal{H}_\infty$ state-feedback controller considering the parameter uncertainties and external disturbances. The aim of the proposed controller is to make the vehicle track the desired path with the active front-wheel steering (AFS), while improving the vehicle handling and stability. The main contributions of this chapter lie on the following aspects: (1) The inevitable delay and data dropout are explicitly considered in the path following model of AGVs, and a generalized delay form is presented to facilitate the controller design; and (2) the path following model combined with vehicle lateral dynamics is established to make the vehicle track the desired path and improve the vehicle stability. In addition, the uncertainties of the tire cornering stiffness and the external disturbances are considered in the controller design to enhance the robustness of the path following ability and vehicle handling stability, simultaneously.

6.2 System Modeling and Problem Formulation

6.2.1 Path Following Model

The path following model of AGV is shown in Fig. 6.1. e represents the lateral offset from the vehicle center of gravity (CG) to the closest point T on the desired path. The heading error ψ is defined as the error between the actual heading angle ψ_h and the desired heading angle ψ_d; therefore, $\psi = \psi_h - \psi_d$. $\dot{\psi}_h = r$, with r being the yaw rate of the vehicle. v_x and v_y are the longitudinal and lateral velocity of the vehicle, respectively. δ_f is the front-wheel steering angle. σ represents the curvilinear coordinate (arc-length) of point T along the path from an initial position predefined, while we know $\sigma \geq 0$ and $\dot{\sigma} = d\sigma/dt$. Finally, $\rho(\sigma)$ represents the curvature of the desired path at the point T. The curvilinear coordinate of points T along the path σ can be given as

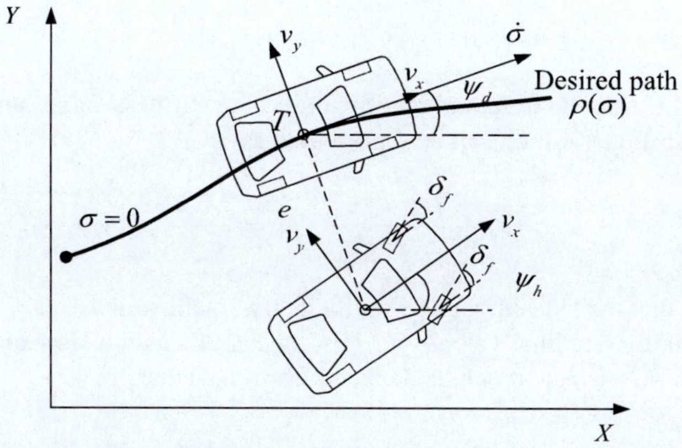

Fig. 6.1 Schematic diagram of path following model

$$\dot{\sigma} = \frac{1}{1 - e \cdot \rho(\sigma)}(v_x \cos \psi - v_y \sin \psi). \tag{6.1}$$

Based on the Serret-Frenet equation in [24], the path following model of an AGV is given as

$$\begin{cases} \dot{e} = v_x \sin \psi + v_y \cos \psi, \\ \dot{\psi} = r - \rho(\sigma) v_x. \end{cases} \tag{6.2}$$

By assuming that the heading error ψ is small, the error e can be rewritten in the linear form as

$$\dot{e} = v_x \psi + v_y + d_1, \tag{6.3}$$

where d_1 represents the modeling error and external disturbance.

6.2.2 Vehicle Model

A schematic diagram of a vehicle model is shown in Fig. 6.2. Assuming the front-wheel steering angle is small, the vehicle's lateral dynamics can be expressed as:

$$\begin{cases} \dot{\beta} = \frac{1}{m}\left(F_{yf} + F_{yr}\right) - r + d_2, \\ \dot{r} = \frac{1}{I_z}\left(l_f F_{yf} - l_r F_{yr}\right) + d_3, \end{cases} \tag{6.4}$$

where m and I_z are the mass and yaw inertia of the vehicle, respectively. d_2 and d_3 are external disturbances. β is the vehicle sideslip angle. l_f and l_r denote the distances between the CG of vehicle to the front and rear wheel axes, respectively. The lateral forces F_{yf} and F_{yr} of the front and rear tires can be modeled as

$$F_{yf} = C_f \alpha_f, \quad F_{yr} = C_r \alpha_r, \tag{6.5}$$

where C_f and C_r are the front and rear tire cornering stiffness values and α_f and α_r are the tire slip angles which can be represented as

$$\alpha_f = \delta_f - \frac{l_f r}{v_x} - \frac{v_y}{v_x}, \quad \alpha_r = \frac{l_r r}{v_x} - \frac{v_y}{v_x}, \tag{6.6}$$

By assuming that the sideslip angle is sufficiently small, we have $\beta = v_y/v_x$. Supposing that the longitudinal velocity v_x is a constant or with a slow changing rate, we can obtain $\dot{\beta} = \dot{v}_y/v_x$. It can be deduced from (6.4) that

$$\begin{cases} \dot{\beta} = a_{11}\beta + a_{12} r + b_1 \delta_f + d_2, \\ \dot{r} = a_{21}\beta + a_{22} r + b_2 \delta_f + d_3, \end{cases} \tag{6.7}$$

with

$$a_{11} = -\frac{(C_f+C_r)}{mv_x^2},\ a_{12} = -(1+\frac{(l_fC_f-l_rC_r)}{mv_x^2}),\ b_1 = \frac{C_f}{mv_x},$$
$$a_{21} = \frac{(l_rC_r-l_fC_f)}{I_z},\ a_{22} = -\frac{\left(l_f^2C_f+l_r^2C_r\right)}{v_xI_z},\ b_2 = \frac{l_fC_f}{I_z}.$$

As $\dot{e} = v_x\psi + v_y + d_1$ with $v_y = \beta v_x$, we have

$$\dot{e} = \psi v_x + \beta v_x + d_1. \tag{6.8}$$

Combined (6.2) with (6.7) and (6.8), we can get

$$\begin{cases} \dot{e} = v_x\beta + v_x\psi + d_1, \\ \dot{\psi} = r - \rho(\sigma)v_x, \\ \dot{\beta} = a_{11}\beta + a_{12}r + b_1\delta_f + d_2, \\ \dot{r} = a_{21}\beta + a_{22}r + b_2\delta_f + d_3. \end{cases} \tag{6.9}$$

Define the state vector $x(t) = [e,\ \psi,\ \beta,\ r]^{\mathrm{T}}$, the control input $u(t) = \delta_f$, and the disturbance $w(t) = [d_1,\ -\rho(\sigma)v_x,\ d_2,\ d_3]^{\mathrm{T}}$, the state-space form of the model can be given as follows

$$\dot{x}(t) = Ax(t) + Bu(t) + w(t), \tag{6.10}$$

where

$$A = \begin{bmatrix} 0 & v_x & v_x & 0 \\ 0 & 0 & 0 & 1 \\ 0 & 0 & a_{11} & a_{12} \\ 0 & 0 & a_{21} & a_{22} \end{bmatrix},\ B = \begin{bmatrix} 0 \\ 0 \\ b_1 \\ b_2 \end{bmatrix}.$$

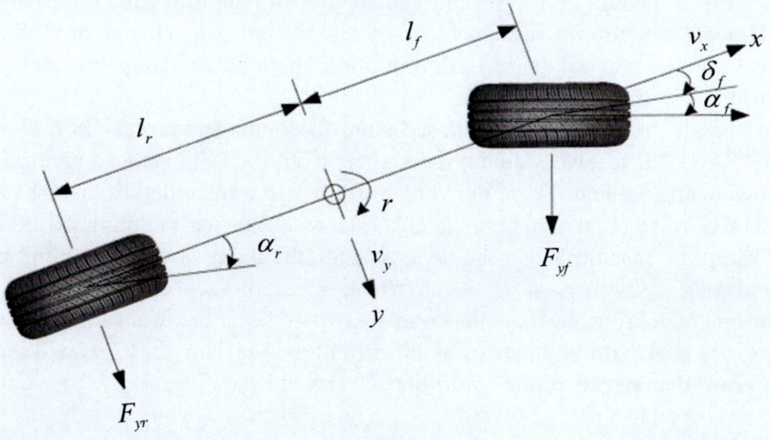

Fig. 6.2 Schematic diagram of a vehicle planar motion model

The tire cornering stiffness, which is time-varying due to the change of the road conditions and the vehicle states, can be represented as

$$C_f = C_{0f} + \lambda_f \tilde{C}_f, \; C_r = C_{0r} + \lambda_r \tilde{C}_r, \tag{6.11}$$

where λ_f and λ_r are time-varying parameters and satisfy $|\lambda_i| \leq 1, (i = f, r)$, and C_{0f} and C_{0r} are the nominal values of C_f and C_r, respectively. So we have

$$A = A_0 + \Delta A, \; B = B_0 + \Delta B, \tag{6.12}$$

where A_0 and B_0 are the nominal values of A and B, respectively. ΔA and ΔB are the variations of A and B, respectively. Based on the definitions of ΔA and ΔB, we have

$$[\Delta A \;\; \Delta B] = H \Lambda_1 [E_1 \;\; E_2], \tag{6.13}$$

with $\Lambda_1 = \text{diag}\left\{\lambda_f, \lambda_f\right\}$. H, E_1, and E_2 are matrices with proper dimensions. Since the road conditions are usually uniform to the front and rear wheels, it is reasonable to assume that $\lambda_f = \lambda_r$ to reduce the possible computational burden and design conservation. Based on (6.12), the system plant (6.10) can be rewritten as

$$\dot{x}(t) = (A_0 + \Delta A)x(t) + (B_0 + \Delta B)u(t) + w(t). \tag{6.14}$$

6.2.3 *Path Following with Delay and Data Packet Dropout*

For the measurements of the vehicle states, the yaw rate r can be obtained by the vehicle on-board sensor such as inertial measurement unit (IMU) while the vehicle lateral offset e, heading error ψ, and vehicle sideslip angle β can be measured by GPS. However, due to the influences of vehicle velocity, the change of GPS signal intensity, and the network bandwidth limitation, there unavoidably exist delays and data dropouts in the control system.

The overall structure of the proposed control schema based on NCS is shown in Fig. 6.3. The GPS receivers obtain the signals from the GPS stations and calculate the states of the vehicle. Then, the vehicle states are transmitted to the zero-order-hold (ZOH) in the control system via CAN. Denote h as the sampling period and t_k as the sampling instant. When the delay occurs, the delay at each sampling period is denoted as τ_k, with $\tau_k = \tau_{sc} + \tau_{ca}$, where τ_{sc} is the sensor-to-controller delay (measurement delay), and τ_{ca} is the controller-to-actuator delay at sampling time t_k. When a data packet dropout occurs at the current time instant, the ZOH will send the last received data packet to the controller. The received data packet in the controller can be described as [19]:

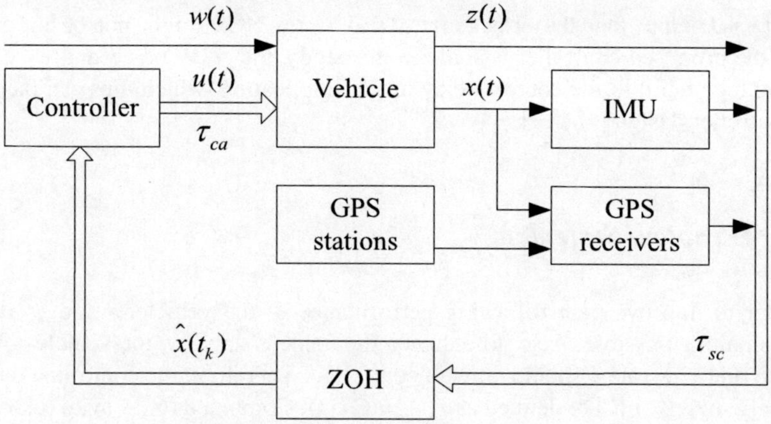

Fig. 6.3 Schematic diagram of vehicle path following control based on NCS

$$\begin{cases} \hat{x}(t_k) = x(t_k), & n(k) = 0; \\ \hat{x}(t_k) = x(t_k - h), & n(k) = 1; \\ \hat{x}(t_k) = x(t_k - d(k)h), & n(k) = d(k). \end{cases} \tag{6.15}$$

where $n(k)$ stands for the number of the data packet dropouts at the time t_k. Then, it can be deduced that

$$\hat{x}(t_k) = x(t_k - d(k)h - \tau_k). \tag{6.16}$$

The delay and data packet dropout can be lumped into a generalized form as

$$\tau(t) = t_k - d(k)h - \tau_k, \tag{6.17}$$

and we can get

$$\hat{x}(t_k) = x(t - \tau(t)). \tag{6.18}$$

Note that the time-varying $\tau(t)$ can be assumed to satisfy

$$0 < \tau(t) \leq \tau_{\max}, \tag{6.19}$$

with $\tau_{\max} > 0$ being the upper bound of the generalized delay $\tau(t)$. Since controller receives the state signal $\hat{x}(t_k)$ at the sampling instant t_k, the state-feedback controller can be designed as follows:

$$\begin{aligned} u(t) &= K\hat{x}(t_k) \\ &= Kx(t - \tau(t)). \end{aligned} \tag{6.20}$$

with K being the desired control gain.

Remark 6.1 Note that the effect of τ_{ca} and τ_{sc} in NCS might not be the same. Since the proposed controller is static in this study, the delay of τ_{ca} and τ_{sc} can be lumped together into the control delay in real application, which thus can facilitate the controller design [16, 19].

6.2.4 Problem Statement

In order to improve path following performance of the vehicle, e and ψ should be as small as possible. Also, to enhance the vehicle stability, the vehicle sideslip angle β and yaw rate r should be well controlled. Therefore, the controlled outputs $z(t) = [e,\ \psi,\ \beta,\ r]^{\mathrm{T}}$ are defined as $z(t) = Cx(t)$. Combined (6.14) with (6.20), the path following vehicle model can be rewritten as

$$\dot{x}(t) = \bar{A}x(t) + \bar{B}Kx(t - \tau(t)) + w(t), \tag{6.21}$$

with $\bar{A} = A_0 + \Delta A,\ \bar{B} = B_0 + \Delta B$. The control objective of the vehicle path following is to design a robust $\mathcal{H}_\infty$ controller, such that the closed-loop system plant in (6.21) is asymptotically stable and has the following $\mathcal{H}_\infty$ disturbance attenuation performance, i.e.,

$$\int_0^t z^{\mathrm{T}}(t)z(t)dt < \gamma^2 \int_0^t w^{\mathrm{T}}(t)w(t)dt, \tag{6.22}$$

where γ is the prescribed attenuation level.

6.3 Robust $\mathcal{H}_\infty$ Controller Design with Delay and Data Dropout

In this section, a robust $\mathcal{H}_\infty$ state-feedback controller with delay and data dropout consideration is proposed for the path following control of AGVs. In order to deal with the uncertainties and data dropout, we firstly introduce the following lemmas:

Lemma 6.1 [16] *For any positive-definite matrix X, matrices (or scalars) a and b with proper dimensions, the following inequality holds:*

$$-2ab \le a^{\mathrm{T}}X^{-1}a + b^{\mathrm{T}}Xb. \tag{6.23}$$

Lemma 6.2 [25][26] *Let $Y = Y^{\mathrm{T}}$, Γ and Ψ be the real matrices with proper dimensions, and Λ satisfies $\Lambda^{\mathrm{T}}\Lambda < I$, then the following conditions:*

$$Y + \Gamma\Lambda\Psi + \Psi^{\mathrm{T}}\Lambda^{\mathrm{T}}\Gamma^{\mathrm{T}} < 0, \tag{6.24}$$

holds if and only if there exists a positive scalar $\epsilon > 0$ *such that*

$$Y + \epsilon \Gamma \Gamma^{\mathrm{T}} + \epsilon^{-1} \Psi^{\mathrm{T}} \Psi < 0. \tag{6.25}$$

Lemma 6.3 [19] *If there exists a matrix* X *and matrix* $W > 0$*, then the following conditions are equivalent:*

$$\begin{array}{ll} (1) & (W - X)^{\mathrm{T}} W^{-1} (W - X) \geq 0, \\ (2) & -XW^{-1}X \leq W - 2X. \end{array}$$

Now, we are in the position to propose the robust $\mathcal{H}_\infty$ *state-feedback controller design method.*

Theorem 6.1 *Given positive constants* γ *and* τ_{max}*, the closed-loop system in (6.21) is asymptotically stable with* $w(t) = 0$ *and can satisfy the* $\mathcal{H}_\infty$ *performance index (6.22) for all* $w(t) \in [0, \infty)$*, if there exists a symmetric positive-definite matrices* $\bar{P}$*,* $\bar{Q}$*,* $\bar{R}$*, general matrix* X*, and scalars* $\epsilon_i > 0$*,* $i = 1, 2$ *such that*

$$\begin{bmatrix} \bar{\Lambda}_{11} & \bar{\Lambda}_{12} \\ * & \bar{\Lambda}_{22} \end{bmatrix} < 0, \tag{6.26}$$

with

$$\bar{\Lambda}_{11} = \begin{bmatrix} \bar{\Xi} & \bar{P} & 0 & I & \bar{P}A_0^{\mathrm{T}} & \bar{P}C^{\mathrm{T}} \\ * & \tau_{\max}^{-1}(\bar{R} - 2\bar{P}) & 0 & 0 & X^{\mathrm{T}}B_0^{\mathrm{T}} & 0 \\ * & * & -\bar{Q} & 0 & 0 & 0 \\ * & * & * & -\gamma^2 I & I & 0 \\ * & * & * & * & -\tau_{max}^{-1}\bar{R} & 0 \\ * & * & * & * & * & -I \end{bmatrix},$$

$$\bar{\Lambda}_{12} = \begin{bmatrix} \epsilon_1 H & 0 & X^{\mathrm{T}}E_2^{\mathrm{T}} + \bar{P}E_1^{\mathrm{T}} & \bar{P}E_1^{\mathrm{T}} \\ 0 & 0 & 0 & X^{\mathrm{T}}E_2^{\mathrm{T}} \\ 0 & 0 & 0 & 0 \\ 0 & 0 & 0 & 0 \\ 0 & \epsilon_2 H & 0 & 0 \\ 0 & 0 & 0 & 0 \end{bmatrix},$$

$$\begin{aligned} \bar{\Lambda}_{22} &= \mathrm{diag}\{-\epsilon_1 \mathrm{I} \ \ -\epsilon_2 \mathrm{I} \ \ -\epsilon_1 \mathrm{I} \ \ -\epsilon_2 \mathrm{I}\}, \\ \bar{\Xi} &= \mathrm{sys}\{\bar{P}A_0^{\mathrm{T}} + B_0 X\} + \bar{Q}, \end{aligned}$$

where $\mathrm{sys}\{\bullet\}$ *denotes* $\bullet + \bullet^{\mathrm{T}}$*. Moreover, the control gain can be calculated as* $K = X\bar{P}^{-1}$*.*

Proof Since

$$x(t) - x(t - \tau(t)) - \int_{t-\tau(t)}^{t} \dot{x}(\alpha) \mathrm{d}\alpha = 0,$$

we have

$$\dot{x}(t) = (\bar{A} + \bar{B}K)x(t) - \bar{B}K \int_{t-\tau(t)}^{t} \dot{x}(\alpha)\mathrm{d}\alpha + \mathrm{w}(\mathrm{t}).$$

Define a Lyapunov function for the system in (6.21) as

$$V(t) = V_1(t) + V_2(t) + V_3(t),$$

where

$$\begin{aligned} V_1(t) &= x^{\mathrm{T}}(t)Px(t), \\ V_2(t) &= \int_{t-\tau_{\max}}^{t} x^{\mathrm{T}}(s)Qx(s)\mathrm{d}s, \\ V_3(t) &= \int_{-\tau_{\max}}^{0}\int_{t+\theta}^{t} \dot{x}^{\mathrm{T}}(s)R\dot{x}(s)\mathrm{d}s\mathrm{d}\theta, \end{aligned}$$

with P, Q, and R being symmetric positive-definite matrices. Computing the time derivative of $V(t)$ along the trajectory of system (6.21) using *Lemma* 6.1, we have

$$\begin{aligned} \dot{V}(t) &= \dot{V}_1(t) + \dot{V}_2(t) + \dot{V}_3(t), \\ &\le x^{\mathrm{T}}(t)[\mathrm{sys}\{P(\bar{A} + \bar{B}K)\} + Q + \tau_{\max}PR^{-1}P]x(t) \\ &\quad - x^{\mathrm{T}}(t-\tau_{\max})Qx(t-\tau_{\max}) + \tau_{\max}\dot{x}^{\mathrm{T}}(t)R\dot{x}(t) \\ &\quad + x^{\mathrm{T}}(t)Pw(t) + w^{\mathrm{T}}(t)Px(t). \end{aligned}$$

Defining $\xi^{\mathrm{T}}(t) = [x^{\mathrm{T}}(t), x^{\mathrm{T}}(t-\tau(t)), x^{\mathrm{T}}(t-\tau_{\max}), w^{\mathrm{T}}(t)]^{\mathrm{T}}$ and adding $z^{\mathrm{T}}(t)z(t) - \gamma^2 w^{\mathrm{T}}(t)w(t)$ to the above inequality, we can get

$$\dot{V}(t) + z^{\mathrm{T}}(t)z(t) - \gamma^2 w^{\mathrm{T}}(t)w(t) \le \xi^{\mathrm{T}}(t)\Phi\xi(t),$$

where

$$\Phi = \begin{bmatrix} \Xi & 0 & 0 & P & \bar{A}^{\mathrm{T}} \\ * & -\tau_{\max}^{-1}R & 0 & 0 & (\bar{B}K)^{\mathrm{T}} \\ * & * & -Q & 0 & 0 \\ * & * & * & -\gamma^2 I & I \\ * & * & * & * & -\tau_{\max}^{-1}R^{-1} \end{bmatrix},$$

$$\Xi = \mathrm{sys}\{P(\bar{A} + \bar{B}K)\} + Q + \tau_{\max}PR^{-1}P + C^{\mathrm{T}}C.$$

It can be observed that the $\mathcal{H}_\infty$ performance of the closed-loop system can be guaranteed if $\Phi < 0$. By using Schur complement, we can get $\Phi < 0$ is equivalent to $\Phi_1 < 0$, where

$$\Phi_1 = \begin{bmatrix} \Xi_1 & P & 0 & P & A_0^{\mathrm{T}} & C^{\mathrm{T}} \\ * & -\tau_{\max}^{-1} R & 0 & 0 & (B_0 K)^{\mathrm{T}} & 0 \\ * & * & -Q & 0 & 0 & 0 \\ * & * & * & -\gamma^2 I & I & 0 \\ * & * & * & * & -\tau_{\max}^{-1} R^{-1} & 0 \\ * & * & * & * & * & -I \end{bmatrix} +$$

$$\mathrm{sys}\left\{ \begin{bmatrix} PH & 0 \\ 0 & 0 \\ 0 & 0 \\ 0 & 0 \\ 0 & H \\ 0 & 0 \end{bmatrix} \begin{bmatrix} \Lambda_1 & 0 \\ 0 & \Lambda_1 \end{bmatrix} \begin{bmatrix} \bar{E}_k & 0 & 0 & 0 & 0 & 0 \\ E_1 & E_2 K & 0 & 0 & 0 & 0 \end{bmatrix} \right\},$$

$$\Xi_1 = \mathrm{sys}\{P(A_0 + B_0 K)\} + Q,$$
$$\bar{E}_k = E_1 + E_2 K.$$

Following *Lemma* 6.2, $\Phi_1 < 0$ is equivalent to

$$\Phi_2 < 0, \tag{6.27}$$

where

$$\Phi_2 = \begin{bmatrix} \Lambda_{11} & \Lambda_{12} \\ * & \Lambda_{22} \end{bmatrix},$$

with

$$\Lambda_{11} = \begin{bmatrix} \Xi_2 & P & 0 & P & A_0^{\mathrm{T}} & C^{\mathrm{T}} \\ * & -\tau_{\max}^{-1} R & 0 & 0 & K^{\mathrm{T}} B_0^{\mathrm{T}} & 0 \\ * & * & -Q & 0 & 0 & 0 \\ * & * & * & -\gamma^2 I & I & 0 \\ * & * & * & * & -\tau_{\max}^{-1} R & 0 \\ * & * & * & * & * & -I \end{bmatrix},$$

$$\Lambda_{12} = \begin{bmatrix} \epsilon_1 PH & 0 & K^{\mathrm{T}} E_2^{\mathrm{T}} + E_1^{\mathrm{T}} & E_1^{\mathrm{T}} \\ 0 & 0 & 0 & K^{\mathrm{T}} E_2^{\mathrm{T}} \\ 0 & 0 & 0 & 0 \\ 0 & 0 & 0 & 0 \\ 0 & \epsilon_2 H & 0 & 0 \\ 0 & 0 & 0 & 0 \end{bmatrix},$$

$$\Lambda_{22} = \mathrm{diag}\{-\epsilon_1 I \quad -\epsilon_2 I \quad -\epsilon_1 I \quad -\epsilon_2 I\},$$
$$\Xi_2 = \mathrm{sys}\{P(A_0 + B_0 K)\} + Q.$$

Defining $\bar{P} = P^{-1}$, $\bar{R} = R^{-1}$, $\bar{Q} = P^{-1} Q P^{-1}$, and $X = K P^{-1}$, by performing a congruence transformation with $\Gamma = \mathrm{diag}\{\bar{P}, \bar{P}, \bar{P}, \underbrace{I, \ldots, I}_{7}\}$ to the above inequality, we have

$$\begin{bmatrix} \tilde{\Lambda}_{11} & \bar{\Lambda}_{12} \\ * & \bar{\Lambda}_{22} \end{bmatrix} < 0, \tag{6.28}$$

with

$$\tilde{\Lambda}_{11} = \begin{bmatrix} \bar{\Xi} & \bar{P} & 0 & I & \bar{P}A_0^{\mathrm{T}} & \bar{P}C^{\mathrm{T}} \\ * & -\tau_{\max}^{-1}\bar{P}R\bar{P} & 0 & 0 & X^{\mathrm{T}}B_0^{\mathrm{T}} & 0 \\ * & * & -\bar{Q} & 0 & 0 & 0 \\ * & * & * & -\gamma^2 I & I & 0 \\ * & * & * & * & -\tau_{\max}^{-1}\bar{R} & 0 \\ * & * & * & * & * & -I \end{bmatrix}.$$

Note that the above inequality cannot be directly implemented due to the existence of the nonlinear term $\bar{P}R\bar{P}$. Since $R > 0$, we have $-\tau_{\max}\bar{P}R\bar{P} \leq 0$. It follows from *Lemma* 6.3 that $-\tau_{\max}^{-1}\bar{P}R\bar{P} \leq \tau_{\max}^{-1}(\bar{R} - 2\bar{P})$, which means that the inequality (6.28) is equivalent to (6.26). This completes the proof. □

To show the advantage of the proposed method, a robust $\mathcal{H}_\infty$ state-feedback controller without delay and data dropout consideration is also given as follows. Suppose the system plant without delay and data dropout can be written as

$$\begin{cases} \dot{x}(t) = \bar{A}x(t) + \bar{B}u(t) + w(t) \\ z(t) = Cx(t) \end{cases}, \tag{6.29}$$

where the state-feedback controller has the form of

$$u(t) = K_{sf}x(t), \tag{6.30}$$

with K_{sf} being the control gain to be designed. Applying (6.30) to (6.29) yields the closed-loop system as

$$\begin{cases} \dot{x}(t) = (\bar{A} + \bar{B}K_{sf})x(t) + w(t) \\ z(t) = Cx(t) \end{cases}. \tag{6.31}$$

Then, the following corollary is given with similar proof of *Theorem* 6.1.

Corollary 6.1 *Given positive constant γ_1, the closed-loop system in (6.31) is asymptotically stable with $w(t) = 0$ and has the $\mathcal{H}_\infty$ performance index (6.22) for all $w(t) \in [0, \infty)$, if there exists a symmetric positive-definite matrix $\bar{P}_1$, general matrix X_1, and scalars $\epsilon_3 > 0$ such that*

$$\begin{bmatrix} \bar{\Xi}_1 & \bar{P}_1 & \bar{P}_1 C^{\mathrm{T}} & \epsilon_3 H & X_1^{\mathrm{T}} E_2^{\mathrm{T}} + \bar{P}_1 E_1^{\mathrm{T}} \\ * & -\gamma_1^2 I & 0 & 0 & 0 \\ * & * & -I & 0 & 0 \\ * & * & * & -\epsilon_3 I & 0 \\ * & * & * & * & -\epsilon_3 I \end{bmatrix} < 0, \tag{6.32}$$

with $\bar{\Xi}_1 = \mathrm{sys}\{\bar{P}_1 A_0^{\mathrm{T}}\} + \mathrm{sys}\{B_0 X_1\}$. □

The control gain of this controller can be described as $K_{sf} = X_1 \bar{P}_1^{-1}$. In the following simulation, we denote the controller design with delay and data dropout consideration as Controller 1 and the controller without delay and data dropout consideration is named as Controller 2.

6.4 Simulation Results

In this section, two simulation cases including the single-lane change and double-lane change maneuvers are presented to validate the effectiveness of the proposed method. The simulation is implemented on CarSim-Simulink joint platform with a full-vehicle model. The vehicle nominal parameters are listed as follows. $m = 1500\,\mathrm{kg}$, $I_z = 2500\,\mathrm{kg \cdot m^2}$, $l_s = 1\,\mathrm{m}$, $l_f = 1.3\,\mathrm{m}$, $l_r = 1.4\,\mathrm{m}$, $C_f = 40{,}000\,\mathrm{N/rad}$, $C_r = 40{,}000\,\mathrm{N/rad}$. The uncertainty of the cornering stiffness is assumed as the 20% of the nominal value. Suppose the maximum delay caused by measurement and the transmission in CAN is 30 ms. The sampling period is 2.5 ms, and the maximum number of data packet dropout is $d(k) = 4$; then, by (6.17) and (6.19) we can identify $\tau_{max} = 30 + 2.5 \times 4 = 40\,\mathrm{ms}$. By using Theorem 6.1 and Corollary 6.1, the gains of Controller 1 and Controller 2 can be obtained, respectively.

6.4.1 *Single-Lane Change Maneuver*

In the simulation, the vehicle is supposed to complete a single-lane change maneuver at the speed of 25 m/s on the low-adherence road ($\mu = 0.2$). The road curvature is plotted in Fig. 6.4, from which one can see that the road curvature changes smoothly in this case.

The longitudinal velocity and control inputs are presented in Fig. 6.4, respectively. The sub-figure for the longitudinal velocity shows that the delay and data dropout investigated in this study do not obviously influence the longitudinal vehicle motion. According to the sub-figure for the steering angle, it is identified that the steering angle is much sensitive to the delay and data dropout. For the given time delay $\tau = 0.04\,\mathrm{s}$ in the simulation, the front-wheel steering angle δ_f generated by the Controller 1 is maintained in reasonable regions with few fluctuations and shows similar changing trend compared to the road curvature. However, with $\tau = 0.04\,\mathrm{s}$ the δ_f generated by

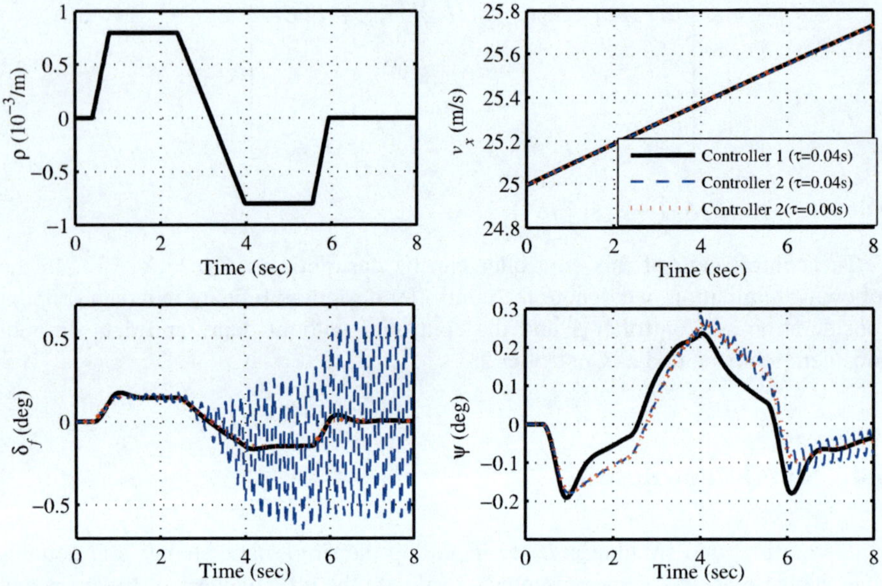

Fig. 6.4 The simulation results for the road curvature, longitudinal velocity, and control inputs in the single-lane change maneuver

the Controller 2 begins to oscillate from 2.5 s, and the oscillation expands sharply until the road curvature returns to zero at the last 2 s. This phenomenon, which is caused by the delay and data dropout, indicates that the vehicle is running near the criticality of instability. Furthermore, we can see that the control effect of the steering angle of the Controller 1 ($\tau = 0.04$ s) is as good as that of Controller 2 ($\tau = 0.00$ s). This verifies that the Controller 1 is effective in handling the delay and data dropout with considering the parametric uncertainties. The sub-figure for the heading error shows that the Controller 1 has more smooth response for the vehicle heading control; however, the curve for Controller 2 ($\tau = 0.04$ s) has several fluctuations.

The simulation results for the sideslip angle, yaw rate, acceleration, and tracking errors in the single-lane change maneuver are presented in Fig. 6.5. From the sub-figures for the sideslip angle and the yaw rate, we can see that the Controller 1 can yield smooth responses and maintain them in the reasonable regions. Their changing trends are much similar as that of the road curvature, while the delay and data dropout cause large fluctuations in the vehicle states and deteriorate the performances of Controller 2. With the same time delay $\tau = 0.04$ s, the tracking errors of the vehicle by Controller 1 are smaller than those of Controller 2 and is even better than those of Controller 2 without time delay. With the given time delay $\tau = 0.04$ s, the vehicle lateral acceleration a_y by Controller 1 is maintained in safe regions. However, with the given time delay $\tau = 0.04$ s, the vehicle lateral acceleration by Controller 2 is oscillating which means that the vehicle is in dangerous state. Figure 6.6 shows the global trajectory results with different controllers. It can be found that the path

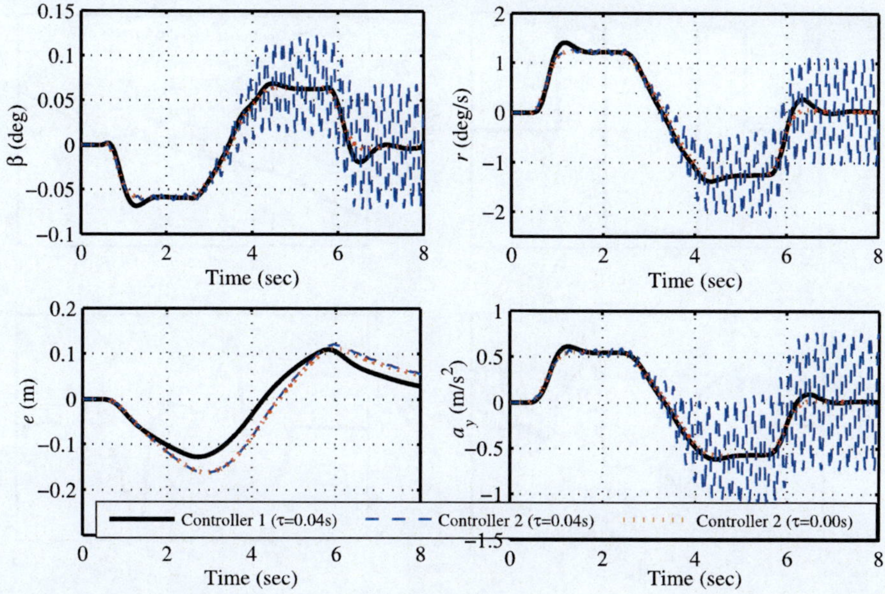

Fig. 6.5 The simulation results for the sideslip angle, yaw rate, acceleration, and tracking errors in the single-lane change maneuver

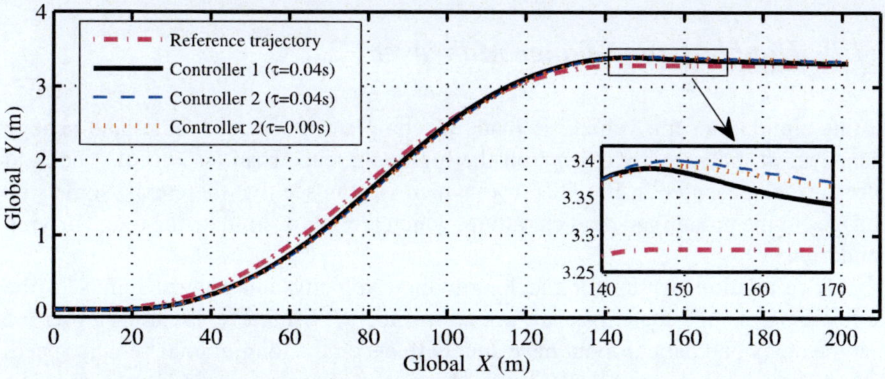

Fig. 6.6 Global trajectories of the AGV in the single-lane change maneuver

following effect is more accurate with fewer fluctuations by the Controller 1 than by the Controller 2. From Figs. 6.5 and 6.6, we can see that the vehicle path following and stability control are simultaneously completed, in the presence of the measurement delay, data dropout, parameters uncertainties, and external disturbances.

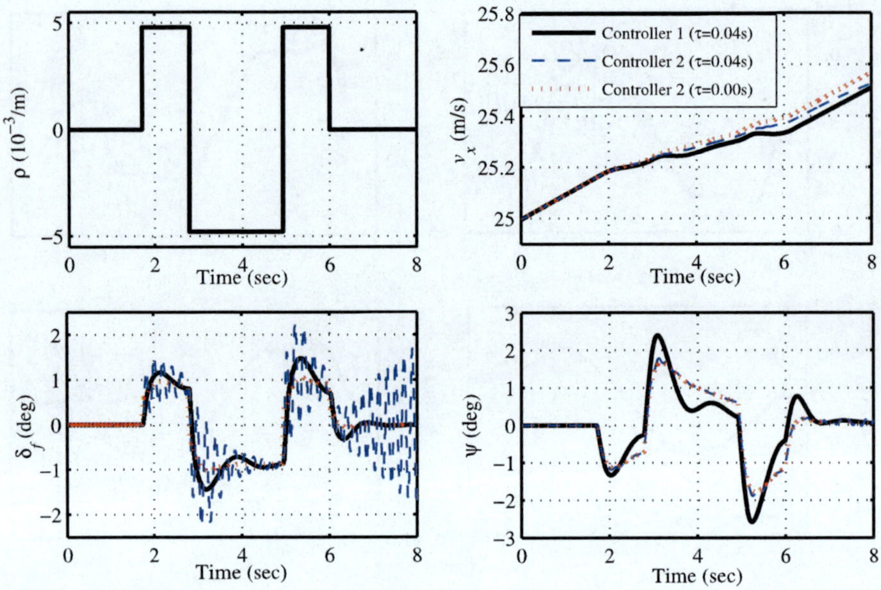

Fig. 6.7 The simulation results for the road curvature, longitudinal velocity, and control inputs in the double-lane change maneuver

6.4.2 Double-Lane Change Maneuver

In this simulation, the vehicle is made to complete a severe double-lane change maneuver at the speed of 25 m/s on the high-adherence road ($\mu = 0.6$). The road curvature is presented in Fig. 6.7, from which we can see that there exist some step changes in the predefined road curvature, which are set to simulate the tough driving conditions.

The simulation results for the longitudinal velocity and control inputs in the double-lane change maneuver are potted in Fig. 6.7. Similarly, it can be observed that the delay and data dropout have small effects on the longitudinal vehicle speed. The steering angle by using Controller 1 has smooth response and remains in the reasonable region, while the delay and data dropout will cause severe instability and oscillations in the steering angle by using Controller 2. In addition, when we introduce some step changes in the road curvature, it can be seen that the steering angle with delay and data dropout oscillates much earlier by using Controller 2 compared with the single-lane change simulation. Controller 2 ($\tau = 0.04$ s) even causes much larger overshoot and finally leads to the divergence of the controlled system. The heading errors using both of the Controller 1 and 2 are maintained in reasonable regions.

The simulation results for the sideslip angle, yaw rate, acceleration, and tracking errors in the single-lane change maneuver are presented in Fig. 6.8. Similarly, oscil-

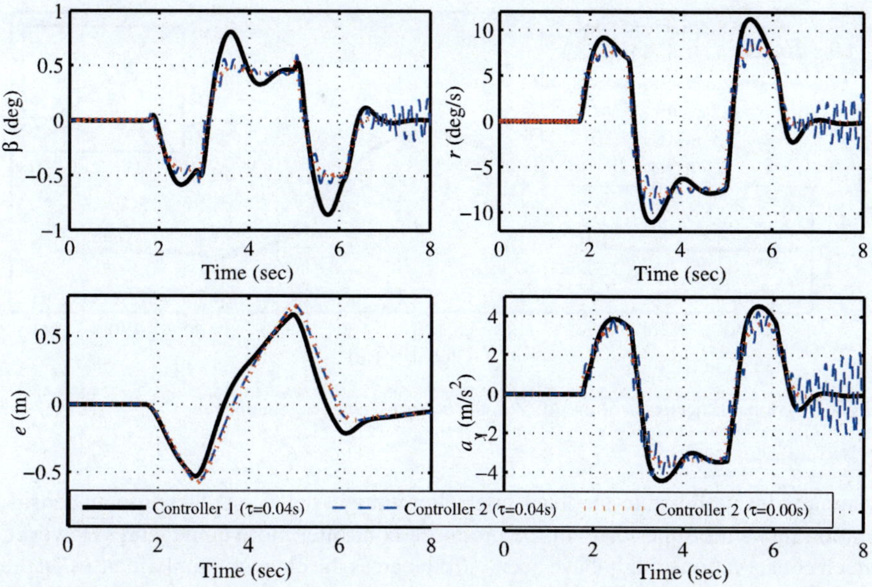

Fig. 6.8 The simulation results for the sideslip angle, yaw rate, acceleration, and tracking errors in the double-lane change maneuver

lations can be seen in vehicle sideslip angle β, yaw rate r, and lateral acceleration a_y, which indicate that the vehicle might be at the criticality of instability when there are delays and data dropouts. The global trajectories are shown in Fig. 6.9, and it is observed that the path following maneuver is completed more accurately by using Controller 1 than by using Controller 2. From the above figures, we can see that the AGV is strictly limited in the traffic lane which is marked by traffic cones. Moreover, the trajectory controlled by Controller 1 has larger distances to the traffic cone at 135 m, compared with that of Controller 2. This means that the vehicle can still be well controlled even on the severe driving conditions. Consequently, it is evident that the delay and data dropout can substantially be attenuated, and the effectiveness and robustness of the proposed controller is verified.

6.5 Conclusion

To handle the delay and data dropout in path following problem of an AGV, a robust $\mathcal{H}_\infty$ state-feedback controller is designed to control the vehicle to track the desired path and regulate the vehicle lateral dynamics. Both the parameter uncertainties and external disturbances are considered in the controller design to enhance the robustness of the proposed controller. Simulation results based on CarSim and Simulink platform verify the effectiveness of the proposed control approach. Further more,

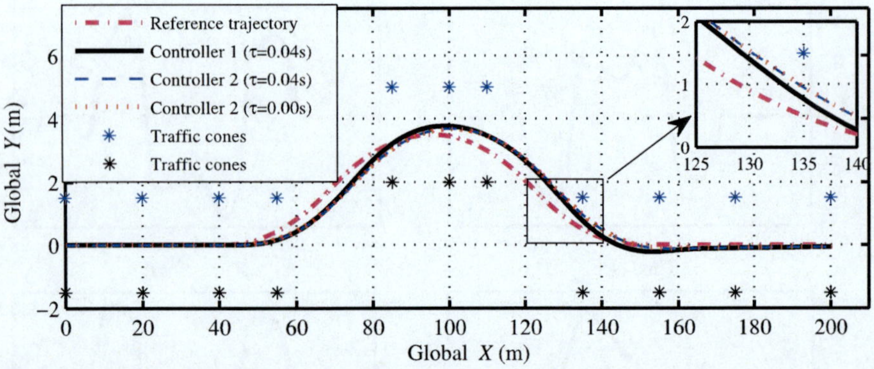

Fig. 6.9 Global trajectories of the AGV in the double-lane change maneuver

comparisons with a state-feedback controller without delay and data dropout consideration make the superiority of the proposed controller more convincing. However, much conservations might have been introduced as the chapter mainly focuses on the maximum time-invariant delays and maximum data dropouts. For the path following of AGVs, the delay and data dropout in real application might be time-varying or unknown, which is left for our the future studies.

Acknowledgements This chapter is from the previous work in [27], and some typos are corrected here.

References

1. M. Zakaria, H. Zamzuri, R. Mamat, S. Mazlan, A path tracking algorithm using future prediction control with spike detection for an autonomous vehicle robot. Int. J. Adv. Robot. Syst. **10**, 309–317 (2013)
2. Y. Chen and L. Li, *Advances in Intelligent Vehicles*, (Academic Press, 2014)
3. L. Consolini, C. Verrelli, Learning control in spatial coordinates for the path-following of autonomous vehicles. Automatica **50**(7), 1867–1874 (2014)
4. S. Yin, Z. Huang, Performance monitoring for vehicle suspension system via fuzzy positivistic C-Means clustering based on accelerometer measurements. IEEE-ASME Trans. Mechatron. **pp**(99), 1–8 (2015)
5. W. Sun, H. Gao, B. Yao, Adaptive robust vibration control of full-car active suspensions with electrohydraulic actuators. IEEE Trans. Control Syst. Technol. **21**(6), 2417–2422 (2013)
6. L. Li, Y. Zhang, C. Yang, X. Jiao, L. Zhanga, J. Song, Hybrid genetic algorithm-based optimization of power train and control parameters of plug-in hybrid electric bus. J. Frankl. Inst.-Eng. Appl. Math. **352**(3), 776–801 (2014)
7. H. Fang, L. Dou, J. Chen, R. Lenain, B. Thuilot, P. Martinet, Robust anti-sliding control of autonomous vehicles in presence of lateral disturbances. Control Eng. Pract. **19**(5), 468–478 (2011)
8. X. Zhu, G. Dong, D. Hu, Unified nonsingular tracking and stabilization controller design for unicycle-type wheeled mobile robots. Adv. Robot. **21**(5–6), 711–728 (2012)

9. S. Aouaouda, M. Chadli, H.R. Karimi, Robust static output-feedback controller design against sensor failure for vehicle dynamics. IET Contr. Theory Appl. **8**(9), 728–737 (2014)
10. M. Choi, S.B. Choi, Model predictive control for vehicle yaw stability with practical concerns. IEEE Trans. Veh. Technol. **63**(8), 3539–3548 (2014)
11. S.A. Arogeti, N. Berman, Path following of autonomous vehicles in the presence of sliding effects. IEEE Trans. Veh. Technol. **61**(4), 1481–1492 (2012)
12. J. Yoon, H. Peng, A cost-effective sideslip estimation method using velocity measurements from two GPS receiver. IEEE Trans. Veh. Technol. **63**(6), 2589–2599 (2014)
13. Z. Shuai, H. Zhang, J. Wang, J. Li, M. Ouyang, Combined AFS and DYC control of four-wheel-independent-drive electric vehicles over CAN network with time-varying delays. IEEE Trans. Veh. Technol. **63**(2), 591–602 (2014)
14. H. Gao, T. Chen, J. Lam, A new delay system approach to network-based control. Automatica **44**(1), 39–52 (2008)
15. H. Li, B. Chen, Q. Zhou, C. Lin, A delay dependent approach to robust $\mathcal{H}_\infty$ control for uncertain stochastic systems with state and input delays. Circuits Syst. Signal Process. **28**(1), 169–183 (2009)
16. M. Yu, L. Wang, T. Chu, F. Hao, An LMI approach to networked control systems with data packet dropout and transmission delays. in *Proceedings of IEEE Conference on Decision and Control* (2004), pp. 3545–3550
17. H. Zhang, Y. Shi, M. Liu, $\mathcal{H}_\infty$ step tracking control for networked discrete-time nonlinear systems with integral and predictive actions. IEEE Trans. Ind. Inform. **9**(1), 337–345 (2013)
18. J. Yu, Z. Deng, M. Yu, Y. Gao, Design of multiple controllers for networked control systems with delays and packet losses. Trans. I. Meas. Control **35**(6), 720–729 (2012)
19. J. Luo, J. Zhao, Robust $\mathcal{H}_\infty$ control for networked switched fuzzy systems with network-induced delays and packet dropout. Circuits Syst. Signal Process. **34**(2), 663–679 (2015)
20. Y. Wang, Q. Han, Quantitative analysis and synthesis for networked control systems with non-uniformly distributed packet dropouts and interval time-varying sampling periods. Int. J. Robust. Nonlinear Control **25**(2), 282–300 (2015)
21. H. Li, C. Wu, Z. Feng, Fuzzy dynamic output-feedback control of non-linear networked discrete-time system with missing measurements. IET Control Theory Appl. **9**(3), 327–335 (2015)
22. S. Srinivasan, R. Ayyagari, Advanced driver assistance system for AHS over communication links with random packet dropouts. Mech. Syst. Signal Pr. **49**(1–2), 53–62 (2014)
23. C. Lozoya, P. Martä, M. Velasco, J. Fuertes, E. Martän, Simulation study of a remote wireless path tracking control with delay estimation for an autonomous guided vehicle. Int. J. Adv. Manuf. Technol. **52**(5–8), 751–761 (2011)
24. R. Skjetne, T. Fossen, Nonlinear maneuvering and control of ships. in *Proc* (MTS/IEEE Oceans, Honolulu, HI, 2001), pp. 1808–1815
25. R. Wang, H. Zhang, J. Wang, Linear parameter-varying controller design for four wheel independently-actuated electric ground vehicles with active steering systems. IEEE Trans. Control Syst. Technol. **22**(4), 1281–1296 (2014)
26. H. Du, N. Zhang, F. Naghdy, Velocity-dependent robust control for improving vehicle lateral dynamics. Transport Res. C-Emer. **19**(3), 454–468 (2011)
27. R. Wang, H. Jing, C. Hu, F. Yan, N. Chen, Robust $\mathcal{H}_\infty$ path following control for autonomous ground vehicles with delay and data dropout. IEEE Trans. Intell. Transp. Syst. **17**(7), 2042–2050 (2016)

Chapter 7
Robust Lateral Motion Control of Four-Wheel Independently Actuated Electric Vehicles with Tire Force Saturation Consideration

Abstract The chapter presents a vehicle lateral-plane motion stability control approach for four-wheel independently actuated (FWIA) electric ground vehicles considering the tire force saturations. In order to deal with the possible modeling inaccuracies and parametric uncertainties, a linear-parameter-varying (LPV)-based robust $\mathcal{H}_\infty$ controller is designed to yield the desired external yaw moment. The lower-level controller operates the four in-wheel (or hub) motors such as the required control efforts can be satisfied. An analytical method without using the numerical optimization-based control-allocation algorithms is given to distribute the higher-level control efforts. The tire force constraints are also explicitly considered in the control-allocation design. Simulation results based on a high-fidelity, CarSim, full-vehicle model show the effectiveness of the proposed control approach.

7.1 Introduction

Compared to the internal combustion engine-powered vehicles, electric vehicles have several advantages in terms of energy efficiency, environmental friendliness, performance benefits, and so on [1, 2]. Among all the types of electric vehicles, the four-wheel independently actuated (FWIA) electric vehicle, in which each wheel is independently actuated with an in-wheel (or hub) motor, is an emerging configuration thanks to the actuation flexibility and fast but still precise torque responses of electric motors [1–4].

The actuator redundancy of FWIA electric vehicle makes the control of this type of vehicle rewarding but challenging. Many studies have been carried out on the vehicle motion control methods for improving the vehicle stability and maneuverability. However, most of these control algorithms are designed for conventional vehicle architectures, but not for the FWIA electric ground vehicles. Recently, a few lateral-plane motion control methods have been proposed for electric vehicles with reductant actuators. For example, a sliding control theory-based control method was proposed for a four-wheel driving and four-wheel steering vehicle in [5]; although an analytic solution of allocating the ground forces was designed, the proposed method relies on the assumption that the tire force model is accurate and the tire-road friction

H. Zhang et al., *Robust Gain-Scheduled Estimation and Control of Electrified Vehicles via LPV Technique*, Key Technologies on New Energy Vehicles,
https://doi.org/10.1007/978-981-19-8509-6_7

coefficient (TRFC) is known. Moreover, the tire force saturation was not considered in the control-allocation design. An online self-evolving fuzzy controller is proposed for autonomous mobile robots to follow a moving object on a desired distance with respect to the leader which is designed in [6], and the proposed controller works well without any pre-training and is capable of adapting the fuzzy rules. A direct yaw-moment control system for an FWIA electric vehicle was proposed in [7], a half-vehicle model which is a linear approximation of vehicle dynamics was used in the controller design, and the tire force saturation was not considered either. A braking control method for electric vehicle which was driven by independent front and rear motors was proposed in [8]; however, the vehicle stability controller design was not reported in the paper. A stability control method for four-wheel-driven hybrid electric vehicle was proposed in [9]. As the required torque split for yaw motion control was generated by the rear motor with an electro-hydraulic brake, the control problem in [9] is thus different from the one considered in this study.

As a typical overactuated system, FWIA electric vehicles usually adopt numerical optimization-based control-allocation algorithms such as daisy chaining, linear programming, nonlinear programming, fixed-point method, to distribute the higher-level control signals to the lower-level motors [10–14]. For example, accelerated fixed-point-based control-allocation methods were proposed in [12], general quadratic programming based on control allocation was proposed in [13], and rule-based control-allocation method was adopted in [14]. It is worthwhile to note that, however, the numerical optimization-based control-allocation methods usually have the drawback of high computational requirements, which may discourage their real-time implementations. Therefore, improving or replacing such methods with low-cost computing methods for ground vehicles is more desirable. Another difficulty encountered in the tire force allocation design for FWIA electric vehicles is that the tire longitudinal and lateral forces are coupled by nonlinear constraints due to the tire-road friction ellipse [15, 16]. For example, the tire force may become saturated if a sufficiently large motor control signal is applied in some extreme cases such as hard brake on a slippery road surface. Once the tire longitudinal force reaches its maximal value, further increasing of slip makes the tire work in the unstable range and the tire longitudinal force will decrease quickly. Moreover, locking/skidding tires no longer provide any grip on the road, and thus, the tire cornering forces transferred from the ground will be significantly reduced and consequently make the vehicle unsteerable or unstable. So the constraints of the tire forces should be explicitly considered in the tire force allocation designs.

This chapter considers the lateral-plane motion control of FWIA electric vehicles to improve the vehicle stability and maneuverability. The proposed control system consists of a higher-level controller and a lower-level controller. A linear-parameter-varying (LPV)-based robust $\mathcal{H}_\infty$ higher-level controller is proposed to attenuate the effects of modeling errors and possible disturbances. The lower-level controller allocates the control signals from the higher-level controller to the four wheels. The main contributions of this chapter lie in the following aspects.

- Uncertainties in both the tire cornering stiffness and vehicle parameters are simultaneously considered in the controller design, and physical limitations on the tire forces and the in-wheel motor power are also considered in the controller design.
- The numerical optimization-based control-allocation algorithms were not used to distribute higher-level control efforts to the four wheels in the lower-level control design. Instead, an analytical solution for allocating the ground forces is given to distribute the higher-level control efforts. The tire force constraints are considered in the control-allocation design, making the tires always work in the stable regions.
- As the wheel speed acceleration cannot be calculated by directly taking the derivative of the wheel speed signal due to the measurement noise, a tire force observer is designed to update the motor control signals such that the desired tire forces can be provided.

The rest of the chapter is organized as follows. Vehicle model considering tire force modeling error and vehicle parameter uncertainties is presented in Sect. 7.2. Controller designs including the robust $\mathcal{H}_\infty$ higher-level controller and lower-level control-allocation designs are described in Sect. 7.3. Simulation results based on a high-fidelity, CarSim, full-vehicle model are provided in Sect. 7.4 followed by conclusive remarks.

7.2 System Modeling

7.2.1 Vehicle Model

A schematic diagram of a vehicle model is shown in Fig. 7.1. Different from the conventional vehicle architectures, each wheel in an FWIA electric vehicle is independently driven by an in-wheel (or hub) motor. Therefore, an external yaw moment can be easily generated to regulate the vehicle yaw and lateral motions thanks to the fast and precise torque responses of electric motors. Ignoring the pitch and roll motions, the vehicle's handling dynamics in the yaw plane can be expressed as

Fig. 7.1 Schematic diagram of a vehicle planar motion model

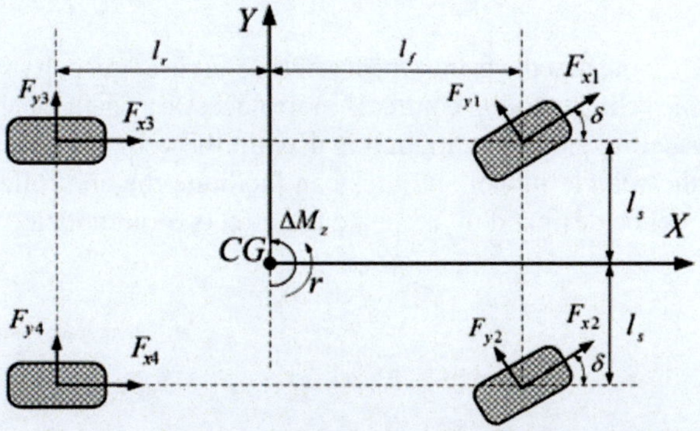

$$\begin{cases} \dot{\beta} = \frac{1}{mv_x}\left(F_{yf} + F_{yr}\right) - r, \\ \dot{r} = \frac{1}{I_z}\left(l_f F_{yf} - l_r F_{yr}\right) + \frac{1}{I_z}\Delta M_z, \end{cases} \tag{7.1}$$

where v_x is the vehicle longitudinal speed, β is the vehicle sideslip angle, r is the yaw rate, m and I_z are the vehicle mass and yaw inertia, respectively. F_{rf} and F_{yr} are the front and rear tire lateral forces, respectively. ΔM_z is the external yaw moment and is used to compensate for the vehicle yaw rate and slip angle tracking errors. As ΔM_z is generated with the longitudinal tire force difference between the left and right-side wheels, we have

$$\Delta M_z = \sum_{1}^{4} (-1)^i F_{\text{xi}} l_s, \tag{7.2}$$

where F_{xi} is the longitudinal force of the ith tire. The front and rear tire lateral forces F_{rf}, F_{yr} are functions of tire slip angles and can be modeled as

$$F_{\text{yf}} = C_f \beta_f, \quad F_{\text{yr}} = C_r \beta_r, \tag{7.3}$$

where C_f and C_r are the front and rear tire cornering stiffness values and β_f and β_r are the wheel slip angles which can be calculated as

$$\beta_f = \delta - \frac{l_f r}{v_x} - \beta, \quad \beta_r = \frac{l_r r}{v_x} - \beta, \tag{7.4}$$

with δ being the front-wheel steering angle. Based on Eqs. (7.3 and 7.4), the vehicle model (7.1) can be further written as

$$\dot{\beta} = -\frac{\left(C_f + C_r\right)\beta}{mv_x} + \frac{\left(C_r l_r - C_f l_f\right) r}{mv_x^2} - r + \frac{C_f}{mv_x}\delta, \tag{7.5}$$

$$\dot{r} = \frac{\left(C_r l_r - C_f l_f\right)\beta}{I_z} - \frac{\left(C_f l_f^2 + C_r l_r^2\right) r}{I_z v_x} + \frac{1}{I_z}\Delta M_z + \frac{C_f l_f}{I_z}\delta. \tag{7.6}$$

Note that the front-wheel steering angle δ is usually small at high vehicle speed. As the vehicle motion control is more necessary when the vehicle speed is high, the small steering angle assumption is used in the vehicle model. This assumption simplifies the vehicle model and thus can facilitate the controller design. A robust controller will be designed to attenuate the effects of unmodeled dynamics and disturbance.

7.2.2 Vehicle Model Considering Parameter Uncertainties

In the above section, the vehicle lateral-plane motions are modeled with the assumption that all the parameters including vehicle mass, yaw inertia, and tire cornering stiffness are precisely known. The parameter uncertainties in the vehicle model are also considered. Denote $x = [\beta, \gamma]^{\mathrm{T}}$, and the vehicle model can be rewritten as

$$\dot{x}(t) = Ax(t) + B_1u(t) + B_2w(t), \tag{7.7}$$

where $A = \begin{bmatrix} -\frac{C_f+C_r}{mv_x} & -1+\frac{C_rl_r-C_fl_f}{mv_x^2} \\ \frac{C_rl_r-C_fl_f}{I_z} & -\frac{C_fl_f^2+C_rl_r^2}{I_zv_x} \end{bmatrix}$, $B_1 = \begin{bmatrix} 0 \\ \frac{1}{I_z} \end{bmatrix}$, $B_2 = \begin{bmatrix} \frac{C_f}{mv_x} \\ \frac{C_fl_f}{I_z} \end{bmatrix}$, $w(t) = \delta(t), u(t) = \Delta M_z$.

Since the vehicle model is a nonlinear function, to facilitate the controller design, we convert the vehicle model into an LPV system. As the vehicle longitudinal speed v_x is time-varying but measurable, the auxiliary time-varying parameters are chosen as $\rho_1(t) = 1/V_x$, $\rho_2(t) = 1/V_x^2$. Denoting $\rho = [\rho_1, \rho_2]^{\mathrm{T}}$, we have

$$\dot{x}(t) = A(\rho)x(t) + B_1u(t) + B_2(\rho)w(t), \tag{7.8}$$

where $A(\rho) = \begin{bmatrix} -\frac{C_f+C_r}{m}\rho_1 & -1+\frac{C_rl_r-C_fl_f}{m}\rho_2 \\ \frac{C_rl_r-C_fl_f}{I_z} & -\frac{C_fl_f^2+C_rl_r^2}{I_z}\rho_1 \end{bmatrix}$, $B_2(\rho) = \begin{bmatrix} \frac{C_f\rho_1}{m} \\ \frac{C_fl_f}{I_z} \end{bmatrix}$.

The tire cornering stiffness can be affected by many factors such as the tire-road friction coefficient (TRFC), wear of the tires, and tire normal load [16]. The uncertainties in the tire cornering stiffness are considered in this chapter. Denote $C_{\max,k}$ and $C_{\min,k}$ as the maximal and minimal values of $C_k (k = f, r)$, respectively, and the tire cornering stiffness can be represented as

$$C_k = C_{k0} + N_k\tilde{C}_k, \tag{7.9}$$

where N_k are time-varying parameters and satisfy $|N_k(t)| < 1$ and $C_{k0}$is the preselected cornering stiffness and can be calculated as $C_{k0} = (C_{\max k} + C_{\min k})/2$ with $\tilde{C}_k = (C_{\max k} - C_{\min k})/2$. As the road conditions are usually uniform to the front and rear wheels, we assume

$$N_f = N_r. \tag{7.10}$$

Due to the payload change, the vehicle mass m is also a varying parameter and can be assumed to be bounded by its minimum value $m_{\min}$ and its maximum value $m_{\max}$. Therefore, the term $1/m$ can be represented by

$$\frac{1}{m} = \frac{1}{m_0} + N_m\frac{1}{\tilde{m}}, \tag{7.11}$$

where N_m is time-varying parameter satisfying $|N_m(t)| < 1,\ 1/m_0 = 2m_{\min}m_{\max}/(m_{\min} + m_{\max}),\ \tilde{m} = 2m_{\min}m_{\max}/(m_{\max} - m_{\min})$. Similarly, the uncertainty in the vehicle yaw inertia can be represented by

$$\frac{1}{I_z} = \frac{1}{I_{z0}} + N_{I_z}\frac{1}{\tilde{I}_z}, \tag{7.12}$$

where N_{I_z} satisfies $\left|N_{I_z}(t)\right| < 1,\ 1/I_{z0} = 2I_{\min}I_{\max}/(I_{\min} + I_{\max}),\ \tilde{I}_z = 2I_{\min}I_{\max}/(I_{\max} - I_{\min})$ with $I_{\min}$ and $I_{\max}$ being the minimum and maximum values of I_z, respectively. Note that we can assume that the following holds for vehicles:

$$N_m = N_{I_z}. \tag{7.13}$$

Based on Eqs. (7.9)–(7.13), C_k/m and C_k/I_z, $(k = f, r)$ can be rewritten as

$$\begin{aligned} \frac{C_k}{m} &= \frac{C_{ko}}{m_0} + \frac{\tilde{C}_k}{\tilde{m}} + N^\dagger\left(\frac{C_{ko}}{\tilde{m}} + \frac{\tilde{C}_k}{m_0}\right), \\ \frac{C_k}{I_z} &= \frac{C_{ko}}{I_{z0}} + \frac{\tilde{C}_k}{\tilde{I}_z} + N^\dagger\left(\frac{C_{ko}}{\tilde{I}_z} + \frac{\tilde{C}_k}{I_{z0}}\right), \end{aligned} \tag{7.14}$$

where $N^\dagger$ satisfies $\left|N^\dagger\right| < 1$. Then, we have

$$\begin{cases} A(\rho) = A^\dagger(\rho) + \tilde{A}(\rho)N_1, \\ B_1 = B_1^\dagger + \tilde{B}_1N_2, \\ B_2(\rho) = B_2^\dagger(\rho) + \tilde{B}_2(\rho)N_3, \end{cases} \tag{7.15}$$

with

$$\begin{aligned} A^\dagger(\rho) &= \begin{bmatrix} -\left(\frac{C_{f0}+C_{r0}}{m_0} + \frac{\tilde{C}_f+\tilde{C}_r}{\tilde{m}}\right)\rho_1 & -1 + \left(\frac{C_{r0}l_r - C_{f0}l_f}{m_0} + \frac{\tilde{C}_rl_r - \tilde{C}_fl_f}{\tilde{m}}\right)\rho_2 \\ \frac{C_{r0}l_r - C_{f0}l_f}{I_{z0}} + \frac{\tilde{C}_rl_r - \tilde{C}_fl_f}{\tilde{I}_z} & -\left(\frac{C_{f0}l_f^2 + C_{r0}l_r^2}{I_{z0}} + \frac{\tilde{C}_fl_f^2 + \tilde{C}_rl_r^2}{\tilde{I}_z}\right)\rho_1 \end{bmatrix}, \\ \tilde{A}(\rho) &= \begin{bmatrix} -\left(\frac{C_{f0}+C_{r0}}{\tilde{m}} + \frac{\tilde{C}_f+\tilde{C}_r}{m_0}\right)\rho_1 & \left(\frac{C_{r0}l_r - C_{f0}l_f}{\tilde{m}} + \frac{\tilde{C}_rl_r - \tilde{C}_fl_f}{m_0}\right)\rho_2 \\ \frac{C_{r0}l_r - C_{f0}l_f}{\tilde{I}_z} + \frac{\tilde{C}_rl_r - \tilde{C}_fl_f}{I_{z0}} & -\left(\frac{C_{f0}l_f^2 + C_{r0}l_r^2}{\tilde{I}_z} + \frac{\tilde{C}_fl_f^2 + \tilde{C}_rl_r^2}{I_{z0}}\right)\rho_1 \end{bmatrix}, \\ N_1 &= \begin{bmatrix} N^\dagger & 0 \\ 0 & N^\dagger \end{bmatrix},\ N_2 = N_m,\ N_3 = N^\dagger, \\ B_1^\dagger &= \begin{bmatrix} 0 \\ \frac{1}{I_{z0}} \end{bmatrix},\quad \tilde{B}_1 = \begin{bmatrix} 0 \\ \frac{1}{\tilde{I}_z} \end{bmatrix},\ B_2^\dagger(\rho) = \begin{bmatrix} \left(\frac{C_{f0}}{m_0} + \frac{\tilde{C}_f}{\tilde{m}}\right)\rho_1 \\ \frac{C_{f0}l_f}{I_{z0}} + \frac{\tilde{C}_fl_f}{\tilde{I}_z} \end{bmatrix}, \\ \tilde{B}_2(\rho) &= \begin{bmatrix} \left(\frac{C_{f0}}{\tilde{m}} + \frac{\tilde{C}_f}{m_0}\right)\rho_1 \\ \frac{C_{f0}l_f}{\tilde{I}_z} + \frac{\tilde{C}_fl_f}{I_{z0}} \end{bmatrix}. \end{aligned} \tag{7.16}$$

Noting that $\left|N^\dagger\right| < 1$ and $|N_m| < 1$, we have $|N_2| < 1$, $|N_3| < 1$, and $N_1^{\mathrm{T}}N_1 < I$.

7.3 Control System Design

Multi-layer control architecture is more effective and flexible in the controller design for overactuated ground vehicles [5, 12]. In this chapter, a hierarchical control structure as Fig. 7.2 shows is designed. In the higher-level controller, a robust $\mathcal{H}_\infty$ controller is proposed to determine the desired yaw moment. In the lower-level controller, a control-allocation method with tire force saturation consideration is implemented to allocate the desired yaw moment to each wheel.

7.3.1 *Higher-Level Controller Design*

Robust controllers based on the fuzzy or LPV systems can effectively deal with the system parametric uncertainties and external disturbances [6, 17–19]. To achieve the desired control performance, we propose the design of a robust gain-scheduling state-feedback higher-level controller based on the LPV method in this chapter. The higher-level controller is designed based on the vehicle model (7.8). Denoting the reference vector as $x_r = [\beta_r, r_r]^{\mathrm{T}}$ with β_r and r_r being the desired vehicle sideslip angle and yaw rate, respectively, the model (7.8) can be rewritten as

$$\begin{cases} \dot{\xi}(t) = A(\rho)\xi(t) + B_1 u(t) + B_2(\rho)w(t) + d(t) \\ z(t) = C\xi(t), \end{cases} \tag{7.17}$$

where $z(t)$ is the controlled output, $\xi(t) = x(t) - x_r(t)$ is the tracking error, $d(t) = -\dot{x}_r$ is the disturbance term. Note that the external disturbances such as the effects of crosswind and tire rolling resistances can also be put into this disturbance term, if the external disturbances are considered, $d(t)$ can be reformulated as $d(t) = d_{\mathrm{ext}}(t) - \dot{x}_r(t)$, where $d_{\mathrm{ext}}(t)$ is the external disturbances. Suppose that the controller can be written as

$$u(t) = K(\rho)\xi(t), \tag{7.18}$$

with $K(\rho)$ being the controller gain to be designed, and the system model (7.17) can be rewritten as

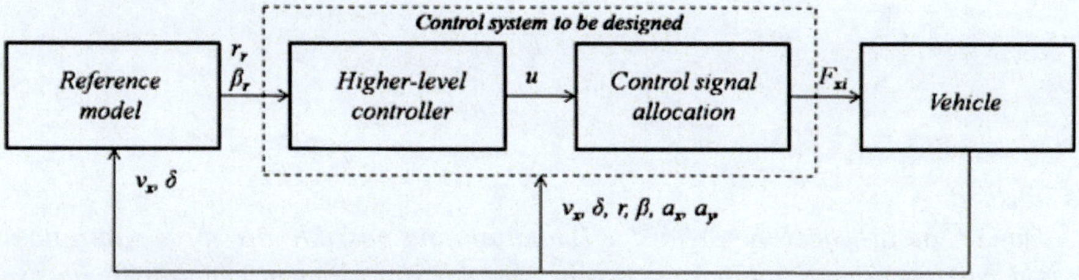

Fig. 7.2 Control structure for an FWIA electric vehicle

$$\begin{cases} \dot{\xi}(t) = A_c(\rho)\xi(t) + B_2(\rho)w(t) + d(t) \\ z(t) = C\xi(t), \end{cases} \tag{7.19}$$

where

$$A_c(\rho) = A(\rho) + B_1 K(\rho). \tag{7.20}$$

Note that there are two disturbance terms $w(t)$ and $d(t)$ involved in Eq. (7.19). These two disturbances are not treated equally here since they are different in physical meanings. The following indices are thus chosen to attenuate the effects of these two disturbances to the controlled outputs

$$\|z(t)\|_2 < \gamma_1 \|d(t)\|, \quad \|z(t)\|_2 < \gamma_2 \|w(t)\|. \tag{7.21}$$

The control objective is to design a gain-scheduled feedback controller such that the indices are satisfied under zero initial conditions. To deal with the uncertainties and external disturbances, we first introduce the following two lemmas.

Lemma 7.1 *[20] Given two positive constants γ_1 and γ_2, considering the closed-loop system in Eq. (7.19), a state-feedback controller exists such that the (extended) bounded real lemma condition holds for some L_2 performance level γ_1 and γ_2 if and only if there exists a symmetric positive-definite matrix P satisfying the following conditions:*

$$\begin{bmatrix} A_c^{\mathrm{T}}(\rho)P + PA_c(\rho) & P & C^{\mathrm{T}} \\ * & -\gamma_1^2 I & 0 \\ * & * & -I \end{bmatrix} < 0, \tag{7.22}$$

$$\begin{bmatrix} A_c^{T}(\rho)P + PA_c(\rho) & PB_2(\rho) & C^{\mathrm{T}} \\ * & -\gamma_2^2 I & 0 \\ * & * & -I \end{bmatrix} < 0. \tag{7.23}$$

Lemma 7.2 *[21] Let $\Theta = \Theta^{\mathrm{T}}$, L and E be real matrices with compatible dimensions, and $N^{\mathrm{T}}(t)$ be time-varying and satisfy $N^{\mathrm{T}}(t)N(t) < I$, then the following condition:*

$$\Theta + LN(t)E + E^{\mathrm{T}}N(t)L^{\mathrm{T}} < 0, \tag{7.24}$$

holds if and only if there exists a positive scalar $\varepsilon > 0$ such that

$$\begin{bmatrix} \Theta & \varepsilon L & E^{\mathrm{T}} \\ * & -\varepsilon I & 0 \\ * & * & -\varepsilon I \end{bmatrix} < 0, \tag{7.25}$$

is satisfied.

Due to the in-wheel motor torque limitation and possible tire force saturation, it is necessary to mention that the actual external yaw moment generated with tire force difference between the two sides of the vehicles is constrained. The control input constraint should also be considered in the controller design. In order to deal with the possible control input saturation, we introduce the following lemma.

Lemma 7.3 *[22]. Consider the closed-loop system in Eq. (7.19), given the positive symmetric matrix P. Then, for all the input constraint, $\|u(t)\|^2 \leq u_{\max}$ can be ensured if there exists a positive scalar ξ satisfying*

$$\begin{bmatrix} -u_{\max}^2 I & H(\rho) \\ H^{\mathrm{T}}(\rho) & -\frac{1}{\xi} Q \end{bmatrix} < 0. \tag{7.26}$$

Now we are in the position to propose the gain-scheduled controller design method. Based on the aforementioned lemmas, we introduce the following theorem.

Theorem 7.1 *Considering the closed-loop system in Eq. (7.19) under the constraint input $\|u(t)\|^2 \leq u_{\max}$, a state-feedback controller exists such that the $\mathcal{H}_\infty$ performance index in (7.21) holds, if there exist scalars $\varepsilon_i(\rho) > 0 (i = 1, \ldots, 5)$, a matrix H, and a symmetric positive-definite matrix Q satisfying the following conditions:*

$$\begin{bmatrix} -u_{\max}^2 I & H(\rho) \\ H^{\mathrm{T}}(\rho) & -\frac{1}{\xi} Q \end{bmatrix} < 0, \tag{7.27}$$

$$\begin{bmatrix} \Pi(\rho) & I & QC^{\mathrm{T}} & \varepsilon_1(\rho)\tilde{B}_1 & H^{\mathrm{T}}(\rho) & \varepsilon_2(\rho)\tilde{A}(\rho) & Q \\ * & -\gamma_1^2 I & 0 & 0 & 0 & 0 & 0 \\ * & * & -I & 0 & 0 & 0 & 0 \\ * & * & * & -\varepsilon_1(\rho)I & 0 & 0 & 0 \\ * & * & * & * & -\varepsilon_1(\rho)I & 0 & 0 \\ * & * & * & * & * & -\varepsilon_2(\rho)I & 0 \\ * & * & * & * & * & * & -\varepsilon_2(\rho)I \end{bmatrix} < 0, \tag{7.28}$$

$$\begin{bmatrix} \Pi(\rho) & B_2^{\dagger}(\rho) & QC^{\mathrm{T}} & \varepsilon_3(\rho)\tilde{B}_1 & H^{\mathrm{T}}(\rho) & \varepsilon_4(\rho)\tilde{A}(\rho) & Q & \varepsilon_5(\rho)\tilde{B}_2(\rho) & 0 \\ * & -\gamma_2^2 I & 0 & 0 & 0 & 0 & 0 & 0 & I \\ * & * & -I & 0 & 0 & 0 & 0 & 0 & 0 \\ * & * & * & -\varepsilon_3(\rho)I & 0 & 0 & 0 & 0 & 0 \\ * & * & * & * & -\varepsilon_3(\rho)I & 0 & 0 & 0 & 0 \\ * & * & * & * & * & -\varepsilon_4(\rho)I & 0 & 0 & 0 \\ * & * & * & * & * & * & -\varepsilon_4(\rho)I & 0 & 0 \\ * & * & * & * & * & * & * & -\varepsilon_5(\rho)I & 0 \\ * & * & * & * & * & * & * & * & -\varepsilon_5(\rho)I \end{bmatrix} < 0, \tag{7.29}$$

where $\Pi(\rho) = QA^{\dagger}(\rho)^{\mathrm{T}} + A^{\dagger}(\rho)Q + (H(\rho)B_1^{\dagger})^{\mathrm{T}} + B_1^{\dagger}H(\rho)$. Moreover, $H(\rho) = K(\rho)Q$.

Proof Pre- and post-multiplying Eq. (7.26) by diag$\{I, Q\}$ and its transpose with $Q = P^{-1}$, respectively, we can readily obtain the equivalent condition (7.27), which ensures the control input constraint. As $A_c(\rho) = A(\rho) + B_1K$, $B_1 = B_1^{\dagger} + \tilde{B}_1 N_2$, we have

$$
\begin{aligned}
&A_c^{\mathrm{T}}(\rho)P + PA_c(\rho)\\
&= (A(\rho) + B_1K)^{\mathrm{T}}P + P(A(\rho) + B_1K)\\
&= \left(A(\rho) + \left(B_1^{\dagger} + \tilde{B}_1N_2\right)K\right)^{\mathrm{T}}P + P\left(A(\rho) + \left(B_1^{\dagger} + \tilde{B}_1N_2\right)K\right)\\
&= A(\rho)^{\mathrm{T}}P + PA(\rho) + PB_1^{\dagger}K + P\tilde{B}_1N_2K + (PB_1^{\dagger}K + P\tilde{B}_1N_2K)^{\mathrm{T}}.
\end{aligned}
\tag{7.30}
$$

Therefore, it follows from Lemma 7.2 that the following condition holds:

$$
\begin{bmatrix}
\begin{array}{l}A(\rho)^{\mathrm{T}}P + PA(\rho)+\\(PB_1^{\dagger}K)^{\mathrm{T}} + PB_1^{\dagger}K\end{array} & P & C^{\mathrm{T}} & \varepsilon_1(\rho)P\tilde{B}_1 & K^{\mathrm{T}}(\rho)\\
* & -\gamma_1^2 I & 0 & 0 & 0\\
* & * & -I & 0 & 0\\
* & * & * & -\varepsilon_1(\rho)I & 0\\
* & * & * & * & -\varepsilon_1(\rho)I
\end{bmatrix} < 0. \tag{7.31}
$$

As $A(\rho) = A^{\dagger}(\rho) + \tilde{A}N_1$, we have

$$
\begin{aligned}
&A(\rho)^{\mathrm{T}}P + PA(\rho) + (PB_1^{\dagger}K)^{\mathrm{T}} + PB_1^{\dagger}K\\
&= \left(A^{\dagger}(\rho) + \tilde{A}(\rho)N_1\right)^{\mathrm{T}}P + P\left(A^{\dagger}(\rho) + \tilde{A}(\rho)N_1\right) + (PB_1^{\dagger}K)^{\mathrm{T}} + PB_1^{\dagger}K\\
&= A^{\dagger}(\rho)^{\mathrm{T}}P + PA^{\dagger}(\rho) + (PB_1^{\dagger}K)^{\mathrm{T}} + PB_1^{\dagger}K + N_1\tilde{A}^{\mathrm{T}}(\rho)P + P\tilde{A}(\rho)N_1
\end{aligned}
\tag{7.32}
$$

Repeating the above process, we have

$$
\begin{bmatrix}
\begin{array}{l}A^{\dagger}(\rho)^{\mathrm{T}}P + PA^{\dagger}(\rho)\\+(PB_1^{\dagger}K)^{\mathrm{T}} + PB_1^{\dagger}K\end{array} & P & C^{\mathrm{T}} & \varepsilon_1(\rho)P\tilde{B}_1 & K^{\mathrm{T}}(\rho) & \varepsilon_2(\rho)P\tilde{A} & I\\
* & -\gamma_1^2 I & 0 & 0 & 0 & 0 & 0\\
* & * & -I & 0 & 0 & 0 & 0\\
* & * & * & -\varepsilon_1(\rho)I & 0 & 0 & 0\\
* & * & * & * & -\varepsilon_1(\rho)I & 0 & 0\\
* & * & * & * & * & -\varepsilon_2(\rho)I & 0\\
* & * & * & * & * & * & -\varepsilon_2(\rho)I
\end{bmatrix} < 0 \tag{7.33}
$$

Performing a congruence transformation with diag$\{Q, I, I, I, I, I, I\}$to the above inequality, we can obtain Eq. (7.28). The inequality in Eq. (7.29) can be proved in a similar way. This finishes the proof. □

As ρ is a time-varying parameter vector, Theorem 7.1 cannot be used directly. Notice that the system matrices $A_C(\rho)$ and $B_2(\rho)$ are linearly dependent on the defined time-varying parameters ρ, which means $A_C(\rho)$ and $B_2(\rho)$ can be written as

$$
A_c(\rho) = \sum_{i=1}^{4}\theta_i(\rho)A_{ci}(\varpi_i), \quad B_2(\rho) = \sum_{i=1}^{4}\theta_i(\rho)B_2(\varpi_i) \tag{7.34}
$$

where ϖ_i are the vertices of the polytope formed by all the extremities of each varying parameter ρ, $and\theta_i$ are defined as

$$\theta_i(\rho) = \frac{|\rho_1 - \varphi(\varpi_i)_1 \| \rho_2 - \varphi(\varpi_i)_2|}{|\rho_{1\max} - \rho_{1\min} \| \rho_{2\max} - \rho_{2\min}|} \tag{7.35}$$

with $\varphi(\varpi_i)_l (l = 1, 2)$ can be calculated as

$$\varphi(\varpi_i)_l = \begin{cases} \rho_{l\max} & \text{if } (\varpi_i)_l = \rho_{l\min} \\ \rho_{l\min} & \text{otherwise} \end{cases} \tag{7.36}$$

Note that $\theta_i(\rho)$ satisfy $\theta_i(\rho) = 1$. Therefore, Theorem 7.1 can be formulated to the following theorem which projects the results in Theorem 7.1 to each vertex of the polytope.

Theorem 7.2 *Considering the closed-loop system in Eq.(7.19) under the constraint input $\|u(t)\|^2 \le u_{\max}$, a state-feedback controller exists such that the $\mathcal{H}_\infty$ performance index in Eq.(7.21) holds, if there exist scalars $\varepsilon_{pi}(\rho) > 0 (i = 1, \ldots, 4)$, matrices $H_i(i = 1, 2, 3, 4)$, and a symmetric positive-definite matrix Q satisfying f:*

$$\begin{cases} \Phi_{1i} < 0 \\ \Phi_{2ij} + \Phi_{2ji} < 0 \\ \Phi_{3ij} + \Phi_{3ji} < 0 \end{cases} \tag{7.37}$$

for $1 \le i < j \le 4$. Here,

$$\Phi_{1i} = \begin{bmatrix} -u_{\max}^2 I & H_i(\rho) \\ H_i^{\mathrm{T}}(\rho) & -\frac{1}{\zeta} Q \end{bmatrix},$$

$$\Phi_{2\mathrm{ij}} = \begin{bmatrix} \Pi_{i,j} & I & QC^{\mathrm{T}} & \varepsilon_{1,i}\tilde{B}_1 & H_i^{\mathrm{T}} & \varepsilon_{2,i}\tilde{A}_j & Q \\ * & -\gamma_1^2 I & 0 & 0 & 0 & 0 & 0 \\ * & * & -I & 0 & 0 & 0 & 0 \\ * & * & * & -\varepsilon_{1,i} I & 0 & 0 & 0 \\ * & * & * & * & -\varepsilon_{1,i} I & 0 & 0 \\ * & * & * & * & * & -\varepsilon_{2,i} I & 0 \\ * & * & * & * & * & * & -\varepsilon_{2,i} I \end{bmatrix},$$

$$\Phi_{3\mathrm{ij}} = \begin{bmatrix} \Pi_{\mathrm{ij}} & B_{2i}^{\dagger} & QC_i^{\mathrm{T}} & \varepsilon_{3i}\tilde{B}_{1j} & H_i^{\mathrm{T}} & \varepsilon_{4i}\tilde{A}_j & Q & \varepsilon_{5i}\tilde{B}_{2j} & 0 \\ * & -\gamma_2^2 I & 0 & 0 & 0 & 0 & 0 & 0 & I \\ * & * & -I & 0 & 0 & 0 & 0 & 0 & 0 \\ * & * & * & -\varepsilon_{3i} I & 0 & 0 & 0 & 0 & 0 \\ * & * & * & * & -\varepsilon_{3i} I & 0 & 0 & 0 & 0 \\ * & * & * & * & * & -\varepsilon_{4i} I & 0 & 0 & 0 \\ * & * & * & * & * & * & -\varepsilon_{4i} I & 0 & 0 \\ * & * & * & * & * & * & * & -\varepsilon_{5i} I & 0 \\ * & * & * & * & * & * & * & * & -\varepsilon_{5i} I \end{bmatrix}.$$

Moreover, the controller gain can be calculated via the following expression:

$$K_i = H_i Q^{-1}. \tag{7.38}$$

Similar proof of this theorem can be found in [23–25], so the proof is omitted here. The final controller gain can be calculated as

$$K(\rho) = \sum_{i=1}^{4} \theta_i K_i. \tag{7.39}$$

In the control gain matrix $K(\rho)$ calculation, the parameter $_i$ is calculated online with Eq.(7.35), while K_i are calculated offline with Eq.(7.38), and thus the computation requirement on the proposed higher-level controller is low. Note that there are two performance indices γ_1 and γ_2 in Eq. (7.21); in the implementation, γ_2 is optimized, while γ_1 is constrained under a prescribed level, and the prescribed value of γ_1 is manually chosen by checking the control performance.

7.3.2 *Lower-Level Controller Design*

In the above section, the higher-level controller which yields the desired external yaw moment is designed. The lower-level controller to be design in this section operates the four in-wheel motors such as the control requirements from the higher-level controller can be satisfied. An analytical solution is given to distribute the higher-level control effort without using the numerical optimization-based control-allocation algorithms.

7.3.2.1 Cost Function Definition

In order to control the vehicle lateral motion, the external yaw moment given by the higher-level control should be generated with the help of the tire force difference between the two sides of the vehicle, that is, Eq. (7.2) should be satisfied. Since the vehicle longitudinal speed is another state that needs to be controlled, the following constraint should be considered in the tire force allocation design as well:

$$\sum_{i=1}^{4} F_{\text{xi}} = F_D \tag{7.40}$$

where F_D is the driver's desired total driving/braking forces which can be calculated from the vehicle accelerator/brake pedal positions. Also, in order to maximize the tire force usage, it is better to make the generated longitudinal tire force be proportional to the tire normal load F_{zi}, i.e.,

$$\frac{F_{x1}}{F_{z1}} = \frac{F_{x2}}{F_{z2}} = \frac{F_{x3}}{F_{z3}} = \frac{F_{x4}}{F_{z4}} \tag{7.41}$$

where the tire normal force F_{zi} can be calculated as

$$\begin{aligned} F_{z1} &= \frac{mgl_r}{2l_r+2l_f} - \frac{mh_{CG}a_x}{2l_r+2l_f} - \frac{mh_{CG}a_yl_r}{2l_r+2l_f}, \\ F_{z2} &= \frac{mgl_r}{2l_r+2l_f} - \frac{mh_{CG}a_x}{2l_r+2l_f} + \frac{mh_{CG}a_yl_r}{2l_r+2l_f}, \\ F_{z3} &= \frac{mgl_r}{2l_r+2l_f} + \frac{mh_{CG}a_x}{2l_r+2l_f} - \frac{mh_{CG}a_yl_r}{2l_r+2l_f}, \\ F_{z4} &= \frac{mgl_r}{2l_r+2l_f} + \frac{mh_{CG}a_x}{2l_r+2l_f} + \frac{mh_{CG}a_yl_r}{2l_r+2l_f}, \end{aligned} \tag{7.42}$$

where h_{CG} is the height of the mass center of gravity and a_x and a_y are the vehicle longitudinal and lateral accelerations, respectively. As a_x and a_y can be accurately measured with an accelerometer on the vehicle, $F_{zi}/\sum_{i=1}^{4} F_{zi}$ are assumed to be known. Note that if Eq.(7.41) holds, the tire longitudinal force can be represented by

$$F_{\mathrm{xi}} = \frac{F_{zi}}{\sum_{i=1}^{4} F_{\mathrm{zi}}} F_D. \tag{7.43}$$

Based on the constraints (7.2), (7.40), and (7.43), the cost function for allocating the four tire forces is defined as

$$J = \frac{1}{2}\left(F_x^{\mathrm{T}} W^{\dagger} F_x + (F_x - cF_D)^{\mathrm{T}} W^{*} (F_x - cF_D) + \left(BF_x - u^{\dagger}\right)^{\mathrm{T}} R \left(BF_x - u^{\dagger}\right)\right) \tag{7.44}$$

where $R = \mathrm{diag}\{r_1, r_2\}$, $W^{\dagger} = \mathrm{diag}\left\{w_1^{\dagger}, w_1^{\dagger}, w_3^{\dagger}, w_4^{\dagger}\right\}$, and $W^{*} = \mathrm{diag}\left\{w_1^{*}, w_1^{*}, w_{\frac{*}{3}}, w_{\frac{1}{4}}^{*}\right\}$ are the weighting factors, $F_x = [F_{x1}, F_{x2}, F_{x3}, F_{x4}]^{\mathrm{T}}$ is the tire force vector, $u^{\dagger} = [F_D u]^{\mathrm{T}}$ with u being the desired yaw moment required by the higher-level controller, $c = \left[c_1\ c_2\ c_3\ c_4\right]^{\mathrm{T}}$ with $c_i = F_{\mathrm{ij}}/\sum_{i=1}^{4} F_{ij}$. B is the control effectiveness matrix, and it can be written as based on Eqs. (7.2) and (7.40) as follows:

$$B = \begin{bmatrix} 1 & 1 & 1 & 1 \\ -l_s & l_s & -l_s & l_s \end{bmatrix}. \tag{7.45}$$

By adding the term $F_x^{\mathrm{T}} W^{\dagger} F_x$ in the cost function, the power consumption of in-wheel motors is also optimized.

7.3.2.2 Tire Force Allocation Considering Tire Force Saturations

It can be observed from the cost function definition (7.44) that

$$\begin{aligned}\frac{\partial J}{\partial F_x} &= F_x^{\mathrm{T}} W^{\dagger} + F_x^{\mathrm{T}} W^{*} + F_x^{\mathrm{T}} B^{\mathrm{T}} R B - B^{\mathrm{T}} R u^{\dagger} + c W^{*} F_D \\ &= F_x^{\mathrm{T}} \left(W^{\dagger} + W^{*} + B^{\mathrm{T}} R B\right) - \left(B^{\mathrm{T}} R u^{\dagger} + c W^{*} F_D\right),\end{aligned} \tag{7.46}$$

and

$$\frac{\partial^2 J}{\partial^2 F_x} = W^{\dagger} + W^{*} + B^{\mathrm{T}} R B. \tag{7.47}$$

As $B^{\mathrm{T}} R B > 0$ and the weighting factors satisfy $W^{\dagger} > 0$ and $W^{*} > 0$, one can claim $\partial^2 J / \partial^2 F_x > 0$, which implies that the objective function J has a global minimum with the minimizing F_{x0} given by

$$F_{x0} = \left(W^{\dagger} + W^{*} + B^{T} R B\right)^{-1} \left(B^{\mathrm{T}} R u^{\dagger} + c W^{*} F_D\right). \tag{7.48}$$

In normal driving conditions, the desired tire force F_{x0} given by the above equation can be generated. However, if a sufficiently large motor control signal is sent to an in-wheel motor or when a hard brake is applied on a low-μ road, the tire longitudinal forces may become saturated. Once a tire longitudinal force reaches its maximal value, further increasing of the motor driving/braking torque of the saturated tire will make the tire spin/skid, and the tire force will decrease quickly. Moreover, the tire longitudinal force saturation will also significantly decrease the tire lateral force, making the vehicle become unsteerable or unstable. So the tire slip ratios should be limited depending on the tire slip angles.

It can be observed from Eq.(7.48) that the constraint violations of a tire force can be discouraged by choosing a larger weighting factor $w_i^{\dagger}$ for this wheel. So one can limit the tire force by operating the weighting factor for this wheel. Inspired by the concept of slip circle [26] which is plotted in Fig. 7.3, the weighting factors wi in the cost function (7.44) can be chosen as

$$w_i^{\dagger} = \frac{w_0}{\psi_i^{\kappa}}, \tag{7.49}$$

where $w_0 > 0$ is preselected initial weighting factor and $\kappa > 0$ is a constant which is used to determine the shape of the weighting factor. Denoting s_i^{peak} as the tire slip ratio where the maximum longitudinal force occurs at zero slip angle and β_i^{peak} as the tire slip angle corresponding to the maximum lateral force at zero slip ratio, ψ_i can be calculated as

$$\psi_i = \sqrt{1 - \bar{\beta}_i^2} - |\bar{s}_i|, \tag{7.50}$$

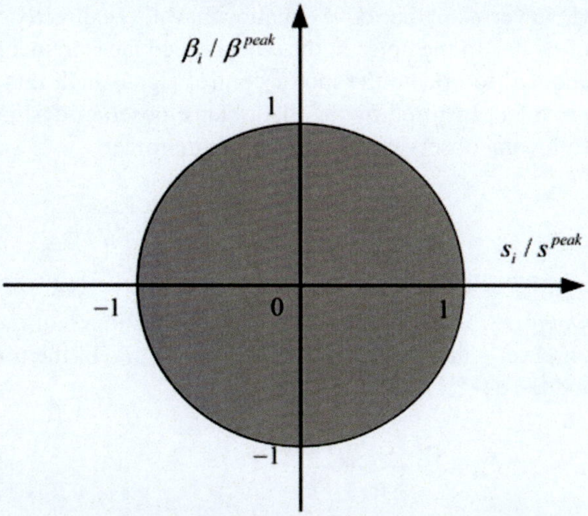

Fig. 7.3 Diagram of the tire slip circle (tire stable region in gray)

where $\bar{s}_i = s_i / s_i^{\text{peak}}$ and $\bar{\beta}_i = \beta_i / \beta_i^{\text{peak}}$ are the normalized tire slip ratios and tire slip angles, respectively. Note that the slip circle can be written as $\bar{s}_i^2 + \bar{\beta}_i^2 = 1$. At a given slip angle, if the normalized tire slip ratio of a wheel moves close to the largest possible values $\sqrt{1 - \bar{\beta}_i^2}$, the weighting factor for this wheel $w_i^{\dagger}$ becomes large. In this way, the tire slip ratio can always be limited within the stable region.

7.3.2.3 In-Wheel Motor Control Signal Generation

An in-wheel motor torque controller is designed in this section such that the desired tire force F_{xi} can be provided. It is known that the in-wheel motor torque T_i can be represented by $T_i = k_i u_{\text{mi}}$, with u_{mi} and k_i being the torque control signal and control gain for the ith in-wheel motor and motor driver pair, respectively [1, 4]. Thus, the rotational dynamics of a wheel with an in-wheel motor can be written as

$$I\dot{\omega}_i = k_i u_{\text{mi}} - F_{\text{xi}} R_T, \tag{7.51}$$

where I is the wheel moment of inertia, R_T is the tire effective rolling radius, and ω_i is the tire rotational speed. Note that the in-wheel motor control gain k_i is a constant which can be calibrated with experimental data [1].

If the wheel rational speed acceleration $\dot{\omega}_i$ is known, the ideal motor control signal for each motor can be directly calculated from the wheel dynamics (7.51) as

$$u_{\text{mi}}^* = \frac{I\dot{\omega}_i + R_T F_{\text{xi}}}{k_i}. \tag{7.52}$$

However, $\dot{\omega}_i$ in the above equation may not be directly calculated by taking derivative of ω_i due to the noise in the wheel speed measurement. A tire force observer is thus adopted to update the motor control signal such that the desired tire force can be provided. Inspired by the disturbance observer design by Do [27], we propose the following observer to estimate the tire force:

$$\begin{cases} \hat{F}_{\mathrm{xi}} = -\frac{I\zeta_i}{R_T} - \frac{I\omega_i\vartheta_i}{R_T} \\ \dot{\zeta}_i = -\zeta_i\vartheta_i - \vartheta_i\left(\frac{u_{mi}k_i}{I} + \vartheta_i\omega_i\right) \end{cases} \tag{7.53}$$

where $\hat{F}_{xi}$ is the longitudinal tire force estimate of the ith wheel and $\vartheta_i > 0$ is the observer gain. Based on Eq.(7.53), dynamic of the tire force estimation $\hat{F}_{xi}$ can be written as

$$\begin{aligned} \dot{\hat{F}}_{\mathrm{xi}} &= -\frac{I\dot{\zeta}_i}{R_T} - \frac{I\dot{\omega}_i\vartheta_i}{R_T} \\ &= -\frac{I}{R_T}\left(-\zeta_i\vartheta_i - \vartheta_i\left(\frac{u_{\mathrm{mi}}k_i}{I} + \vartheta_i\omega_i\right)\right) - \frac{I\dot{\omega}_i\vartheta_i}{R_T} \\ &= -\frac{I}{R_T}\left(-\vartheta_i\left(-\frac{\hat{F}_{\mathrm{xi}}R_T}{I} - \omega_i\vartheta_i\right) - \vartheta_i\left(\frac{v_ik_i}{I} + \vartheta_i\omega_i\right)\right) \\ &\quad - \frac{\vartheta_i\left(k_iv_i - F_{\mathrm{xi}}R_T\right)}{R_T} \\ &= -\frac{I}{R_T}\left(\frac{\vartheta_i\hat{F}_{\mathrm{xi}}R_T + \vartheta_iv_ik_i}{I}\right) - \frac{\vartheta_ik_iv_i}{R_T} + \vartheta_iF_{\mathrm{xi}} \\ &= -\vartheta_i\left(\hat{F}_{\mathrm{xi}} - F_{\mathrm{xi}}\right). \end{aligned} \tag{7.54}$$

Choose a Lyapunov function candidate as

$$V = \frac{1}{2}\left(F_{\mathrm{xi}} - \hat{F}_{\mathrm{xi}}\right)^2 \tag{7.55}$$

Evaluating the time derivative of the above Lyapunov function, we obtain

$$\begin{aligned} \dot{V}_1 &= \left(F_{\mathrm{xi}} - \hat{F}_{\mathrm{xi}}\right)\left(\dot{F}_{\mathrm{xi}} - \dot{\hat{F}}_{\mathrm{xi}}\right) \\ &= \left(F_{\mathrm{xi}} - \hat{F}_{\mathrm{xi}}\right)\left(\dot{F}_{\mathrm{xi}} + \vartheta_i\left(\hat{F}_{\mathrm{xi}} - F_{\mathrm{xi}}\right)\right) \\ &= -\vartheta_i\left(\hat{F}_{\mathrm{xi}} - F_{\mathrm{xi}}\right)^2 - \left(\hat{F}_{\mathrm{xi}} - F_{\mathrm{xi}}\right)\dot{F}_{\mathrm{xi}} \\ &\le -\vartheta_i\left(\hat{F}_{\mathrm{xi}} - F_{\mathrm{xi}}\right)^2 + \tau_i\left(\hat{F}_{\mathrm{xi}} - F_{\mathrm{xi}}\right)^2 + \frac{\dot{F}_{\mathrm{xi}}^2}{4\tau_i} \\ &= -\left(\vartheta_i - \tau_i\right)\left(\hat{F}_{\mathrm{xi}} - F_{\mathrm{xi}}\right)^2 + \frac{\dot{F}_{\mathrm{xi}}^2}{4\tau_i}, \end{aligned} \tag{7.56}$$

where $\tau_i > 0$. So the tire force estimation error $F_{xi} - \hat{F}_{xi}$ will be bounded as

$$\left|F_{\mathrm{xi}} - \hat{F}_{\mathrm{xi}}\right| \leq \frac{\left|\dot{F}_{xi}\right|_{\max}}{2\sqrt{(\vartheta_i - \tau_i)\,\tau_i}} \tag{7.57}$$

where $\left|\dot{F}_{\mathrm{xi}}\right|_{\max}$ is the maximal value of $\left|\dot{F}_{\mathrm{xi}}\right|$. Defining a new variable as $\tau_i = \vartheta_i/2$, $F_{\mathrm{xi}} - \hat{F}_{\mathrm{xi}}$ will finally be bounded as

$$\left|F_{\mathrm{xi}} - \hat{F}_{\mathrm{xi}}\right| \leq \frac{\left|\dot{F}_{\mathrm{xi}}\right|_{\max}}{\vartheta_i} \tag{7.58}$$

Since ϑ_i can be chosen arbitrarily large, the boundary of the estimation error can be arbitrarily small. Based on the observed tire forces, a control law can be designed to adjust the motor control signals. Define a Lyapunov function candidate as

$$V_2 = \frac{1}{2}\left(F_{\mathrm{xi}} - \hat{F}_{\mathrm{xi}}\right)^2 \tag{7.59}$$

Evaluating the time derivative of the above Lyapunov function, we obtain

$$\begin{aligned}
\dot{V}_2 &= \left(\hat{F}_{\mathrm{xi}} - F_{x0i}\right)\left(\dot{\hat{F}}_{\mathrm{xi}} - \dot{F}_{x0i}\right) \\
&= \left(\hat{F}_{\mathrm{xi}} - F_{x0i}\right)\left(-\vartheta_i\left(\hat{F}_{\mathrm{xi}} - F_{\mathrm{xi}}\right) - \dot{F}_{x0i}\right) \\
&= -\left(\hat{F}_{\mathrm{xi}} - F_{x0i}\right)\left(\vartheta_i\left(\hat{F}_{\mathrm{xi}} - \frac{k_i v_i - I\dot{\omega}_i}{R_T}\right) + \dot{F}_{x0i}\right)
\end{aligned} \tag{7.60}$$

If the motor torque control law can be designed as

$$v_i = \frac{R_T \hat{F}_{\mathrm{xi}} - L_i \operatorname{sign}\left(\hat{F}_{\mathrm{xi}} - F_{x0i}\right)}{k_i} \tag{7.61}$$

with L_i and ϑ_{i} satisfying

$$\begin{cases} \vartheta_i I\left(L_i - I\left|\dot{\omega}_i\right|\right) > \dot{F}_{x0i} \\ L_i > I\left|\dot{\omega}_i\right| \end{cases} \tag{7.62}$$

the following condition holds:

$$\operatorname{sign}\left(\hat{F}_{\mathrm{xi}} - F_{0\mathrm{xi}}\right) = \operatorname{sign}\left(\vartheta_i\left(\hat{F}_{xi} - \frac{k_i u_{\mathrm{mi}} - I\dot{\omega}_i}{R_T}\right) + \dot{F}_{0\mathrm{xi}}\right), \tag{7.63}$$

which means $\dot{V}_2 < 0$, and the tire force can be controlled to track the desired value. In order to eliminate the chatting effects, the sign function in the motor control signal law (7.63) can be replaced with a saturation function

Table 7.1 Vehicle parameter in the simulation

Parameters	Nominal values
m	900 kg
I_z	1200 kg·m^2
l_s	0.8 m
l_f	1.0 m
l_r	1.0 m
I	1.1 kg·m^2
R_T	0.3 m
C_f	22,000 N/rad
C_r	25,000 N/rad

7.4 Simulation Studies

Two simulation cases based on a high-fidelity full-vehicle model constructed in CarSim were conducted. The vehicle parameters are listed in Table 7.1. As the vehicle sideslip angle should be constrained to improve the vehicle stability [28, 29], the desired slip angle β_r is given as zero. The desired vehicle yaw rate can be generated from the driver's steering angle and vehicle longitudinal speed as [29]

$$r_r = \frac{v_x}{l\left(1 + k_{\text{us}} v_x^2\right)} \delta(t), \tag{7.64}$$

where $l = l_f + l_s$ is the distance between the front and rear axles and k_{us} is the stability factor. To better show the performance of the proposed controller, the states of an uncontrolled vehicle with the same hand-wheel steering input are also given in the simulation results.

7.4.1 Single Lane Changing

In this simulation, a single lane-changing maneuver with a single sinusoidal steering which is plotted in Fig. 7.4 is carried out to evaluate the performance of proposed method. The vehicle ran at a high speed range, and the TRFC was chosen as 0.6. The actual vehicle mass in the CarSim model was set to 1.2 times of the nominal value in this simulation. The vehicle speed is plotted in Fig. 7.5. It is assumed that the driver noticed a roadblock in the middle of the lane, so a hard braking was also applied when the driver started to steer the hand wheel.

Vehicle accelerations in the longitudinal and lateral directions are plotted in Fig. 7.6, where we can see that the lateral acceleration ay goes larger with the increase of the steering angle, while the longitudinal acceleration ax became small when the

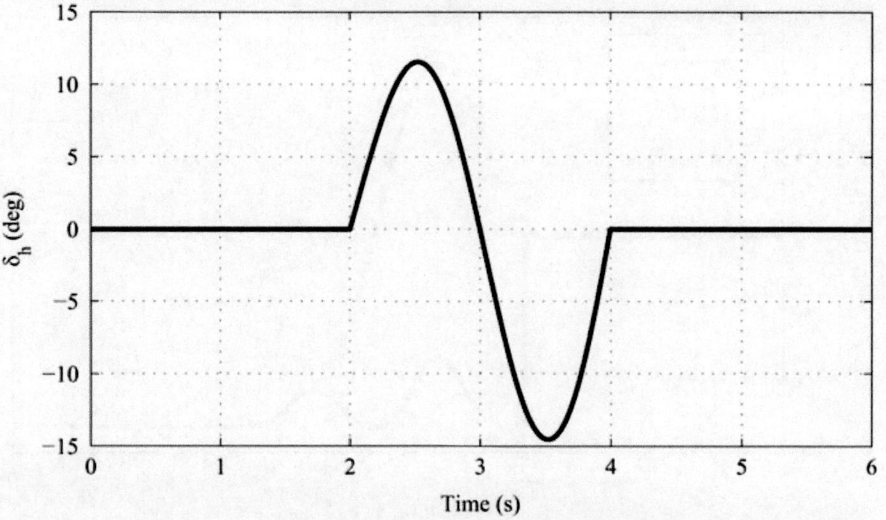

Fig. 7.4 Hand-wheel steering signal in the single lane-changing simulation

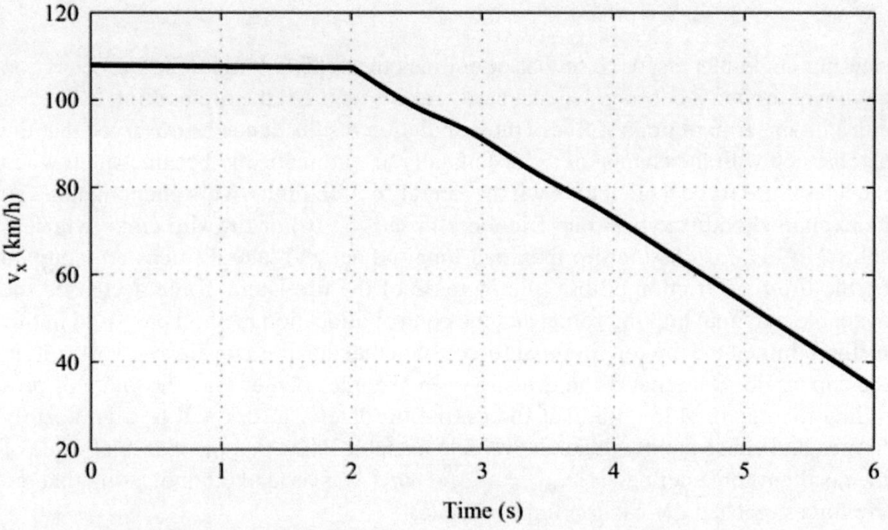

Fig. 7.5 Vehicle longitudinal speeds in the single lane-changing simulation

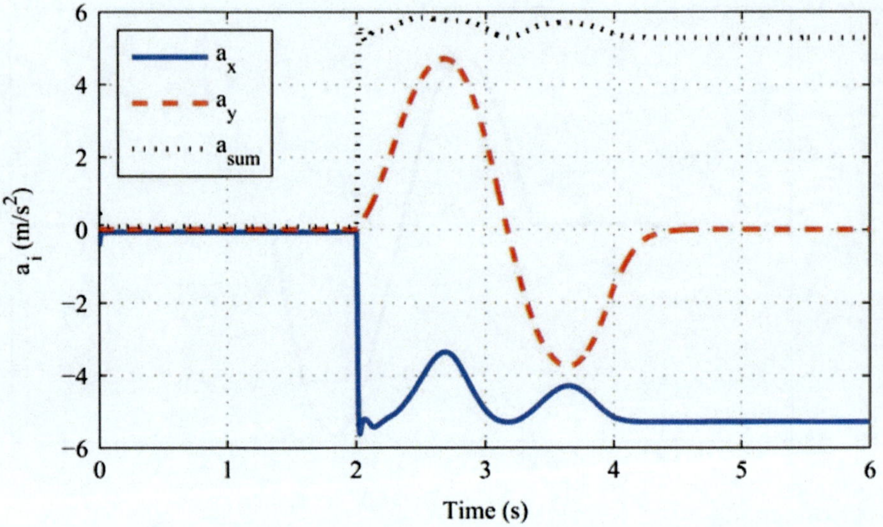

Fig. 7.6 Vehicle accelerations in the single lane-changing simulation

steering angle increased. Note that oftentimes the vehicle longitudinal acceleration ax moves around a constant when a hard brake is applied (the tire braking forces are constant under hard brake). But in our simulation results, it can be observed that the ax changed with the change of ay, specifically ax automatically became small when the lateral tire force tended to reach the lateral friction limits. This phenomenon can be explained with the tire-road friction ellipse [15, 16] or tire slip circle which is shown in Fig. 7.3. As the tire maximal longitudinal and lateral forces are coupled by the tire-road friction ellipse, the increase of the tire lateral force decreased the available maximal braking force, and the control-allocation method proposed in this method limited the tire longitudinal forces such that the tires can always work within the slip circle. Note that if the tire slip ratio becomes bigger than the value of providing the maximal longitudinal forces, the tire lateral forces will be significantly limited and consequently make the vehicle unstable. Note that the maximal value of the resultant force defined as $a_{\text{sum}} = \sqrt{{a_x}^2 + {a_y}^2}$ was around 0.6 indicating that the tire forces reached the friction limit in the test.

Vehicle yaw rates are plotted in Fig. 7.7 where we can see that the yaw rate of the controlled vehicle could track the reference even though a small tracking error existed, where as for the uncontrolled vehicle, the yaw rate deviated from the desired value fast when the hard brake was applied. The small tracking error in the yaw rate of the controlled vehicle was caused by the limited tire forces, i.e., the desired external yaw moment required by higher-level controller cannot be fully provided as the tires reached the friction limit in the test. Note that the vehicle could still be stabilized even if the tire forces reached the friction limits in the test. Figure 7.8 compares the vehicle sideslip angles, and it can be observed from this figure that

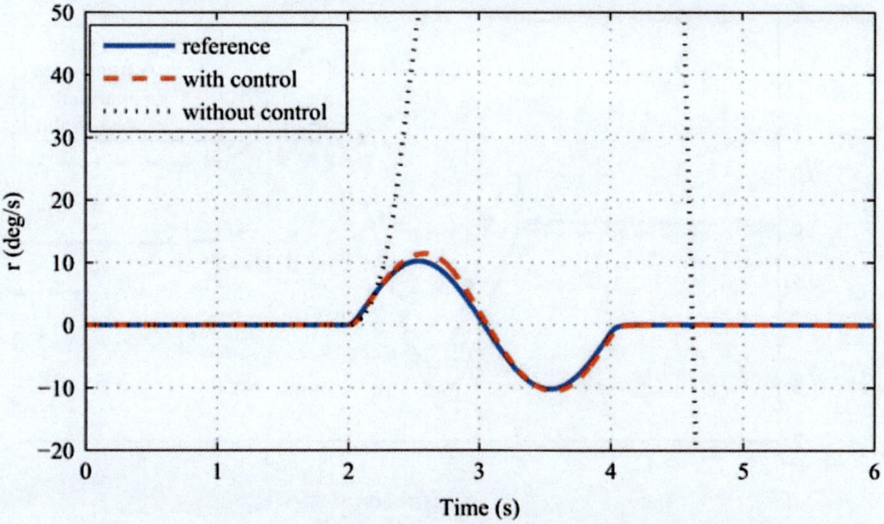

Fig. 7.7 Vehicle yaw rates in the single lane-changing simulation

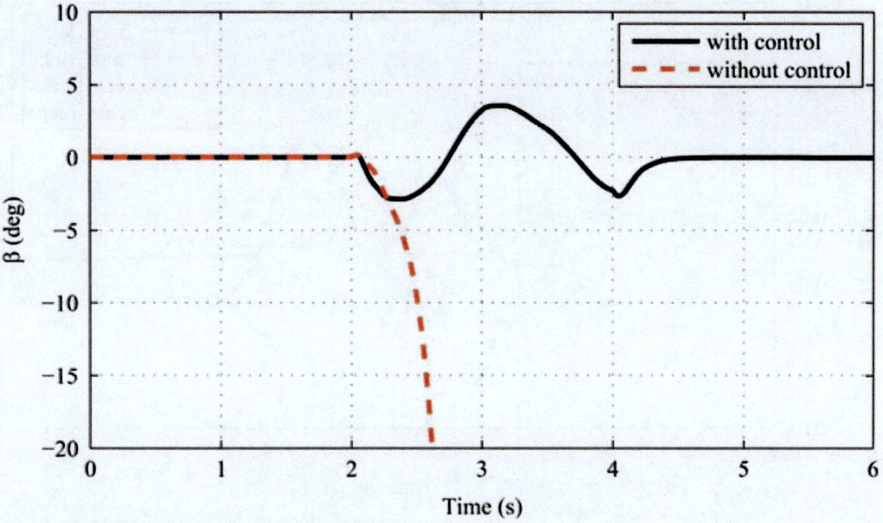

Fig. 7.8 Vehicle slip angles in the single lane-changing simulation

the sideslip angle of the controlled vehicle could always be restrained around zero, indicating that the vehicle was stabilized. Due to the lose of tire lateral forces during the single lane-changing maneuver, the sideslip angle of the uncontrolled vehicle with the same hand-wheel steering input and braking forces increased fast as soon as the hard brake was applied, which proves the effectiveness of the proposed control method.

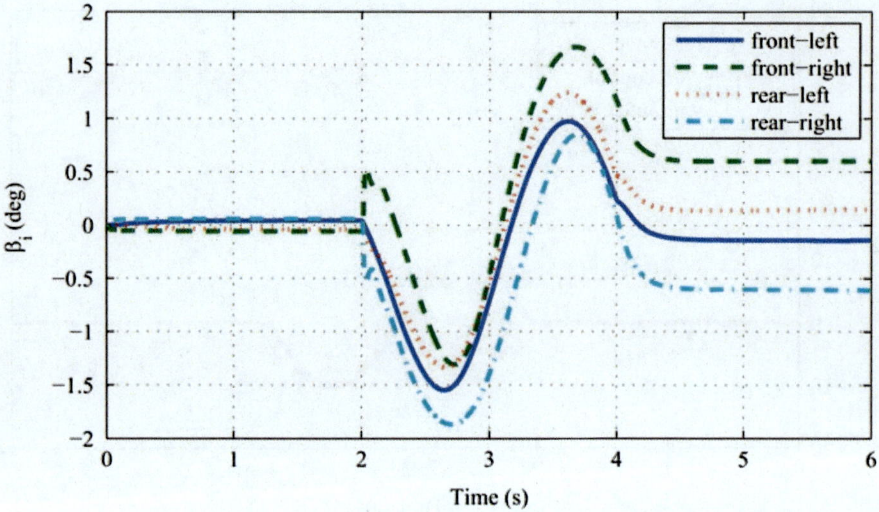

Fig. 7.9 Tire slip angles in the single lane-changing simulation

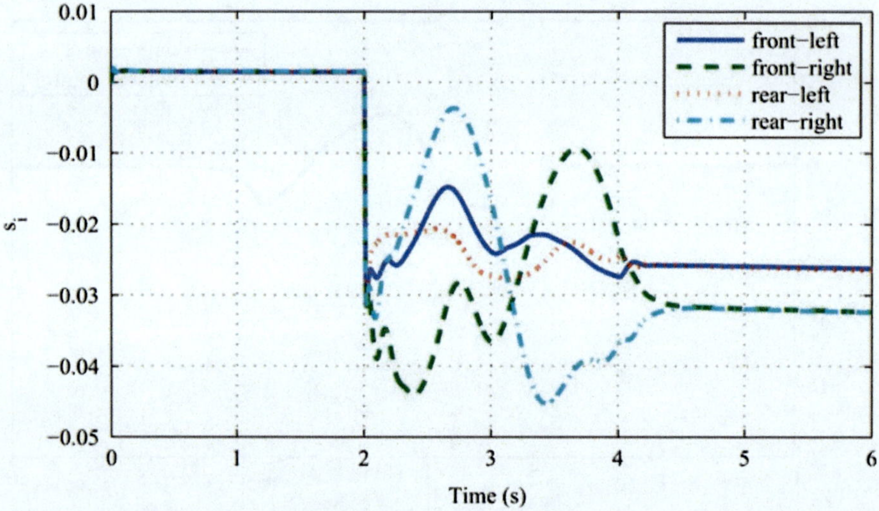

Fig. 7.10 Tire slip ratios in the single lane-changing simulation

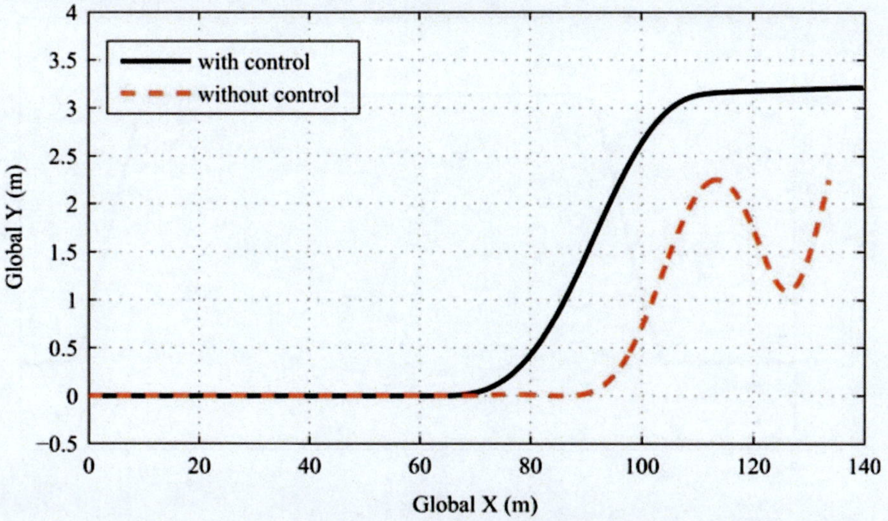

Fig. 7.11 Comparison of the vehicle trajectories in the single lane-changing simulation

Tire slip angles and slip ratios are plotted in Figs. 7.9 and 7.10, respectively. We can see that the tire slip ratios and slip angles were always limited even when a hard brake and a large steering angle were applied. Figure 7.11 compares the vehicle trajectories in the single lane-changing simulation. We can see that the controlled vehicle could track the desired trajectory expected by the driver, while the uncontrolled vehicle failed accomplish the single lane-changing task due to the hard brake and large steering input.

7.5 J-Turn Simulation

In this simulation, the vehicle was controlled to make a J-turn at a low speed. The TRFC was set to 0.2, and the actual vehicle mass in the CarSim model was set to 1.2 times of the nominal value. A counterclockwise turn was introduced at 2 s, and Fig. 7.12 shows the hand-wheel steering angle manipulated by the driver in the J-turn simulation. The initial vehicle speed was chosen 20 km/h, and the driver pressed the accelerator to gather speed at 5 s.

Vehicle states in this maneuver are plotted in Figs. 7.13, 7.14, 7.15, and 7.16. It is obvious that the yaw rate of the controlled vehicle could precisely track the desired reference expected by the driver, and the vehicle sideslip angle could also be restrained in a narrow scope which is close to zero. Figure 7.15 shows the vehicle accelerations in the J-turn maneuver. We can come to a conclusion similar to that in the single lane-changing simulation, i.e., when the lateral acceleration became large, the tire driving forces of the controlled vehicle decreased depending on the tire slip

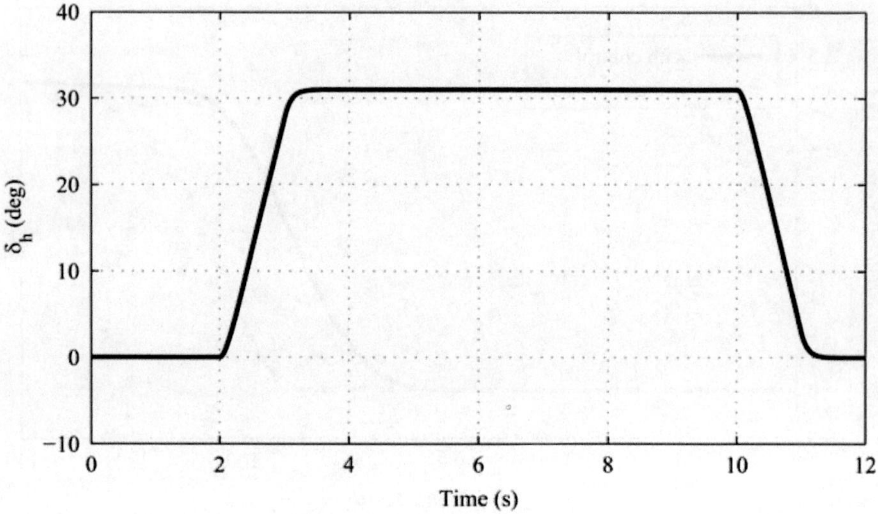

Fig. 7.12 Hand-wheel steering signal in the J-turn simulation

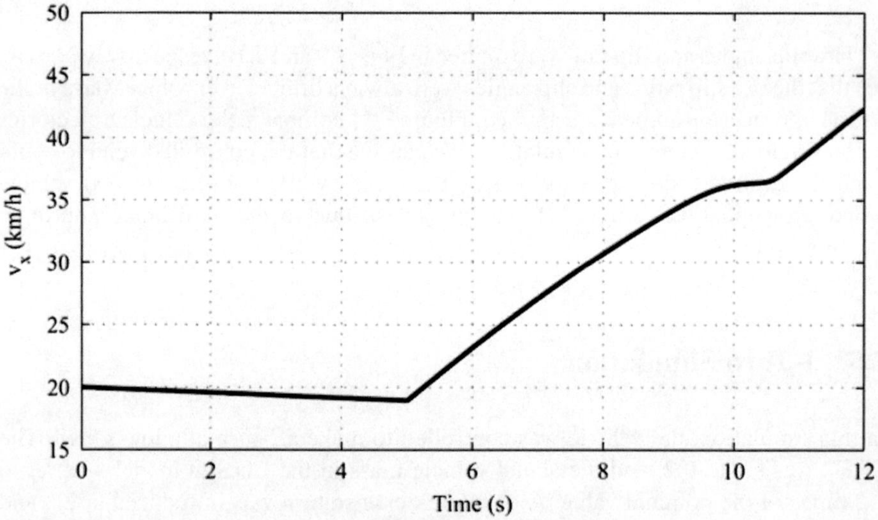

Fig. 7.13 Vehicle longitudinal speeds in the J-turn simulation

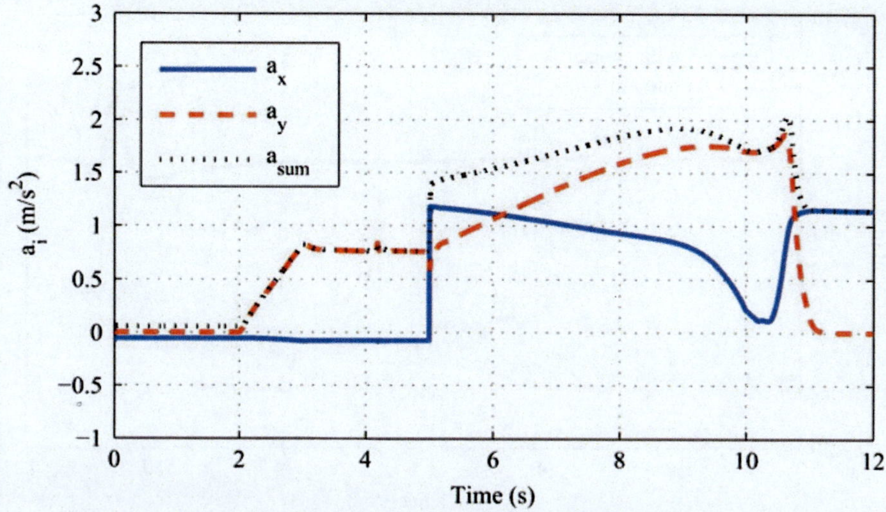

Fig. 7.14 Vehicle accelerations in the J-turn simulation

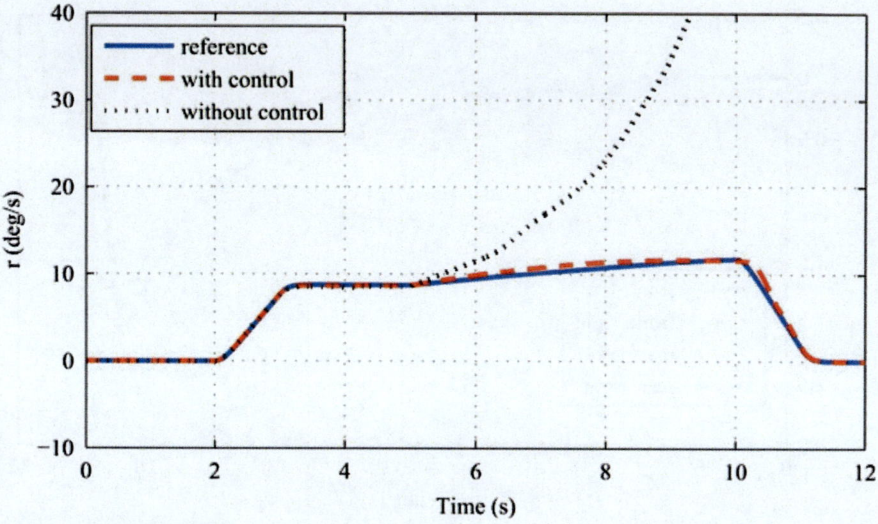

Fig. 7.15 Vehicle yaw rates in the J-turn simulation

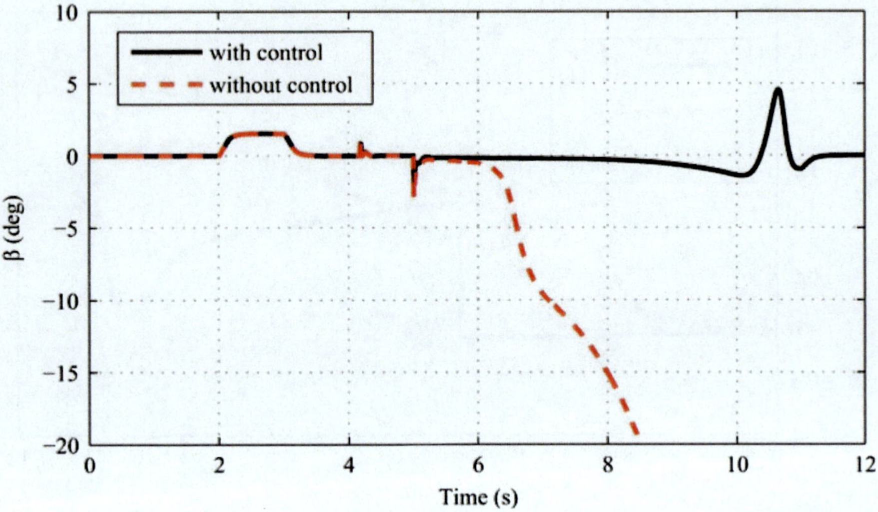

Fig. 7.16 Vehicle slip angles in the J-turn simulation

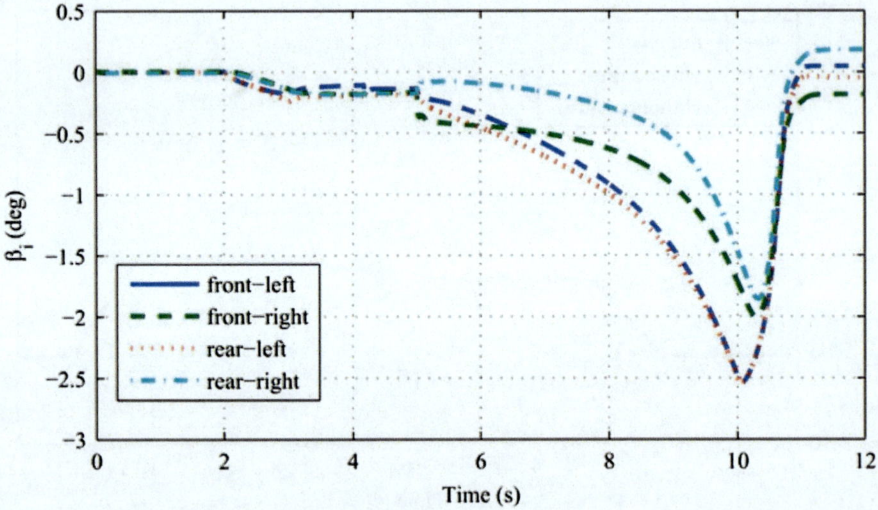

Fig. 7.17 Tire slip angles in the J-turn simulation

angles to provide the greatest possible lateral tire forces and stabilize the vehicle. Tire slip angles and slip ratios are plotted in Figs. 7.17 and 7.18. We can see again that both tire slip angles and slip ratios were always restricted in the stable regions which prove the effectiveness of the proposed controller. The vehicle global trajectories are shown in Fig. 7.19. One can observe that the proposed controller ensured the vehicle trajectory tracking performance even when the vehicle runs on a low-μ road.

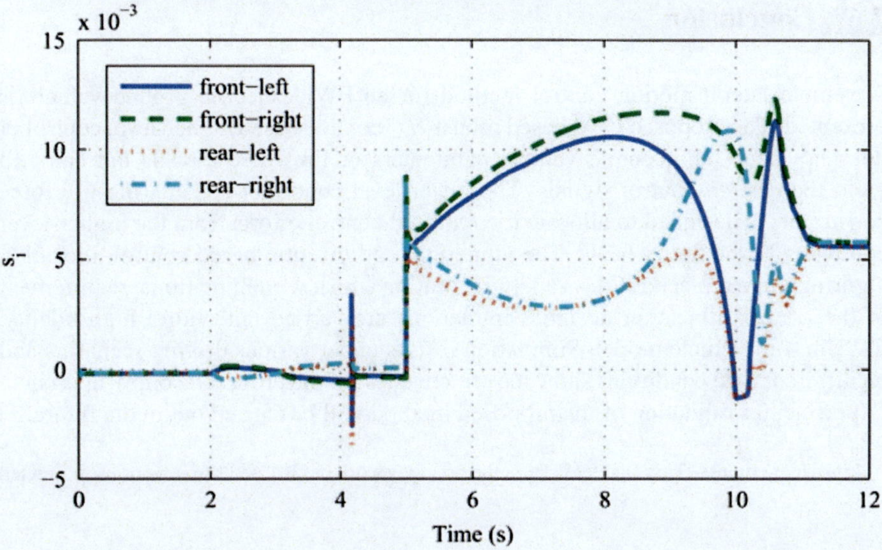

Fig. 7.18 Tire slip ratios in the J-turn simulation

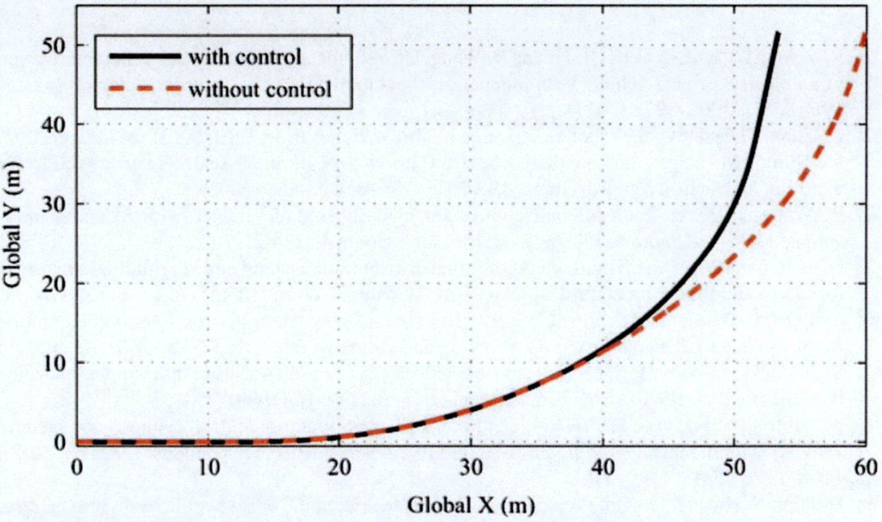

Fig. 7.19 Comparison of the vehicle trajectories in the J-turn simulation

7.6 Conclusion

A vehicle lateral motion control method for an FWIA electric ground vehicle is presented. The proposed LPV-based robust $\mathcal{H}_\infty$ control-based higher-level controller does not need the accurate vehicle parameters or tire force models but can still yield the desired control signals. The lower-level controller considering tire force constraints is designed to allocate the required control efforts from the higher-lever controller to the four wheels. The numerical optimization-based control-allocation algorithms are replaced with an analytic solution with low computational requirement in the control-allocation design. Simulations are carried out with a high-fidelity, CarSim, full-vehicle model. Simulation results under various driving scenarios and on different road conditions show the effectiveness of the proposed control approach. Experimental validation of the proposed method will be carried out in the future.

Acknowledgements This chapter is from the previous work in [30], and some typos are corrected here.

References

1. R. Wang, Y. Chen, D. Feng, X. Huang, J. Wang, Development and performance characterization of an electric ground vehicle with independently actuated in-wheel motors. J. Power Sources **196** (8, SI), 3962–3971 (2011)
2. C. Chan, The state of the art of electric and hybrid vehicles. Proc. IEEE **90**(2), 247–275 (2002)
3. M. Shino, M. Nagai, Independent wheel torque control of small-scale electric vehicle for handling and stability improvement. JSAE Rev. **24**(4), 449–456 (2003)
4. R. Wang, J. Wang, Fault-tolerant control for electric ground vehicles with independently-actuated in-wheel motors. J. Dyn. Syst. Measure. Control., 134(2)
5. O. Mokhiamar, M. Abe, Simultaneous optimal distribution of lateral and longitudinal tire forces for the model following control. J. Dyn. Syst. Measure. Control **126**(4), 753–763 (2004)
6. p. Sadeghi-Tehran, P. Angelov, Online self-evolving fuzzy controller for autonomous mobile robots, in 2011 IEEE Workshop on Evolving and Adaptive Intelligent Systems (EAIS) (2011)
7. S. Sakai, H. Sado, Y. Hori, Motion control in an electric vehicle with four independently driven in-wheel motors. IEEE/ASME Trans. Mechatron. **4**(1), 9–16 (1999)
8. N. Mutoh, Y. Hayano, H. Yahagi, K. Takita, Electric braking control methods for electric vehicles with independently driven front and rear wheels. IEEE Trans. Indus. Electron. **54**(2), 1168–1176 (2007)
9. D. Kim, S. Hwang, H. Kim, Vehicle stability enhancement of four-wheel-drive hybrid electric vehicle using rear motor control. IEEE Trans. Vehicular Technol. **57**(2), 727–735 (2008)
10. M. Bodson, Evaluation of optimization methods for control allocation. J. Guidance Control Dyn. **25**(4), 703–711 (2002)
11. T.A. Johansen, T.I. Fossen, Control allocation A-survey. Automatica **49**(5), 1087–1103 (2013)
12. J. Wang, R.G. Longoria, Coordinated and reconfigurable vehicle dynamics control. IEEE Trans. Control Syst. Technol. **17**(3), 723–732 (2009)
13. J. Plumlee, D. Bevly, A. Hodel, Control of a ground vehicle using quadratic programming based control allocation techniques, in Proceedings of the American Control Conference, pp. 4704–4709 (2004)
14. S. Sakai, H. Sado, Y. Hori, Dynamic driving/braking force distribution in electric vehicles with independently driven four wheels. Electric. Eng. Jpn. **138**(1), 79–89 (2002)

15. J. Wang, Coordinated and Reconfigurable Vehicle Dynamics Control(Ph.D. dissertation), Department of Mechanical Engineering (The University of Texas at Austin, Austin, TX, USA, 2007)
16. J. Wong, C. Chiang, A general theory for skid steering of tracked vehicles on firm ground. Proc. Inst. Mech. Eng. Part D-J. Automobile Eng. **215**(D3), 343–355 (2001)
17. D. Filev, P. Angelov, Fuzzy optimal control. Fuzzy Sets Syst. **47**(2), 151–156 (1992)
18. R. Precup, S. Preitl, Popov-type stability analysis method for fuzzy control systems. Proc. Fifth EUFIT **97**, 1306–1310 (1997)
19. E. Joelianto, D.C. Anura, M.P. Priyanto, Anfis-hybrid reference control for improving transient response of controlled systems using pid controller. Int. J. Artific. Intell. **10**(13), 88–111 (2013)
20. W. Xie, An equivalent LMI representation of bounded real lemma for continuous-time systems. J. Inequalities Appl.
21. H. Zhang, Y. Shi, A.S. Mehr, Robust energy-to-peak filtering for networked systems with time-varying delays and randomly missing data. IET Control Theory Appl. **4**(12), 2921–2936 (2010)
22. H. Gao, X. Yang, P. Shi, Multi-Objective Robust $\mathcal{H}_\infty$ Control of Spacecraft Rendezvous. IEEE Trans. Control Syst. Technol. **17**(4), 794–802 (2009)
23. H. Zhang, Y. Shi, A.S. Mehr, Parameter-dependent mixed $\mathcal{H}_2/\mathcal{H}_\infty$ filtering for linear parameter-varying systems. IET Signal Process. **6**(7), 697–703 (2012)
24. H. Zhang, X. Zhang, J. Wang, Robust gain-scheduled energy-to-peak control of vehicle lateral dynamics stabilisation. Vehicle Syst. Dyn. **52**(3), 309–340 (2014)
25. H. Zhang, Y. Shi, J. Wang, On energy-to-peak filtering for nonuniformly sampled nonlinear systems: a Markovian Jump system approach. IEEE Trans. Fuzzy Syst. **22**(1), 212–222 (2014)
26. K. Kritayakirana, Autonomous vehicle control at the limits of handling (Ph.D. dissertation), Department of Mechanical Engineering (The Stanford University, USA, 2012)
27. K.D. Do, Control of nonlinear systems with output tracking error constraints and its application to magnetic bearings. Int. J. Control **83**(6), 1199–1216 (2010)
28. J. Ahmadi, A.K. Sedigh, M. Kabganian, Adaptive vehicle lateral-plane motion control using optimal tire friction forces with saturation limits consideration. IEEE Trans. Vehicular Technol. **58**(8), 4098–4107 (2009)
29. H. Du, N. Zhang, G. Dong, Stabilizing vehicle lateral dynamics with considerations of parameter uncertainties and control saturation through robust yaw control. IEEE Trans. Vehicular Technol. **59**(5), 2593–2597 (2010)
30. R. Wang, H. Zhang, J. Wang, F. Yan, N. Chen, Robust lateral motion control of four-wheel independently actuated electric vehicles with tire force saturation consideration. J. Franklin Inst. **352**(2), 645–668 (2015)

Appendix: Fundamentals of Robust $\mathcal{H}_\infty$ Control

A.1 Definition of $\mathcal{H}_\infty$ in the Frequency Domain

Consider a stable linear time-invariant (LTI) system as follows:

$$\begin{aligned} \dot{x}(t) &= Ax(t) + Bv(t) \\ z(t) &= Cx(t) + Dv(t) \end{aligned} \tag{A.1}$$

where $x(t)$ is the state vector, $v(t)$ is the external input, and $z(t)$ is the controlled output, which is related to the performance requirement. A, B, C, and D are the system matrix, input matrix, output matrix, and given matrix with appropriate dimension, respectively.

For the linear time-invariant system, the corresponding transfer function could be obtained

$$T(s) = \frac{Z(s)}{V(s)} = C(sI - A)^{-1}B + D \tag{A.2}$$

The definition of $\mathcal{H}_\infty$ norm in frequency domain could be expressed as:

$$\|T(s)\|_\infty = \sup_{w} \sigma_{\max}(T(jw)) \tag{A.3}$$

Here, w denotes the frequency, and the $\mathcal{H}_\infty$ norm of an LTI system refers to the peak of largest singular value of the corresponding transfer function.

A.2 Definition of $\mathcal{H}_\infty$ in the Time Domain

Consider an LPV system expressed as follows:

H. Zhang et al., *Robust Gain-Scheduled Estimation and Control of Electrified Vehicles via LPV Technique*, Key Technologies on New Energy Vehicles,
https://doi.org/10.1007/978-981-19-8509-6

$$\begin{aligned}
\dot{x}(t) &= A(\rho)x(t) + B(\rho)u(t) + E(\rho)v(t) \\
&= \sum_{i=1}^{n} \theta_i(\rho(t))(A_i x(t) + B_i u(t) + E_i v(t)) \\
z(t) &= C(\rho)x(t) + D(\rho)v(t) \\
&= \sum_{i=1}^{n} \theta_i(\rho(t))(C_i x(t) + D_i v(t))
\end{aligned} \tag{A.4}$$

where $x(t)$, $u(t)$, and $v(t)$ stand for the state vector, control input, and external disturbance, respectively. $z(t)$ is the controlled output which is selected in terms of the design objective.

Note that only the LTI systems have the transfer functions. Therefore, the $\mathcal{H}_\infty$ norm of an LPV system with a zero initial condition cannot be defined in the frequency domain and is defined in the time domain as:

$$\|z(t)\|_2 < \gamma \|v(t)\|_2, \tag{A.5}$$

for all nonzero $v(t) \in l_2[0, \infty)$, where

$$\|z(t)\|_2 = \sqrt{\int_{t=0}^{\infty} z^{\mathrm{T}}(t)z(t)\mathrm{d}t},$$

$$\|v(t)\|_2 = \sqrt{\int_{t=0}^{\infty} v^{\mathrm{T}}(t)v(t)\mathrm{d}t},$$

and γ is the $\mathcal{H}_\infty$ performance index.